Web to the Stars

A History of The University of Tennessee Space Institute

Weldon Payne
University of Tennessee Space Institute

KENDALL/HUNT PUBLISHING COMPANY
2460 Kerper Boulevard P.O. Box 539 Dubuque, Iowa 52004-0539

Library of Congress Catalog Card Number 91-75635

ISBN 0-8403-7088-1

Printed in the United States of America
10 9 8 7 6 5 4 3 2 1

CONTENTS

Dr. Bernhard H. Goethert
October 20, 1907–March 29, 1988
Dedicated to Doc's memory and to all those who have
helped to make his dream come true.

PREFACE

No one can scoop up the waves of the sea in their entirety, but each vanishing handful is teeming with life, and the surf, echoing, has a story to tell. Neither is it possible to testify, truly, to all that has passed in the night, to pen a history and claim wholeness, for Time shifts the sands and dims the roar. I have sought the movement and the foam and the essence.

I set out to write not so much a formal history as to tell a true story. In lieu of footnotes, I have freely attributed within the text, and a partial accounting of research sources is included as Appendix C.

Since late 1988 when Gary Smith suggested the task of tracking the Space Institute's first twenty-five years, I have especially looked for the prints of individuals who have made UTSI happen. Even so, I have not been able to include every deserving person or to acknowledge their contributions to the Space Institute's story. I beg their pardon.

I have placed a high value on accuracy, but in attempting to reconstruct the fragile past, not every piece will square with all remembrances and perceptions.

The book contains photographs from UTSI's archives, many of which were taken by U.S. Air Force and AEDC photographers.

To all who have supported this effort, I am grateful. Special thanks must go to Art and Helen Mason, Bob Young, Milton Klein, Shirley and Bob Kamm, Judy Rudder, Boyd Stubblefield, Joyce Moore and Dave Hiebert, and to Ken Harwell, Dick Roberds, and Wes Harris for their sanctioning of this project.

the author

Chapter One

A TIME FOR DREAMS

". . . even a spider refuses to lie down and die if a rope can still be spun to a star."

Loren Eiseley

PROUDLY, now, she stands in quiet dignity, embodiment of the dream to weld graduate study and aeronautical research into a unique institution for training the best scientific minds. White, semicircular brilliance gleams in the sunlight while Canada geese skim the pooled waters of Elk River two miles inside Franklin County, Tennessee. The **University of Tennessee Space Institute,** tucked away from the busy world, belies the international reputation that it has achieved since it opened classes in temporary quarters on September 24, 1964.

The Space Institute reflects the indomitable spirit of an individual who would not accept "no," and it also is a witness to the spirit of Middle Tennesseans who caught his vision and helped him to realize his dream.

It was, perhaps, the time and place for dreams to come true. Man had put his foot into space, and America's young president had promised to land a man on the moon within the decade. A few miles south, the National Aeronautics and Space Administration had established the Marshall Space Flight Center. **John H. Glenn Jr.** had made history as the first American to orbit the earth. The United States had teetered on the brink of war during the Cuban missile crisis as **President John F. Kennedy** called fourteen thousand Air Force reservists to active duty.

Then America lost her young dreamer to an assassin's bullet in Dallas, but flint had been struck. America had her dream.

Having ignored **Charles Lindbergh's** warnings in the late thirties, the United States had twice found herself lagging in technological expertise, and the mood of the country reflected a determination to put the nation out front. Technology's abrupt intrusion had forced a new awareness of the urgency to link academic excellence with research and development opportunities; the scientific community was impatient to follow the example of the medical profession in tying discovery to application. The college experience was no longer limited to the elite, and graduate studies had emerged as necessities to survival in science and engineering careers. The time indeed was ripe.

Closer home, lightning had, as one writer put it, "struck twice," first bringing Camp Forrest to the area, infusing the sleepy rural psyche with a new awareness and establishing a special link between military and civilian life. Then, after World War II, came the Air Force's Arnold Engineering Development Center, a facility unequaled in the free world, and with it came the opportunity to develop as an adjunct just such an institution as the Space Institute.

But dreams require dreamers. Minds had been at work for years, exploring, experimenting, laying the foundation for what would eventually come. One man, finally, would become obsessed with the dream and, assisted by a cadre of key players who shared his vision, he would bulldoze his dream into reality.

In 1945, those players had not been touched by the dream that would bring them together in Middle Tennessee, nor did they know each other, but each would wind up, eventually, in the right place at the right time.

✧ ✧ ✧

In the spring of that year, **Bernhard Hermann Goethert's** most urgent goal was survival, to dodge Russian soldiers as he trudged two hundred miles westward from Berlin, sleeping in barns and swimming across the River Mulde, on his way to a reunion with his wife and four children, including the baby boy that he had not seen. It took him more than two months, and the tough little doctor—one of the foremost aeronautical experts in the world at that time—almost did not make it.

Goethert had earned his Ph.D. in aeronautical engineering at the Technical University of Berlin in 1938 at the age of thirty-one. As department chief of high-speed aerodynamics in Berlin during World War II, he spearheaded wind tunnel testing for development of the world's first jet and rocket combat airplanes and of the V-1 and V-2 guided missiles. (The first V-1 rocket fell on London on June 13, 1944.) This research was conducted at the German's research institute—the Deutsche Versuchsansralt Luftfaht (German Experimental Establishment for Aeronautics). **Dr. Hans Doetsch** was head of low-speed aerodynamic research.

Goethert and his family—**Mrs. Hertha Goethert** (''Mutti'' to her husband and children), **Hella,** seven, and the five-year-old twins, **Wolfhart** and **Winfried**—lived in Bohnsdorf, a suburb of Berlin, until 1943 when allied bombers began hitting closer. Mrs. Goethert and the children went to Hertha's birthplace, Diemarden, a farming village near Gottingen. Dr. Goethert frequently visited them on weekends, and for Christmas that year, the wife and children returned to Berlin. Heavy bombing forced Mrs. Goethert and the children into a shelter for the night, and on Christmas morning, 1943, they returned with Dr. Goethert to their home in Bohnsdorf amidst flames and debris and, for the children's sake, opened presents in their home, which was undamaged except for shattered panes. Then Hertha, who was pregnant, and the three children took a crowded train back to Diemarden. This was the last time the mother and children saw their comfortable home. **Reinhard** was born the next year in HahanKlee in the Harz Mountains, but his father would not see him until late in 1945.

Early that year, Allied armies pushed the German soldiers on both the east and west sides of Berlin, and Goethert later recalled that the Russians had Berlin totally encircled by about April 25. Employees of the institute and remaining families were quartered in a shelter on the institute's grounds. It was on the evening before the Russians took over the research institute—probably April 25—when Goethert and his friend, **Gunter Hacker,** debated what they should do. Hacker felt that he must join the German soldiers as they retreated and help them in the futile defense of the city. Goethert knew the war was lost, and he argued that further fighting would probably result in both of their deaths. They each had small children, and their obligation, he felt, was to try to reach them and to support them in the trying times that were sure to come.

Hacker retreated with the soldiers to the inner city of Berlin, and Goethert, who stayed in the shelter, never heard from him again.

On May 1, 1945, **Admiral Karl Doenitz** announced that he had been appointed as successor to Adolph Hitler who was rumored to have killed himself. On May 7, Doenitz agreed to an unconditional German surrender. By this time, the Russian soldiers had forced Goethert and other personnel at the institute onto the streets to work in cleaning up debris from bombings. Goethert and Doetsch slipped away and hid in an empty apartment. From there, Goethert evaded soldiers, making his way to his own house. A few days later, he got in touch with Doetsch, and they agreed to strike out on bicycles to Gottingen, about 250 kilometers away. They did not get far.

Russian soldiers confiscated their bicycles and ordered the two scientists to walk into the woods. The soldiers, with guns drawn, clomped behind them. The scientists had seen dead civilians lying in the woods; they had heard stories of Russian soldiers robbing and shooting civilians. Goethert figured it was the end.

Slowly and quietly, they walked along the narrow trail, deeper into the woods, all hope of escape gone. Goethert's thoughts were on his wife and children. He wondered how they could survive. With little identification, he speculated that no one would ever know what had happened to him. He waited for the shots.

At a small clearing, the Russians jammed their guns into the backs of their captives and signaled for them to empty their knapsacks. After looking at the sparse belongings, the soldiers pushed Goethert and Doetsch and motioned for them to get to moving.

They walked back to Berlin.

Shortly afterwards, the president of the research institute, Dr. Bock, accompanied by Russian officials, appeared at Goethert's house. They took Goethert to a villa in Schoneweide where Dr. Doetsch later joined them.

The scientists were treated well, for the Russians wanted them to join the Russians in developing what they hoped would be the most advanced Air Force in the world. They promised the men that they would live very well in Russia. Separately, the three scientists were interrogated often and at odd hours. While they stalled, confident that the Russians would not accept no to their offer, the Germans surrendered, and an armistice went into effect. Amid the jubilant celebration, Goethert and Doetsch walked away from the villa.

Shortly thereafter, Goethert again started his long trek toward Diemarden, which was occupied by Americans, and although he did not know it, toward

Middle Tennessee where, some day, he would insist that a space institute should be built on the shores of **Woods Reservoir,** which in 1945 did not even exist.

The British government also coveted the expertise of the scientists, but the Americans had priority and so Goethert was one of eighty-eight German scientists who got their first glimpse of the Statue of Liberty from aboard the crowded steamship **Argentine,** which also was loaded with returning American servicemen.

About sixty of the Germans—men who had worked at Peenemunde with **Dr. Wernher von Braun** on various missiles—were sent to El Paso to work with the U.S. Army. A few were sent to work with the Navy near Washington. Goethert and ten other engineers and scientists were escorted by train to New York and thence to Boston. They boarded a small boat that took them to a tiny island, Fort Strong. It was here that Goethert participated in his first Thanksgiving celebration.

On December 10, they were flown to Dayton, Ohio, where Goethert was assigned as a consultant for the U.S. Air Force's test facilities. It would be two years before his wife and children would join him, and it would be seven years before they would move to **1703 Sycamore Circle,** Manchester, Tennessee.

Frank Goad Clement was a corporal, serving with the Military Police battalion at Camp Bullis, Texas, in 1945. The young lawyer from Dickson, Tennessee, had been an FBI agent for more than a year before entering the military at Fort Oglethorpe in November, 1943. In a quarter of a century, he had accumulated an impressive resume, entering Cumberland University in 1937 and Vanderbilt in 1939. He took a bar exam in 1940 before graduating from law school in 1942. Bigger things lay ahead for the personable Clement who had an unusual gift with words.

In October, 1947, Clement became general counsel for the Public Service Commission in Tennessee, and he was state commander of the American Legion in 1949 before re-entering the army the next year after the Korean conflict broke out. Along the way, the Jaycees chose Clement as "Outstanding Young Man" in Tennessee. His real destiny, though, awaited him in Nashville where, in January, 1953, he would be inaugurated for his first of three terms as governor of Tennessee—the latter two being four-year terms. In seeking his third term in 1962, Clement would pledge his support for the dreamed-of University of Tennessee Space Institute, and soon after his inauguration in January, 1963, he would come through on that promise with flying colors.

Ernest Crouch, a native of Estill Springs, Tennessee, was on the road as a pharmaceutical representative for the UpJohn Company in 1945, but he dreamed of something more, and the next year he bought a drug store in Columbia, Tennessee, and moved it to his hometown of McMinnville. A graduate of Atlanta Crowe's School of Pharmacy with a Doctor of Pharmacy degree, Crouch also held a degree from Mercer University and would receive the Honorary Doctor of Humane Letters from Southern College of Optometry. A distinguished career in the Tennessee Legislature was in his future—he would establish a record of thirty years of consecutive service as representative and/or senator. His long career would include serving as majority leader and chairman of the Finance Ways and Means Committee, and it was as a state senator that Crouch would be a driving force to make the Space Institute a reality. He would battle opponents to get state funds approved to open the Institute, and he would become the first chairman of the UTSI Support Council, receive the UT Distinguished Service Award, and remain an ally through the years.

In April, 1945, while Americans on Okinawa were fighting the last naval battle of the war, the **USS Mono Island,** a repair ship, lay at anchor in Buckner Bay. As a typhoon approached, the ship attempted to put out to sea, but instead ran aground on coral. A young Tennessee sailor from Midland in Rutherford County was aboard that ship with dreams of marrying his high school sweetheart, **Ann Beachbord** of Bell Buckle, and of becoming a dentist. Both dreams would come true, but there was no way that **Ewing J. Threet,** yeoman second class, could suspect that his neighbor and most famous patient would be **Dr. B. H. Goethert,** and that he would put forty crowns on the hard-driving scientist's teeth over the next forty years. Nor could he suspect that in less than twenty years, he and the German doctor would be partners in selling to the state of Tennessee the idea of a graduate school that would become known worldwide.

Morris L. Simon was on duty as acting managing editor of the Knoxville News-Sentinel on the afternoon of April 12, 1945, when the wires went crazy with news that President **Franklin D. Roosevelt** had died. Simon ramrodded the publication of an **"Extra"** edition and then worked all night, shaping in-depth coverage of the tragedy for the next day's publications. Justifiably proud, the young editor was surprised later to read a memo from a superior, commending other staffers with no mention of Simon's hard work and dedication. The incident only confirmed his conviction that it was time for him to pursue his dream. **Simon wanted his own newspaper.**

Born in Bristol on June 2, 1911, he had dropped out of classes at the University of Tennessee, abandoning hopes of becoming a lawyer, after his dad's store was hit hard by the depression. Walking into the newspaper office in Knoxville, he announced that he wanted to become a reporter, only to be told that they were not hiring.

"You don't understand," Simon said. "I said I want to **learn** to become a reporter." He began work with no pay.

In 1946, Simon and **J. Ralph Harris** started a twice-weekly newspaper in Tullahoma. Thus he moved into position to meet a kindred spirit, Dr. Goethert, and as publisher of The Tullahoma News and Guardian, Simon would be a catalyst in the fight to bring the space institute to reality.

Folks would tell him that his newspaper would never make it in Tullahoma, but **Morris L. Simon,** like **Bernhard H. Goethert,** sometimes had deaf ears.

Marcel Joseph K. Newman of Hannover, Germany, proposed marriage to **Mia (Mitzi) Ervens** of Aachen on their second date soon after he met her in 1929, but she wasn't sure she loved him. Besides, he was going to America for a year. She rode with him to the depot, but her mother sat between them, so they couldn't **really** talk. As they said their hurried farewells—the train was moving—Marcel pressed his ring into her hand, and Mia gave him her bracelet. The young man's year in America turned out to be a lifetime, but despite his busy schedule as a research engineer with Westinghouse Research Laboratory and his continued studies, he did not forget Mitzi. Their long-distance plans to marry were hampered, however, by the Great Depression, which had prompted the United States to ban immigrants. But Newman persisted. Before going with Westinghouse in Pittsburgh, he had lived in Kentucky—his uncle there had sponsored him for his visit to America—and State Senator **Albert (Happy) Chandler,** who three years later would become Kentucky's forty-fourth governor, came to the assistance of the young sweethearts. Mitzi arrived in New York on May 10, 1932. The next day she married Marcel Newman.

Always a scholar, eager to master new things, Marcel had a bright future ahead that would include a prestigious position with the University of Notre Dame, where he taught for eleven years, was head of the Mechanical Engineering and Industrial Engineering Departments, and director of Nuclear Engineering. Earlier—from 1936 to 1944—Newman taught at Columbia University, but by 1945 he had gone back to industry as a research engineer with Walter Kidde, Inc. He received his Ph.D. from Columbia in 1950, the year that he began his tenure at Notre Dame where he would create and guide the development of a nuclear engineering program. From 1934 until he started teaching at Columbia in 1936, Newman worked as a development engineer in the research laboratory of the Gulf Development Corporation in Pennsylvania.

Born April 3, 1903, Newman earned his diploma in engineering from the University of Hannover shortly before his mother gave him a year's trip to America. Once in the U.S., he earned his bachelor's degree in mechanical engineering in 1930 from the University of Kentucky and his master's in 1934 from the University of Pittsburgh, where he had designed a turbine nozzle.

It was 1960 before Dr. Newman met Dr. Goethert and soon worked shoulder to shoulder with him in pursuit of a space institute. He played a pivotal role in getting the Tennessee Society of Professional Engineers to throw their support behind the dream. And Newman would be there, in an administrative role and as one of the first faculty members, when finally the Institute became a reality.

Bob Young was twenty years old, studying mechanical engineering at Northwestern University in 1945, the year he took a class in heat, ventilation and air conditioning under Assistant Professor **Joel Bailey. Robert Lyle Young,** upon graduation from high school in 1943, had been picked for special officers training in the Navy's V-12 program and had attended the University of Illinois for eight months before entering Northwestern. He had never heard of Tullahoma and had no special interest in the University of Tennessee. He would stay on at Northwestern for his master's and Ph.D. degrees and to teach. But he and Bailey would share a destiny a few years later that would have them both making vital contributions to the establishment of the space institute and that would lead to Dr. Young's long career at UTSI. And in the process, Bob Young would wind up only a few miles from the birthplace of his great-great-grandfather **James Dryden.**

In 1945, **Robert W. (Bob) Kamm** was working for the National Advisory Committee for Aeronautics (NACA), forerunner of the National Aeronautics and

Space Administration (NASA), at Langley Laboratory in Hampton, Virginia. The Minneapolis native had gone with NACA in 1940, starting the long career for which he had prepared by earning his degree in aeronautical engineering from New York University in 1939. In the interim, he had worked fifteen hundred feet underground for twenty-five dollars a week on the "Delaware Aqueduct," a project to construct a tunnel to deliver water to New York City. That job required him to be the first to inspect the tunnel after each blast to make sure all dynamite had exploded.

Within a short time, Kamm would become deeply involved in studies that would have far-reaching consequences on Middle Tennessee and a direct influence on the Space Institute, where eventually he would serve for twenty years. Exciting roles lay ahead for **Robert W. Kamm** that year, and in each, he would be the quiet, unassuming man behind the scenes, ever avoiding the spotlight, checking to see that things went off right.

Shirley Rogers also worked for NACA that year (1945) in Cleveland, Ohio, totally unsuspecting that in a few years she, too, would be in Middle Tennessee, tangling with Dr. Goethert to prevent a planned space institute from infringing on a Girl Scout camp. One day, too, she would become **Mrs. Robert W. Kamm.**

The war years were not without intrigue, even for civilians, and the man who later would be called the "godfather of the Space Institute" experienced this mystique through his association with another dentist who lived a double life.

Dr. Frank P. Bowyer and his wife, **Doris,** were waiting at the depot in Knoxville for **Dr. Don Clawsen** to board a train for one of his frequent trips to Washington, D.C. Suddenly, Doris announced: "I've figured out what they're doing out at Oak Ridge. They're splitting the atom!"

Clawsen's expression did not change, but moments later he asked Frank to step aboard the train with him. When the two were alone, Clawsen suggested that Frank might advise his wife to refrain from spreading such wild and obviously false rumors about the **"Manhatten Project"** at Oak Ridge. Crazy stories were rampant, he said, and while Doris' comment was ridiculous, he felt that it might be better for both Doris and Frank if she wouldn't say anything else about that.

Bowyer did not know what was going on at Oak Ridge. He knew that Clawsen had experience with the Secret Service, and that dentistry had been a front for the former dean of MeHarry College in Nashville when he served in Beirut. He knew that Clawsen's presence at Oak Ridge involved more than dental care for the people working there and their families.

But these were not normal times. Bowyer himself was practicing in Knoxville not by choice but because that was where the government thought he should be. After graduating from UT Medical College in Memphis in 1939, he had spent three years in Nashville, taking advanced training as an orthodontist. He was associated there with **Dr. Oren Oliver,** one of the top men in the field. In 1942, even while Bowyer's application to join the navy ostensibly was being considered, **Paul McNutt,** head of the governmental agency on procurement and assignment, sent Bowyer to Knoxville to take over the practice of an individual who had been called to active duty. Bowyer also was designated as an assistant to Dr. Clawsen, specifically to help him set up a dental care program at Oak Ridge. (Scientists from all parts of the country were at Oak Ridge, and security officials did not want children of these employees going into Knoxville for dental work.)

A native of Tampa, Bowyer had studied predentistry at the University of Florida before entering the dental college in Memphis in 1935. His only other choice in the South at that time was in Atlanta, but that school was not affiliated with a college until 1942, when it became associated with Emory. Bowyer's association with Oak Ridge continued until the war ended in 1945. He had no direct military service during the war, but he entered the Air Force in 1952 as a major and served for two years before returning to Knoxville.

Governor Frank Clement's last official act before inaugurating **Buford Ellington** as his successor in January, 1959, was to appoint Bowyer and **Attorney Charles Lockett** to the UT Board of Trustees. It was a stroke of good luck for Goethert and his friends who envisioned a space institute. Bowyer served on the board for the next twenty-one years, and he did more than any other person to convince his fellow trustees to endorse the idea of a space institute. On February 5, 1962, it was **Frank Bowyer** who made the motion to approve locating the institute near Arnold Center.

From the time that a fellow dentist, **Ewing J. Threet,** then a state senator, introduced him to the vision of an institute and to Dr. Goethert, Bowyer burned with enthusiasm and optimism for the project. He became fast friends with Goethert, who in later years often referred to Bowyer as the "godfather of the Space Institute."

✧ ✧ ✧

In 1945, things were beginning to happen that would bring these and other key players together at a place and a time for the fulfillment of a dream that some said could never come true. The next few years would see some of the most powerful and charismatic men in the world involved in bringing America to the front in technological research and development. Bright minds in the political, scientific and military worlds would join in clearing the way for an aeronautical research center second to none and, in the not too distant future, a graduate institution unique in its mission and scope. It all would happen midway of Nashville and Chattanooga, on the Highland Rim of the Cumberland Mountains. But it would not be easy.

✦

Chapter Two

LIGHTNING STRIKES AGAIN

First, there was **Henry Harley (Hap) Arnold,** a pioneer army pilot who took flying lessons in 1911 from **Orville Wright** on his way to becoming one of the first to qualify as a military aviator. Arnold had graduated from the U.S. Military Academy in 1907 and started his military career as an infantry officer before switching to aviation. During World War I, he served as executive officer to the chief of air service; he was the first to report movements of ground troops from the air by radio. In 1938, Arnold became chief of the Army Air Corps, which Congress had established on July 2, 1926. Two years later, he was named U.S. Army deputy chief of staff for air.

On May 16, 1940, as war came closer, **President Roosevelt** authorized fifty thousand airplanes for the Army Air Corps. On June 20 of the next year, Congress set up the Army Air Forces, and Hap Arnold became chief of this new outfit. Then came the war, and in 1942, Arnold became commanding general of the Army Air Forces. In 1944, he was promoted to General of the Army. The U.S. Air Force was created as a separate department on September 18, 1947. Two years later, Arnold was named General of the Air Force, becoming the only man in American history to hold both five-star ranks. He is regarded as the "Father of the U.S. Air Force," and efforts to pinpoint the beginning of the idea that led to the Space Institute invariably lead back to this aeronautical trail-blazer.

Writing in the U.S. Air Force News Review shortly after the Institute opened, **F. Harrie Richardson,** retired Air Force colonel and the first assistant to the director of UTSI, said the concept was "born early in World War II," and that "the man with the idea" was General Arnold. Some say that **Dr. Frank Wattendorf** actually made the first proposal for such an institution; others think of it as originating with **Dr. Theodore von Karman,** and The Tullahoma News once editorially called the institute the "brainchild" of **General Leif J. Sverdrup.**

Certainly, it was General Arnold who got leaders to see that this country was lagging far behind Germany in technological achievements, something that **Charles Lindbergh** had tried to do in the late thirties. After spending some time in Germany, that famous flier tried to warn the U.S. that the Germans were re-arming, and that they were technologically ahead of the United States. Lindbergh was called "unpatriotic" and "Nazi." After the "buzz bombs" had done their damage in London during World War II, General Arnold had more success in getting his country's attention. It was at his urging that **President Harry S Truman** pushed for the "wind tunnel" bill that eventually authorized establishing the center named for General Arnold.

In November, 1944, Arnold established the Army Air Forces Scientific Advisory Group ("Group" later became "Board") to consider the future of Army Air Forces research and development. **Dr. von Karman,** one of the world's leading aeronautical scientists, was chairman. Born in Budapest, Hungary, von Karman had become a U.S. citizen in 1936. During his early years, following studies under **Dr. Ludwig Prandtl** in Goettingen, he had served as professor at the Technical University of Aachen, Germany.

In a report to General Arnold in 1945, von Karman recommended as the only way for America to catch up technologically that Arnold establish a new center for superior research and development in the field of supersonic flight. This recommendation led eventually to establishment of what became the Arnold Engineering Development Center (AEDC). This same report contained the first suggestion for an **adjunct academic university program.** Von Karman said the proposed center should be associated with a civilian academic university program for advanced education and research.

Bob Kamm, who later would work closely with von Karman in bringing the engineering center into existence, said there was one simple reason for the (AEDC) installation: "The Germans were technologically superior to us in World War II. Hap Arnold was wise enough to convince the Air Force to take drastic measures to make sure we never again were technologically inferior. He had von Karman send a team into Germany right at the end of the war and follow the front lines as they moved into Germany to capture installations and find out what they were doing. As a result, the von Karman committee came out with two very valuable reports."

The first report, **"Where We Stand,"** analyzed American and German capabilities in the astronautics field and demonstrated that America was "very far behind." In citing the technological superiority of the Germans during the war, von Karman drew particular attention to the development of the first operational jet-propelled aircraft with swept wings and of the first large high-performance rockets such as V-2. The other report, **"Toward New Horizons,"** recommended steps to take to avoid ever being caught short again, including the establishment of a new **center for superior research and development in supersonic flight.**

Kamm said it was actually Dr. Wattendorf, von Karman's chief scientific advisor, who first proposed that the new center be adjacent to a higher education institution so that employees could continue their studies at the graduate level, and faculty and staff of the academic institution could cooperate with research projects at the new center. Von Karman immediately endorsed the idea, and thus the first "blueprint" for the Space Institute surfaced.

Wattendorf, an Air Technical Service Command member of the Scientific Advisory Group, also was a member of the team that went into Germany at the end of the war, but he was called home by the death of his father. While flying over the ocean in the middle of the night, he scribbled a memorandum to his boss at Wright Field, saying that something must be done to close the technological gap. He proposed two possibilities. One was for the U.S. to take a number of German facilities, disassemble them, ship them to the United States, and then reassemble them as a nucleus for the proposed center. Next, he said, the U.S. should entice as many German scientists who had worked with the equipment as possible to help put it back together in the United States.

Both of Wattendorf's suggestions were followed. The first operating facility at AEDC—the Engine Test Facility—was made up chiefly (90 percent) of German components. The nucleus of what would become the Gas Dynamics Facility at AEDC—a set of hypersonic wind tunnels—was transported from Germany. And, as Bob Kamm noted, a number of German scientists, including **Dr. B.H. Goethert,** were "talked into coming to this country."

The foundation for what would become the Space Institute was being laid with the proposal for an Air Force center, but that center was a long way from reality, and it was anyone's guess as to where it would be built. There also was opposition to the proposal.

"The Air Force met a lot of opposition to the idea of a center," Kamm recalled. "NACA thought they were the experts and that the Air Force didn't know what it was doing." Other branches of the service, especially the Navy, felt they were better qualified to operate such a facility.

In June, 1946, after taking competitive bids, the Air Force awarded a contract to Sverdrup & Parcel of St. Louis (headed by General Sverdrup) to "develop a master plan and conduct site surveys" for the proposed center. Oregon and Washington quickly emerged as possible sites, and S&P eventually recommended that the air center be located at Moses Lake, Washington—a long way from Tullahoma, Tennessee.

In March, 1948, **Stuart Symington,** the first Secretary of the Air Force, and **General Carl Spaatz,** Air Force chief of staff, were briefed on plans for the installation. They rejected the Moses Lake site because of "strategic reasons"—it was too near the Pacific, too vulnerable, so it was back to the drawing board as a search began for an inland site.

"This is the first time that Camp Forrest was seriously considered," said Kamm, who in 1948 had begun working in the Pentagon to review the Air Force proposal for the center. He was director, panel on facilities, committee on aeronautics, of the Research and Development Board, a part of the Department of Defense. This was when Kamm first met Dr. von Karman; later, he would become the doctor's staff assistant.

Camp Forrest was more than a place; it was the **event** that brought World War II to south central Tennessee and transformed Tullahoma in 1940 from a town of about four thousand people to a crowded city of seventy-five thousand—a place crawling with construction workers, soldiers, and their dependents. It was where **George S. Patton,** then a major general, perfected many tactical operations of his armored units that later would prove successful in North Africa and Europe. (In June, 1941, Central High School in Manchester served as headquarters for the 2nd Army's training maneuvers. General Patton and **Henry Stimson,** Secretary of War, both reportedly spent at least one night there during this time.)

While Camp Forrest was being built on one side of Tullahoma, the U.S. government opened a five million dollar base for training B-24 bomber crews on the opposite side, at Northern Field, so that Tullahomans found themselves surrounded by the military. Even earlier, the Camp Forrest site—or a portion of it—had accommodated 2,500 troops. In 1926, the state had established a training camp for Tennessee National Guard units on 1,040 acres outside Tullahoma provided by the Nashville, Chattanooga, and St. Louis Railway Company. Named Camp Peay in honor of the late **Governor Austin Peay,** the site was used for the Guard's annual maneuvers.

Late in 1940, as the United States began to take note of the warning signs, the government picked Camp Peay as location for a training center. The original site was eventually expanded to 43,662 acres, bringing to about 85,000 acres, counting nearby artillery ranges, the acreage in the area used for military training. In 1941, the camp was renamed Camp Forrest in honor of **General Nathan Bedford Forrest.**

Troops began pouring into the area in 1941 as reserve and National Guard units were called to active duty. **Don B. Campbell** suddenly found himself as mayor of a city (Tullahoma) bursting at the seams, with thousands seeking places to live. Manchester, where **G. Bowlin Morton** was mayor, Winchester and other neighboring communities also were affected by the influx of soldiers from throughout the United States.

Al Jolson was among entertainers who performed for the Camp Forrest troops, and **President Franklin D. Roosevelt** included the facility in his tour of military installations. It would be a few years before another president would come to the area on a different mission.

Initially, Camp Forrest served as an infantry division training center with soldiers getting as near to real combat as possible before being shipped overseas. Late in the war, it was used for training by the 17th Airborne Division. Starting in May, 1942, it also was used as a center for interning "alien" civilians who had been rounded up at the start of the war. Before the intern center was phased out in May, 1943, the number of aliens interned at Camp Forrest rose from one hundred twenty-four to eight hundred sixty-eight.

In late July, 1943, a prisoner of war center was established at Camp Forrest, and more than twenty thousand prisoners were confined there before the war ended. Many were put to work on various construction jobs; prisoners of war built what became the Arnold Center golf course. The POW station at Camp Forrest also administered thirteen other branch POW camps located elsewhere in Tennessee and in Alabama, Georgia and North and South Carolina. Finally, the station hospital at Camp Forrest was unexpectedly made one of the two general hospitals for prisoners of war in the United States, and its patient load jumped to more than two thousand.

More than seventy thousand soldiers passed through Camp Forrest before the war's close in 1945. In September of that year, a memorandum prepared by the Office of the Chief of Engineers for the War Department concluded that Camp Forrest might be kept open for training activities, but it estimated that this would involve spending about $18 million to replace barracks with permanent buildings. Most of the reservation was leased from the state of Tennessee, the memo noted, and only a small part was owned by the federal government. The decision was made to shut the facility down. Northern Field—named for **Lt. William L. (Billy) Northern, a World War II pilot and the first Tennessean in the Army Air Corps to lose his life in that war**—also was closed as a military installation.

John W. Harton, a prominent Tullahoma businessman who at twenty-seven had been Tullahoma's youngest mayor, borrowed money and bought the field, and a civilian airport began operations there. Harton had served as state campaign director for one of **Prentice Cooper's** successful campaigns for governor and also served as state treasurer from 1939 to 1943. Harton was a close personal friend of **Senator Kenneth D. McKellar,** one of the most powerful congressmen of that era. Harton also would make the barracks at Northern Field available as their headquarters when personnel moved into Tullahoma a few years later to build the Air Force's aeronautical research and development center.

Nineteen hundred and forty-nine was the year that lightning would strike the area for the second time. **Harry S Truman,** who had been thrust into the presidency at the death of **Franklin D. Roosevelt** on April 12, 1945, began a new term in the White House in January. **Gordon Browning,** in 1948, with assistance from **Estes Kefauver,** had broken the powerful alliance between Memphis boss **E.H. Crump** and **Senator Kenneth D. McKellar.** Now Browning was in his second term as governor. He would win one more two-year term before falling to the young silver-tongued orator from Dickson, **Frank G. Clement,** in 1952.

McKellar, who in 1916 had become the state's first popularly elected senator, was one of the strongest members of the United States Congress. Eighty years old, he had a reputation of getting what he wanted. In his sixth six-year term, he was chairman of the Senate Appropriations Committee. He would try for a seventh term in 1952, but **Albert Gore Sr.,** congressman from Carthage (and father of the future U.S. senator who ran for the Democratic nomination for president in 1988), would send him into retirement. In 1949, though, the crusty old politician, McKellar, carried plenty of clout as he would demonstrate when things got down to the wire about where to locate the Air Force's new center.

On September 21, 1949, a special committee of the Scientific Advisory Board, chaired by **Louis N. Ridenour,** completed a report on Air Force research and development for **General Hoyt S. Vandenberg,** Air Force chief of staff. The Ridenour committee, of which Dr. Wattendorf was a member, recommended a speed-up in making the center—which had first been proposed in 1945—a reality. The committee also recommended that the Air Force seek legislation empowering it **to award a modest number of pre-doctoral and post-doctoral fellowships to highly qualified students.** It also proposed that a small fraction of the Air Force's research and development budget be consistently assigned for making **contracts with educational institutions** for fundamental research.

On October 27, 1949, the 81st Congress authorized establishment of the new Air Force center, called then the Air Engineering Development Center, by passing the Unitary Wind Tunnel Bill. President Truman quickly signed the measure. Bob Kamm said that even after this legislation was passed, the Air Force still would not say where the center should be built, but it was decided that the best location would be in Huntsville, Alabama, not in Tennessee.

W.R. Davidson, McKellar's administrative assistant, remembered that Hap Arnold told Truman "we won this war, but we will lose the next one because we are so far behind." A major aeronautical testing facility was needed to correct the situation, Arnold said. Truman then had a ("wind tunnel") bill drawn, but "nobody paid any attention to it" for awhile, Davidson said. One day **Lou Barringer,** a political savvy cotton baron from Memphis, was visiting with Truman's secretary, when the President entered the room and asked Lou to "see if you can get Mac to sponsor this bill."

Lou said McKellar's first question would be "Where?"

"Truman said 'Anywhere he wants it,' " Davidson said, and the senator successfully sponsored the bill.

Davidson, a later resident of Tullahoma, had sold his dry-cleaning business in Lawrenceburg when he thought he was entering the military service during World War II, but doctors turned him down and he went to Washington, D.C. Davidson was working as a doorman in the Senate chambers and as a night clerk at the Willard Hotel. One day **Representative Tom Stewart** of Winchester told Davidson that McKellar wanted to talk with him.

"The senator told me I didn't need to work two jobs, that he would give me one that would make this unnecessary," Davidson said. "He didn't tell me I would be working day **and** night as his administrative assistant." Davidson took over the job that had been held by McKellar's late brother, **Don.**

In November, 1988, **Bob Kamm** would still remember a day in the fall of 1949 when he attended a meeting in Senator McKellar's Washington office. **Davidson** also was there. Here is how Kamm remembered it:

"The Air Force told McKellar that Huntsville was the best location. TVA provided power for the Air Force at lower rates; transmission lines were already in place at Huntsville, they said. **McKellar went through the roof! He said, 'If you want a center, it will be in Tennessee, not in Huntsville!' "**

And so it was.

(Davidson remembered another day when McKellar confronted TVA Chairman **Gordon Clapp** about TVA's claims that electrical power could be provided more cheaply in Alabama. Angry, McKellar grabbed his cane from a mantel in his office and directed it toward Clapp's head, according to Davidson, who "pushed him (Clapp) aside" just before the cane found its target.

Rumors and speculation had buzzed for months in the Tullahoma area, but it was in late fall of 1949 before everyone knew for sure. The announcement came on Wednesday, November 9. **Stuart Symington** and **Senators McKellar** and **Kefauver** said the Air Engineering Center would be constructed on the Camp Forrest site. There was rejoicing throughout the area. In Tullahoma, whistles blew, bells rang, and an impromptu celebration was held at City Hall. Two days later, on Armistice Day, The Tullahoma News (which was published twice a week then), in a two-line, box-car streamer headline, reported the news: **"WORK ON AIR CENTER HERE WILL START IN TWO MONTHS."**

Smaller headlines advised that John Harton was offering free use of the airport by the air center officials, the City Board would plan expansion of city services, and Tullahoma Mayor **Jack T. Farrar,** a medical doctor, pledged that the city would "meet the needs."

"It will put a burden on existing facilities," he said, "but the board will make plans to take care of the need, whatever it may be."

An Air Force spokesman said major construction work would start within two or three months. Expected to cost more than a billion dollars, the center would be used to test supersonic planes and guided missiles and to conduct research on advanced aircraft. The Air Force said the center would employ about 1,000 to 1,500 permanent employees initially, gradually increasing to about 35,000 when in full operation.

Farrar, Harton, and **A.H. Sanders,** superintendent of the city utilities system, had camped out in Washington before the announcement, keeping in daily touch with the Tennessee congressmen, eagerly waiting and watching for the decision that would change things forever back home.

Major General Franklin O. Carroll was appointed as the first commander of the center. He assumed his duties on January 1, 1950. He asked **Bob Kamm** to join his staff with his initial assignment to work as a staff assistant to von Karman in the Pentagon. Kamm preferred working in Washington. He had been to Tullahoma, but he did not want to move at that time. He quickly became involved in coordinating plans and specifications for the new center with NACA, the Air Force, the Research and Development Board, and other agencies. He was doing what Kamm always seemed to do best—working behind the scenes to get the ducks lined up neat and orderly. He would move to Manchester the next year and, in 1956, to **Sycamore Circle.**

While plans moved ahead on the center, the next few years also would find various efforts being made to establish some sort of **academic institution** to accommodate personnel of the center. These included ambitious efforts to establish a **von Karman Institute.** Success of sorts would come in six years with the start of the University of Tennessee's graduate program at AEDC.

Several significant things happened in 1950. On June 1, **Dr. Wattendorf,** in his capacity as deputy scientific advisor of the Air Engineering Development Division, signed a "memorandum for the record" recommending that AEDC have a complementary **educational institution** to attract scientific personnel. He suggested the name "Arnold Research Institute" and recommended that it be near the center. He visualized this institute being organized and operated by an association of universities similar to the Institute of Nuclear Study at Oak Ridge, site of the Atomic Energy Commission's laboratories. Wattendorf said such an institute could be sponsored and promoted by the Air Force, or it might be proposed and promoted independently by a private or state foundation with the Air Force's encouragement.

Air Force historians say that it was less than two months after this proposal that the Air Force Chief of Staff established the Industry and Educational Advisory Board. Initial members representing the Aircraft Industries Association were **Dr. William Bollay,** North American Aviation Company, Inc., **John Buckwalter,** Douglass Aircraft Co., Inc., **Major General E.M. Powers,** Curtiss-Wright Corp., **W.A. Parkins,** Pratt & Whitney Aircraft Corp., and **Archie Colwell,** vice president of Thompson Products Inc. in Cleveland. Von Karman nominated the first two university representatives: **Professor John R. Markham,** Massachusetts Institute of Technology, and **Dr. C.E. Brehm,** president, University of Tennessee.

Kamm said creation of the Industry and Educational Advisory Board was one of several steps taken to demonstrate the Air Force's capability to handle the job of getting the new center into operation. Kamm, who was executive secretary of the board, also remembered two very positive things that came from it. Members advised dropping a proposal for a large, expensive dynamometer for propeller development in view of the development of jet engines that would make propellers less important to the military. This saved money. Also, within two years after the board was formed, two members—Archie Colwell and **Dr. Cy Ramo,** vice president of Hughes Aircraft—formed the Thompson-Ramo company. A year later, it became the Thompson-Ramo-Woolridge Company, one of the major aerospace companies in the 1980's.

The Air Force also assigned its brightest young technical officers to the center. Kamm remembered among this "cream of the crop" such men as **Barney Marschner, Jack Dodge, Kit Carson, Walter Moe** and **Dan Boone.** A few years later, Second Lieutenant **Bob Leland** (later to become president of Calspan Corporation) would work for Marschner.

Another bold step by the Air Force was its decision to go to an operating contractor. A big reason for doing this was that it was felt that an operating contractor would be more flexible regarding salaries and other benefits and thus could attract competent people, which at that time was considered to be a major challenge.

Kamm vividly remembered that when von Karman's Scientific Advisory Board first visited Tullahoma, members reported that **"this is an intellectual desert** . . . one of the most difficult jobs facing the Air Force will be to assemble a competent crew and convince them to live in this part of the country."

While it is not widely known, the Arnold Research Organization Inc., or "ARO" as it became known during the years that it operated the center, came into being through pressure and pleading by the Secretary of the Air Force. According to Kamm, General Carroll approached General Sverdrup in St. Louis about organizing such a company. Sverdrup balked. He didn't want to take on the responsibility, but **Stuart Symington** called Sverdrup in and pleaded with him, arguing that it was his "patriotic duty." Sverdrup acquiesced, and ARO, Inc. was created as a subsidiary of S&P for the specific task of operating the Air Force's new center. Sverdrup was the first president of ARO.

General "Hap" Arnold died in 1950, and the Air Force announced that the center would be named: **Arnold Engineering Development Center.** ARO was to operate AEDC under contract to the Air Force on a cost-plus-fixed-fee basis. This plan hit a snag in 1952 when **Albert Gore Sr.,** running for the Senate, made the fee an issue, and a rider was attached to the 1953 military appropriations bill, barring any payments to ARO. S&P operated the center on a non-fee basis for a few months while a congressional investigation was under way, but ARO resumed operation of the facility on August 1, 1953.

In the early days of preparing to open Arnold Center, the Air Force got the Army to establish a separate Corps of Engineers district solely for the mammoth construction project. Kamm said the contractor also did an excellent job of recruiting some competent young people and trained them for a year or two in St. Louis and at governmental laboratories such as those at NACA before AEDC facilities were operational. (One of these young men was **Robert E. Smith Jr.** who was among the small group of "recruits" who went to work at Cleveland, Ohio, in preparation for a career at AEDC. Along the way, too, he earned a master's degree in the UT graduate program at Arnold Center.)

It was while working with von Karman in 1950 that Bob Kamm first met Dr. Goethert—a meeting that marked the start of a long friendship and professional association.

"Von Karman asked me to organize a big conference on transonic wind tunnel testing techniques," Kamm recalled in 1988. "We knew one of the major facilities would be a large transonic tunnel. This was a new state-of-the-art concept. He said, **'Be sure to get that German by the name of Gerdit** or something. I think he is working at Wright Field.' I tracked him (Goethert) down and got him on the telephone. After not being able to communicate very clearly, I finally got him to say that he would be there the next Tuesday afternoon at two o'clock, so I went up there (to Wright Field), sat down with him and explained face to face what I wanted. He did attend the meeting at St. Louis. This is how we met, and we got to be very good friends and worked together until he died."

Dr. Goethert's expertise in wind tunnels proved very helpful at that point and likely figured into a change in plans that allowed the Air Force to build smaller wind tunnels than earlier planned. At that time, there were two concepts as to how to design transonic wind tunnels. The one held at NACA involved slotted walls with suction. Goethert's concept did the same, but he used porous walls with drilled holes. Either method allowed for smaller test sections, and plans were changed to build them 16 by 16 feet in size rather than 40 by 40 feet.

Thousands of people swarmed into Tullahoma on June 25, 1951, and packed the spectator sections on the Arnold Center reservation. It was hot and humid, but history was in the making.

(It was the first anniversary of the invasion of South Korea by North Koreans. On this same day in June nine years before, the British Air Force had staged a thousand-bomb raid on Breman, Germany. And on this very day in 1951, the first commercial color telecast took place as CBS transmitted a one-hour special from New York to four other cities.)

But for Tullahomans, it was an historic day because **President Harry S Truman** had come to dedicate Arnold Engineering Development Center. Governor Gordon Browning, Tullahoma Mayor Jack T. Farrar, General Carroll and other dignitaries met Truman's plane at Northern Field and traveled in a motorcade through Tullahoma and on to AEDC. Folks waved and cheered, and a Tullahoma musician, **Mrs. Alan Gray Campbell,** wrote a special song of welcome for the president.

At the shipping and receiving warehouse, a speaker's platform was set up; tables were in place for the press, and areas were roped off for spectators to stand for the program. Governor Browning formally presented title to the former Camp Forrest property to President Truman, who—with General Hap Arnold's widow seated next to him—formally dedicated the center.

Americans were dying in Korea in the conflict that had started exactly a year before when communist troops overran the border into South Korea, and the United Nations Security Council, meeting in emergency session, issued a cease-fire call that went unheeded. The UN authorized the United States to organize armed forces—the first time a world organization's troops had acted as police. On June 27, 1950, Truman had ordered U.S. forces to protect South Korea, and a couple of days later, he ordered American troops in, the start of thirty-seven months of what has been called "one of the bloodiest" periods of fighting ever.

During his dedicatory speech, President Truman repeated the United States' willingness to join in a meaningful peaceful settlement of the Korean conflict: "In Korea and in the rest of the world, we must be ready to take any steps which truly advance toward world peace. But we must avoid like the plague rash actions which would take unnecessary risks of world war or weak actions which would reward aggression."

Actually, the truce would come on July 27, 1953, after the president from Missouri had fired **General Douglas MacArthur** as UN commander in Korea, and after another general, **Dwight D. Eisenhower,** had become president of the United States.

It was on June 25, 1951, too, that **Bob Smith,** newly married and fresh out of Vanderbilt University with a bachelor's degree in mechanical engineering, began working for ARO, Inc., as a trainee in Cleveland.

"I was very fortunate," Smith recalled on May 31, 1989, in a brief interview in front of the Manchester post office on the morning after he had returned from an AGARD meeting in Portugal. At this time he was vice president and chief scientist of Sverdrup Technology, Inc. and a representative to NATO.

Smith had grown up in Pulaski, and he met **Beverly Patterson** (who had left Pulaski at age four) when she returned to attend Martin College there. They were married in 1951. After a year in Cleveland, Smith returned to Middle Tennessee, residing in Manchester and working at AEDC.

"When I came to ARO, we (the group that had been farmed out by ARO) had fifty years experience," he said, explaining that fifty people had been trained with industry for a year. "They took me in, and treated me well from the start."

When UT began its graduate program at AEDC in 1956, Smith signed up immediately, attending classes part-time while working full time, and six years later received a master's degree in mechanical engineering. He praised the quality of instruction he received, particularly singling out one of his professors: "Dr. (Robert L.) Young had a knack for phasing questions on his tests so that if you knew the principle, you could work it (the problem) in a couple of hours," he said. "If you didn't, you could never work it."

Smith became a member of the Manchester School Board in 1960, serving for eleven years, much of this time as chairman.

The government eventually acquired more land in addition to the Camp Forrest property on which to locate wind tunnels and other test chambers and to make room for a four-thousand-acre lake that was impounded on Elk River to provide a source for cooling water for the wind tunnels. The lake was named "Woods Reservoir" in honor of **Colonel Lebbeus B. Woods,** a graduate of the U.S. Military Academy, Class of 1925, and AEDC's first deputy chief of staff, materiel. Woods died on May 22, 1955.

The first major construction project, the shipping and receiving warehouse, was formally transferred to ARO in the fall of 1951. **General Thomas E. Farrell,** a key official on the Atomic Energy Commission, was named as ARO's first managing director.

In October, 1952, ARO first achieved supersonic flow in the transonic model tunnel of the propulsion wind tunnel facility at AEDC. The first unit of the gas dynamic facility, which (at the urging of **Bob Kamm**) later would be named for **Dr. von Karman,** went into operation in 1953, the year that shakedown testing began in the engine test facility.

Before Arnold Center was operating, efforts were being made to **affiliate AEDC with universities.** The Industry and Educational Advisory Board, in meetings on May 11 and 12, 1951, considered drafts of a proposed contract with the University of Tennessee and recommended that a program be developed to provide qualified AEDC personnel university level courses leading to advanced degrees in physical and engineering sciences and "instruction in such other fields as circumstances may justify."

On June 19, 1951, the Air Force awarded a letter contract to UT to study the board's recommendations. Under direction of **Dr. Wiley Thomas,** UT completed the study and submitted a final report on December 1, 1952. Included in this report was a recommendation by the Graduate Study Committee that a program of **residence graduate courses and degrees** for AEDC employees be established. The study also recommended lecture and symposia programs to put center personnel in contact with leaders of aviation, industry and science. Most significantly, the UT study recommended that an **Institute of Flight Sciences** be established at AEDC. The institute should "engage the human and institutional resources of the entire free world and should foster especially close ties with neighboring Southern educational institutions." The report noted that such an institute would be a "very natural extension of the proposed graduate degree program," and projected that it "might well develop in an organic way from these programs." Citing programs at Princeton and the University of California, the proposal also stressed that "Germany owed much of its pre-Hitlerian eminence in science and technology to the various institutes in which teaching and research were combined."

The proposal rested on the expectation that activities of the institute would become a valuable factor in attracting and retaining scientific personnel required for the full development of AEDC.

Dr. Richard Courant, director of the Institute for Mathematics and Mechanics at New York University and a member of the ad hoc Research Study Committee of the UT study, suggested that the graduate degree program being considered might well have an **institute for advanced study** in the **flight sciences** as its ultimate goal.

Such efforts would continue for the next three or four years, and it would be 1956 before the University of Tennessee would offer graduate classes at AEDC.

Meanwhile, in 1952, **Dr. B.H. Goethert** moved to Manchester and became chief of the propulsion wind tunnel at AEDC. He had been a consultant for the U.S. Air Force's test facilities in Dayton from 1945 to 1949, and from 1949 until he moved to Tennessee, he was chief of wind tunnel test activities at Dayton.

Also, in 1952, **Frank G. Clement,** a blue-eyed, scripture-quoting lawyer from Dickson, polled 302,491 votes in the Democratic primary on August 7 to **Gordon Browning's** 245,166, on his way to becoming at thirty-two the youngest governor in Tennessee's history. Congressman **Albert Gore Sr.** also soundly defeated **Senator McKellar,** sending the old warhorse into retirement. This also was the year that the Tennessee senator with the coonskin cap—**Estes Kefauver** from Chatanooga—made a serious bid for the Democratic presidential nomination. Kefauver, a UT graduate, had entered the Senate in 1948, unseating **Tom Stewart,** a former district attorney general from **Winchester.** At the Chicago convention, Kefauver lost his bid for the nomination to **Adlai Stevenson,** and on November 4, **Dwight D. Eisenhower** won the election for president.

While the recommended institute was being scrutinized in 1953 at the Air Research and Development Command Headquarters and in the Pentagon, **Colonel Harrie Richardson,** at that time Arnold Center chief of staff, was looking into the possibility of establishing a local institute for advanced studies in the propulsion sciences. Directing this effort was **Lieutenant General Donald C. Putt,** commanding general of the Air Research and Development Command.

Richardson's efforts continued into 1955 when a formal proposal was drawn up for the Air Force to establish adjacent to AEDC a **research institute** in the "furtherance of the work of Theodore von Karman." Under this proposal, the institute would offer no degrees; activity would be at the post-doctorate level even though the Ph.D. would not be a necessary requisite. AEDC test units would be available for specific research programs. The Air Force would have granted two hundred and forty acres, and the site would have bordered Woods Reservoir. The estimated cost was five million dollars with a yearly operating cost of about six hundred thousand dollars, and von Karman would have been the first director. On August 1, 1955, at a meeting at the Sverdrup & Parcel home office in St. Louis, General Sverdrup agreed to support the proposal. Others at that meeting were Colonel Richardson, **Major General Samuel R. Harris,** then AEDC commander, and **W.E. Moser,** secretary and general counsel for ARO, Inc.

That fall, **Dr. J. Robert Oppenheimer,** director of the Princeton Institute for Advanced Study, was queried on the proposal for what now was called the von Karman Institute. He threw cold water on the proposal, saying that AEDC lacked adequate library facilities, that the center was too far from the main center of learning to attract the steady flow of visitors needed, and that the area around AEDC did not have suitable family facilities. He recommended either Pasadena, California, or Cambridge, Massachusetts, as sites for the proposed institute. Richardson quickly responded that Tullahoma was only about thirty minutes further by air from Washington than Princeton was and that, because of AEDC, Tullahoma was already the world's most important center for aeronautical engineering.

On February 27, 1956, **Dr. von Karman, General Harris,** and **General Sverdrup** met at Pasadena, and von Karman agreed to locating the institute at Arnold Center. But by April 4, when Harris, Sverdrup, Wattendorf and von Karman met in a Washington hotel, von Karman obviously had cooled toward the idea. Efforts to get financing from such private sources as the Guggenheim and Ford Foundations failed.

Dr. Goethert, writing about the early days, referred to UT's in-depth study on the educational needs of AEDC: "With participation of several universities, the Air Force and contractor organizations, a number of committees were established to study various opportunities and develop specific recommendations.'' Sixty-seven people served on these committees. Dr. Goethert called attention specifically to **Dr. E.A. Waters,** dean of the UT Graduate School, **Dr. John Wild,** representing ARO and AEDC, **Colonel Richardson**, chief of staff at AEDC, and **Bob Kamm,** representing the Air Force and AEDC.

When **UT President C.E. Brehm** submitted the final report on the AEDC university affiliation study on December 1, 1952, he prefaced it with excerpts from the Ridenour Report that stressed the significance of technical scientific education. The report's key recommendation was to establish a **graduate education** program for military and civilian personnel of Arnold Center. Other recommendations called for close cooperation between the center and universities in research and in establishing a public lecture program. Efforts to organize a consortium of universities to participate failed so that when the graduate program actually began in 1956, UT was the only participating university.

On June 14, 1956, the Air Force signed a contract with UT for a graduate study program at AEDC, staffed and administered by UT. It would offer a master of science degree in selected fields of engineering. **Dr. Joel F. Bailey,** head of mechanical engineering at UT, was put in charge of the program when it began in September under the overall supervision of the UT Graduate School in Knoxville with **Dr. Dale Wantling** as its head. Wantling soon was succeeded by **Dr. Hilton A. Smith.**

Dr. Robert L. Young, who would be lured to the area because of the graduate program, believed that the initial emphasis for the graduate study came from ARO through **General Sverdrup.**

''Clearly, the Air Force was also interested because initially there were very big plans for AEDC, which would have involved moving most of the U.S. Air Force research activities to AEDC,'' Young said.

During the five-year effort to get the graduate program established, backers sought to have several southeastern universities—including UT, Vanderbilt, Georgia Tech, Kentucky, Alabama and others—make resident programs available at AEDC. **Dr. Wiley Thomas,** director of the UT Engineering College Co-Op program, took up residence in Estill Springs in Franklin County while attempting to get the multi-university concept established. The universities did not reach common ground, according to Young, ''perhaps because of differing academic requirements and maybe lack of interest.''

Kamm agreed that the Air Force was interested in establishing a relationship with the university community and at first a consortium of neighboring universities was thought to be an ideal approach. However, within six months or a year after the study, Kamm said, it became obvious that a consortium was not feasible, and ''we decided to ask UT to establish a master's level program solely for AEDC employees at AEDC.''

Major General Samuel R. Harris was AEDC commander when negotiations first started for the graduate school. Persons involved in the study worked out of a big hangar at Northern Field. The main administration building and warehouse were open, but most of the buildings at Arnold Center were only under construction.

The UT program to award graduate degrees to engineers on the staff of Air Force and ARO at AEDC was approved in Washington on May 3, 1956. (The next week, one hundred scientists and engineers from the nation's top laboratories and industrial establishments attended a two-day meeting at Arnold Center of the American Ordnance Association, guided missiles branch.)

Twenty-two master's degrees would be earned through the program over the next eight years, but the real significance of the graduate program was that it would form the nucleus for the realization of the dream for a space institute. And opportunity was about to knock on the door of an unsuspecting assistant professor at Northwestern University and entice him to move center stage where he would play a major role in making the Space Institute work for the next twenty-five years.

✦

Chapter Three

SOMETHING TO BUILD ON

BOB YOUNG was teaching at Northwestern in 1956. His friend, **Joel Bailey,** had left around 1950 to join UT's staff, and **Dr. Pietro (Pete) Pasqua,** who had been a fellow graduate student with Young at Northwestern, had taken a teaching position in Dr. Bailey's Mechanical Engineering Department at UT two years later.

Early in the fifties, Young and his wife **Phyllis** visited Pasqua and Bailey in Knoxville, golfed, caught fish, and enjoyed the excellent spring weather. The Youngs then drove to Atlanta and came back down a "tremendous mountain" (Monteagle) and spent the night at the Tennessee Motel in Manchester without suspecting what lay in store for them nearby a few years later.

At Bailey's suggestion, **Dr. Nathan W. Dougherty,** engineering dean at UT, called Young in June of 1956 and discussed the possibility of Young's heading up the new graduate program coming up at AEDC in the fall.

"I did not know where Tullahoma was," Young recalled many years later, "but after looking it up on the map, I told him that I was not interested."

Later that summer, Dougherty asked that either Bailey or Pasqua start and head up the program at AEDC. Both men had young children in school at Knoxville and did not wish to move, but they asked the dean to make a choice and agreed that the one chosen would go. Young explained it this way: "Joel thought that Pete would be picked because he felt that Dean Dougherty would not want to lose the services of his ME department head for at least a year. Pete thought that Joel would be picked because Dean Dougherty would want the best. So they both volunteered, and Joel lost, and Dean Dougherty appointed Pete as acting head of mechanical engineering."

Bailey employed **Miss Virginia Richardson** as chief secretary and registrar, secured the services of **Dr. Charles Bruton** of the University of the South at Sewanee to teach math, and got **J. Leith Potter,** an employee of ARO at AEDC, to teach aerodynamics. Classes opened with Bailey teaching thermodynamics, Bruton teaching advanced calculus, and Potter aerodynamics. The Air Force had predicted an enrollment of about four hundred; UT had expected less than forty. Both were wrong. Seventy-one students registered for those first classes, including several ARO employees who would gain high positions at AEDC.

By the spring of 1957, Bob Young was restless in his post as assistant professor of mechanical engineering at Northwestern and advisor of all of the mechanical and industrial engineering students in the department. He had received his undergraduate and master's degrees there and, in 1953, his Ph.D., and he felt that he was still looked upon largely as a student. He went job hunting, visiting Michigan State, Wisconsin and Illinois. The University of Illinois offered him a job as assistant professor at a salary of $8,000 for nine months, and he was tempted. He knew **Dr. Norman Parker,** head of the department, and other faculty members there, and it was close to home. Then his former professor and friend, **Joel Bailey,** called, urging Young to visit Knoxville to talk about the opening that existed due to Bailey's being at AEDC.

"I visited Knoxville in March, 1957," Young said. "The weather was beautiful." At Bailey's insistence, the next month Young flew to Nashville and drove with Bailey to Tullahoma for "a very interesting" visit at Arnold Center. Soon afterward, UT offered Young a choice of either going to the mechanical engineering department in Knoxville or replacing Bailey at AEDC. Either offer involved more pay than Young had been considering at Illinois, and the AEDC post paid better than the one on the

Knoxville campus. Young was fascinated by the new air center and accepted that offer for "two years." ("When I returned to Chicago, it had snowed at Midway Airport, and I barely made it home up beautiful Cicero Avenue.")

Then they learned that Phyllis was pregnant.

Young did some soul-searching about moving. Phyllis had a fine doctor at Evanston who had delivered their first son, **Ronnie.** But Young felt it would be unfair to back down on his promise to UT.

"So in early September, I and my family showed up in Tullahoma in our 1955 Studebaker and rented a house—Joel had helped us find it, and it was a beautiful place. Phyllis soon found a good doctor (**Dr. Winfred Wiser,** who practiced with **Dr. Ralph Brickell**) , and he delivered **Scott Allen** in February."

Young assumed duties as director of the graduate program for the fall quarter of 1957, and Bailey stayed to help him get it under way. In December, Bailey returned to Knoxville, leaving **Young, Miss Richardson, Potter** and **Dr. Victor Mizell,** a math professor. **Dr. Martin Grabau** also joined the program that quarter.

Officials at first saw no need for electrical engineering classes, Young remembered, "but then immediately received a petition from AEDC electrical engineers to put in electrical engineering. I was lucky to find Dr. Grabau to teach a course in control theory that fall quarter."

Young taught and did research in hypersonics and low density gas dynamics. The space chamber "business" came in shortly afterwards, and Young "was able to get in early on that." He also did "a little" administration, which he said was never pressing. "I would visit Knoxville maybe once a quarter. I had a most pleasant relationship with **Dean Armour T. Granger,** who had succeeded Dougherty as dean of engineering, and with **Dean Wantling,** who was dean at the graduate school, succeeding **Dr. Waters** shortly after Joel started the program. Wantling was an education major and not a particularly forceful person; I got along great with him and got a lot of assistance from **Jane Winter** in the graduate office (later in the College of Liberal Arts). The program grew fairly rapidly, and we granted the first degree in 1959, and by 1961, we had a large enrollment."

Actually, **two** master's degrees in mechanical engineering were awarded in March, 1959, and two in the fall of that year. First degrees went to **Max Kinslow** of Manchester and to **Clark Lewis** of Tullahoma. In the fall, **Michael Pindzola** and **Kenneth Tempelmeyer** also received master's degrees in mechanical engineering. (Lewis, in June, 1968, would become the first UTSI student to earn a doctorate in **engineering science,** with Dr. Young as his professor.)

Bailey "much enjoyed his fifteen months in Tullahoma and got the program off to an excellent start," Young said, and Dr. Waters was "most instrumental" in setting up the program. Young believed one thing that largely accounted for the early success of the graduate program was that no residency on the Knoxville campus was required of students working on master's degrees in certain areas at AEDC.

"This was a very progressive move," he said, contrasting it with Huntsville where the University of Alabama for many years **did** require a period of residence on the main campus at Tuscaloosa. Although Huntsville, with NASA and the Army closeby, had the potential for many students, Young believed that their graduate program did not prosper because of the residency rule. Getting the residency requirement waived for the program was a long hard fight, according to Bob Kamm. **Dr. John Wild,** director of engineering at AEDC, representing ARO, and Kamm, representing the Air Force, worked with UT in preparing for the graduate program. Kamm said plans "got serious" in 1954, but it took nearly two years to get the residency requirement waived. ARO and the Air Force took the position that they couldn't afford to send their employees to Knoxville for a quarter. Ultimately, it was Dr. Waters who authorized the waiver. Residency requirements would return to haunt Space Institute officials in the Ph.D. program a few years later.

With **Bob Young** on deck as director of the program in the fall of 1957, **Joel Bailey** returned to UT in December, and **Pasqua** became head of the new Nuclear Engineering Department. Bailey and Pasqua would build distinguished careers at UT and retire there. And Bob Young would stay with the graduate program not two years as he had promised himself, but until it would finally merge into what became the University of Tennessee Space Institute, and then he would stay with the Institute for the next twenty-five years, retiring on June 1, 1990.

In moving to Tullahoma, Bob Young had come close to retracing the steps of his great-great-grandfather, **James Dryden,** who was born in **Blue Stocking Hollow** in southern Bedford County (Tennessee) in 1825. Dryden's wife, **Hannah,** died a few months after their marriage, and when he was about twenty-five years old, Dryden left Tennessee. He walked four hundred miles and settled in central Illinois, near Mattoon and Neoga. On April 3, 1925, **Robert Lyle (Bob) Young** was born in a farm house four miles northeast of Neoga.

Young found the graduate program exceptionally challenging, with some students in those early days bringing a lot of experience and expertise with them. Late in the fifties, AEDC got the job of establishing the feasibility of a large space chamber for upper atmosphere research. Young had been teaching radiation heat transfer, and he worked closely on this project.

"Those were nice days when almost anything you did was new, and it was very fascinating," he remembered. "The whole space chamber project went extremely well, and the results continually exceeded our expectations. It was a very stimulating environment in which to work and teach. I remember a heat transfer class in which I was discussing transition briefly and forced convection, and **Jack Whitfield** raised his hand and said that possibly some of my remarks were too simple. He then gave me some papers that he and **Leith Potter** had written on the subject, and I realized, as I had before, that Jack was a great expert in that area at that time, widely known for his work in this area. (**Whitfield in 1989 was president of both Sverdrup Corporation and of Sverdrup Technology Inc.**) Similar things were true of many people in my class . . . but it was all a very harmonious relationship, and the students worked hard. We've never had any better students than the first ones. . ."

Young was a demanding instructor and a tough grader, and several students who later became his good friends, failed his classes. Looking back, Young speculated that had it not been for the heavy competition with outstanding classmates, many of those who failed might have satisfactorily completed the course.

The formal proposal in 1955 for a von Karman Institute included the following definition of its basic purpose: ". . . to permit advanced study and conduct of research in the fields of engineering and science that directly or indirectly support the advancement in aeronautics, and to provide the necessary interconnection between these fields." It would "engage the human and institutional resources of the entire free world and would foster close ties with the AEDC and Southern educational institutions." Under this proposal, the fields of aerodynamics, flight propulsion, and aeronautical structures would comprise the principal activities of the institute, but also included would be such associated pure sciences as mathematics, physics and chemistry and such applied sciences as electronics and materials.

In response to a letter from **Colonel Harrie Richardson** on June 22, 1955, **Rowan Gaither Jr.**, president of the Ford Foundation, replied that the Foundation was not interested in supporting the proposed institute. On August 1 of that year, when Richardson, AEDC Commander Harris, W.E. Moser and General Sverdrup met in St. Louis, Sverdrup agreed to support the proposal in three steps, according to Air Force historians. First, the institute would be incorporated on a non-profit basis. Then preliminary architectural sketches and a brochure would be prepared. Finally, a campaign would be conducted to raise money.

General Sverdrup, assisted by General Harris after Harris retired, would in fact make a strong effort to raise seven million dollars to bring the von Karman institute into being. This last drive would end, unsuccessfully, in late 1957, but from its ashes would come a new initiative to secure state funding for an institute, and this effort would attract the support of a man who carried within his head blueprints for an institute unlike any yet proposed. That man was **Dr. Bernhard H. Goethert.**

However, in 1956, efforts were still being made to secure money from private foundations for the proposed von Karman facility. On April 18, **General Donald Putt,** commanding general of the Air Research and Development Command, informed General Harris that officials of the Guggenheim Foundation had given the impression that they would be more interested if the institute were to bear the Guggenheim name. Later that summer, **General Jimmy Doolittle,** the famous aviator who had led the successful bombing raid on Tokyo during World War II and from 1956 to 1958 was chairman of the National Advisory Committee for Aeronautics, volunteered to try to get the Guggenheim and Rockefeller Foundations to put up some or all of the money for the institute, but nothing came of this.

Meanwhile, General Harris, who lived in Manchester, retired on July 31, 1956, as commander of Arnold Center, and on December 26, he was employed by Sverdrup and Parcel, Inc. In this capacity, he attempted to raise money for the von Karman Institute from aviation companies around the country. General Sverdrup was still optimistic.

Notes kept by the retired commander (**Harris**) reflected his feeling that the Pentagon had expressed no official interest in the project. **Donald A. Quarles,** Secretary of the Air Force, was willing to provide AEDC land for such an institute, but he did not want Air Force contractors to feel that they were being pressured to support it. And, Harris noted, neither the Air Force nor contractors had any real authority to spend time or money on the project.

Slowly, the stage was being set for the cast of players who would take a different approach to the idea of an institute. One of these was **Morris L. Simon,** newspaper publisher. Air Force historians say that General Harris, as a personal representative of General Sverdrup, contacted Simon on January 3, 1957, briefed him on the institute proposal, and asked him not to publish anything on it at that time. According to this account, Simon suggested that if funding could not be obtained from foundations or other institutions, then contributions from local and state governments might provide the funds. The publisher also indicated that he had already contacted local members of the Tennessee legislature as to the possibility of transferring land for an institute. General **Sverdrup** reportedly had also contacted Governor **Clement.** The Air Force history shows that Sverdrup and Parcel had withdrawn from the project by September, 1957, when the von Karman institute idea was finally laid to rest.

In an interview in December, 1988, **Simon** remembered that it was General **Sverdrup** himself, not Harris, who first discussed with him plans for an institute and showed him a brochure with a sketch of the building, which was to be erected in about the same place as the Space Institute ultimately was located. He remembered that this had happened in 1956, in St. Louis, and the following information seems to support this conclusion:

On May 11, 1956, Simon's newspaper's top headlines reported that Tullahoma merchants had asked the Chamber of Commerce to obtain a big department store to replace Wilson's, which had closed. Recommendations were to be made by the Retail Merchants Committee, chaired by **Jack D. Walker.** On July 18, the Tullahoma News reported that **Goldstein's** would establish a major department store in Tullahoma. **Mayor Jack Farrar** would raise a building for the store on South Jackson Street, adjoining the Coop Building, and stock would be owned by a number of business and civic leaders.

In announcing plans for the store to the Chamber of Commerce on July 17, Walker said that **General Leif J. Sverdrup** had aided the committee's efforts and would be one of the stockholders in the new business.

"We are especially grateful to General Sverdrup," Walker said, adding that the committee had visited Sverdrup to enlist his assistance. "General Sverdrup told the committee that we could count on his full support in this endeavor . . . because of his deep interest in the welfare of Tullahoma. He said he was for anything that would be for the benefit of personnel of ARO and AEDC."

Shortly afterwards, the mayor and Walker were named vice presidents in Goldstein's, and **Attorney James H. Henry** was named secretary. Plans were to open the store in February.

Simon remembered that he, Jim Henry, and Walker went to St. Louis while organizing Goldstein's and sold General Sverdrup ten thousand dollars worth of stock in the proposed store. During that visit, Simon said, Sverdrup "pulled out a brochure (of a proposed institute) and said he wanted to show it to me, even though I couldn't publish anything on it then. I got a copy of it."

Later, Sverdrup and Harris visited Simon in his office where the general explained that Harris, then working for S&P, was trying to raise money for the erection of and support for the von Karman institute, as it was then called. Simon later learned—perhaps in 1957—that Harris was no longer working on the project, and the effort had "practically ceased because, as I understood it, von Karman had joined an institute overseas and would not be coming here even if they built an institute. At that time, I was aware of the fact that the **University of Tennessee had joined hands** at Oak Ridge with the **Oak Ridge Institute,** and I called **Jack Shea** (public affairs official at Arnold Center and a strong supporter of efforts to get an institute) and asked him to contact Sverdrup and ask him if he would object to an effort being made to get the state of Tennessee to be interested in helping to finance an institute here similar to the one at Oak Ridge, but dealing with aerospace sciences. And also one that would be connected with the University of Tennessee. That being so, there would be some reason for the state to help finance it. **Jack Shea** contacted me and said that General Sverdrup, in essence, gave me the green light to do whatever we could."

Simon then talked with **Robert M. (Bob) Williams,** who had succeeded **General Thomas F. Farrell** in 1957 as managing director of ARO, and to Dr. Goethert, who had become chief of engine and rocket testing at Arnold in 1956, and director of engineering at the center in 1959. Simon said this is when **Goethert first became interested** in the project although Goethert's role is generally associated with the early 1960's, and he became especially vocal in support of a space institute in 1961. (Goethert was promoted to Research Vice President and Chief Scientist of ARO in 1963).

Simon met Goethert soon after the scientist moved to Coffee County. Alike in temperament, both men of vision and determination, they became good friends. Like others who worked to bring dreams of an institute to fruition, Simon attributed the success of those efforts to Goethert.

"Dr. Goethert did what I would say was a yeoman's job of moving the thing along because he was vitally interested in it," Simon said. "He probably knew more than anybody else the value of this. Since he was at AEDC at the time, he was, I'm sure, thinking of it in conjunction with AEDC and the value it would be to AEDC to have this kind of an institute here where employees could further their education and could specialize—to have a sort of graduate institute here for the people at AEDC. The whole tenor of it at the time was that the employees could further their education and still live in the small town of Tullahoma and work at AEDC.''

When Simon and Goethert became involved in the project, they began an association that would extend over a quarter of a century and often involve a ''keystone'' type of combination play as they worked to get the institute and then to support it in many ways. As vice chairman of the UTSI Support Council, Simon often made motions and introduced issues because Goethert asked him to do so.

Given the go-ahead by Sverdrup, Simon ''arranged to get a meeting set up with Clement.'' He remembered that **Bob Williams** made the call to the governor and scheduled the meeting. Among those calling on the governor were **Williams, Shea, Goethert,** and several others from Arnold Center as well as **Jack Walker.** Simon didn't go.

''The day before we were going,'' he said, ''Jack Walker called and asked if I minded if he took my place. He was in real estate and a friend of **Herschel Greer,** a Nashville mortgage company executive who was a friend of the governor. I said, 'Go on, if you want to.' I didn't feel that I had any particular influence. I just wanted to see something done.''

The publisher recalled that the governor indicated some interest and suggested that studies be made and that another meeting be held. Simon thought that another meeting probably was held, but the project seems to have hit a snag.

''I was never invited,'' Simon said about other possible meetings with Clement. ''I was out of the picture. From there on out, we (at The News) would frequently call Goethert to see if he was in contact with anybody, to see what they were doing, because of Goethert's predominate interest in having the Space Institute and doing something about it. He began to spearhead the thing, and it began to move. He was speaking. He spoke at Manchester, Tullahoma . . . at clubs all over the area about the possibility of a space institute and what it would mean to this section. The office he held at that time certainly made him a welcome speaker to many groups.''

Air Force historians say the idea for an institute was resurrected in 1961 when Goethert, then director of engineering for ARO, began discussions of a state-sponsored aerospace institute with UT officials. Air Force records show that AEDC officials completed the first draft of a **''Proposal for a Tennessee Space Institute''** on November 21, 1961.

Bob Young remembered when Goethert first began talking about a ''space institute'' that would combine features of the previously planned von Karman institute with the graduate program at AEDC. Dean Dougherty, who was a consultant to ARO, spent several days a month in the graduate program offices in the A&E building. One day, either in late 1960 or early 1961, Goethert called some of the ARO officials in and unfolded his idea that involved moving to a separate campus and having both a research institute and an educational program and to continue serving AEDC students while also possibly admitting outside students.

''Dougherty was not too much attracted to the concept,'' Young recalled, ''and I was somewhat flabbergasted by his (Goethert's) big ideas. Dean Dougherty had too limited a view, and he immediately thought of a building right across from AEDC. He felt that such a program would primarily service AEDC. He also calculated that it would be extremely expensive and doubted the state would ever fund it. He really didn't know the drive that Goethert had because Doc stumped the state himself and got the Tennessee Society of Professional Engineers behind it. He got Dr. Threet, Morris Simon and other influential community leaders behind the idea, and **soon I was reading in newspapers some amazing things about a space institute** to be established at Tullahoma.''

Dr. Hilton Smith took over as dean of the UT graduate school and thus toured the AEDC facilities during frequent visits, and he recognized the potential of AEDC for the type program that Goethert had in mind. Young remembered that he and **Dr. Marcel Newman** (who had joined the faculty of the graduate program at AEDC and was especially instrumental in securing support of professional societies) met with Dr. Smith at the Cumberland Motel in Manchester ''where he explained to us that a space institute was probably a pretty good concept, but the University had many problems to concern itself with at that time, and it probably should be delayed until the University was able to get some other things under way. Dr. Smith, too, very much underestimated Dr. Goethert's drive and determination and the tremendous fascination with space at that time. People in the local area became extremely enthusiastic supporters of the Space Institute as Goethert and his group pointed out the educational and research benefits and potential industrial benefits for the state.''

A brochure, entitled ''Tennessee Space Institute,'' was produced, and it helped tell Goethert's story of opportunities in academics, research and industry. Goethert frequently consulted Young about the project, and Young wrote some of the material on academics. A committee was formed to investigate the feasibility of Goethert's proposal. **Heinrich Ramm** of Manchester represented ARO and Goethert on the committee; **Colonel Charles Alexander** represented the Air Force; **Leith Potter** also represented ARO; **Dr. Smith, Dean Granger** and **Joel Bailey,**

UT, and **Dr. Newman** and **Dr. Young** represented the graduate program at AEDC. Many meetings were held, both in Tullahoma and in Knoxville, and academic plans were well developed.

"Granger labored mightily on administrative plans," Young said, "and funding and research plans were always quite hazy. While Granger diligently pursued his duties, I feel that he was not tremendously optimistic or confident."

In his **"Highlights of UTSI History,"** Dr. Goethert cited the momentum of the graduate program at AEDC and its effect on solidifying thinking for an institute. He said cooperation between resident UT professors and AEDC and ARO became "particularly close" after Goethert was promoted to the position of Director of Engineering and Chief Scientist for ARO.

"Many discussions took place" between him and Dr. Young on how to improve the graduate program, Goethert noted, and Young agreed that Goethert took a large interest in the AEDC graduate study program.

"He, **Dr. Marcel Newman,** Dean Dougherty and I had many conversations about it," Young said. "I found this to be remarkable, for Dr. Goethert was extremely busy as director of engineering for ARO. At first, it was not completely clear what his concept was, and I early suggested that we expand study opportunities for the several co-op undergraduate students who spent alternate periods of work at AEDC with the other quarters in school at their home institution." Young suggested that this program be expanded and these students allowed to enroll in some of the existing graduate courses.

"This idea did not come up to Goethert's expectations," Young remembered, "and it was little considered. However, he did like the work-study implications of the plan" and the Aerospace Age Work Study Program was established in the summer of 1961 with thirty-five students registered from fifteen colleges. Goethert regarded this as preparation for the coming Space Institute's academic-research program, Young said, and he hoped that several of the people who lectured in that program would eventually become Space Institute faculty. Some did, including **Newman, Bill Snyder, John Dicks, Barney Marschner, Goethert and me."**

Further illustrating the lack of recognition of Goethert's vision, Young recalled that "most of us university folk felt that the Aerospace Age Work Study Program was a one-shot deal, lots of work, and would go away so that we could get on with the normal graduate program. We did not understand Goethert's long-range plan, and history has proved him right."

Additional insight into Goethert's vision for an institute is evidenced in his writing about "pre-institute activities":

"The impressive success of the AEDC-UT Graduate program generated extensive thought about the opportunities which an extension of the program would offer." Such an extension would include opening the program to all qualified students, not just AEDC employees, and the curriculum would be extended to admit Ph.D. students. Authority to grant doctorates as well as M.S. degrees would be granted. This extension also would call for a "drastic increase of the faculty beyond the five resident professors of the existing graduate program (and) would result in a greatly extended curriculum with considerably more courses than were possible with the small number of resident professors in the AEDC-UT program."

It was recognized that support of such a program would require a new sponsor in addition to the Air Force at Arnold Center and UT. Following "good experiences" with the work-study program, Goethert visualized the enlarged program being concentrated on graduate education and offering not only classroom studies but also extensive research work conducted jointly by professor and student teams.

"The complementary nature of the classroom studies and research was recognized as most effective for educating superior graduates on the M.S., Ph.D. and post-Ph.D. level," Goethert noted. "Consequently, the guiding principle was to be that both professors and students would divide about evenly their time between classroom course work and research." (Such a concept was true to examples set by von Karman and Ludwig Prandtl.)

Goethert recognized the "tremendous advantage" that such a graduate institute would have in the close cooperation with AEDC: "Its professors could serve as consultants with the AEDC and thus keep abreast of the latest developments in aerospace and associated fields. Likewise, the students would enjoy the same advantages by assisting their professors in their consultant work or by working directly as technical assistants to the AEDC staff. In this manner, the test facilities and laboratories of the AEDC, the largest and highest-performance complex of this type in the world, would become accessible to the new institute. This is an advantage which can not readily be matched by any other educational institution."

Goethert saw that such an institute also would greatly benefit the Air Force, UT, the state of Tennessee, and neighboring counties in Middle Tennessee. He wrote that these and other possibilities were topics of "intense discussions" between personnel of ARO, Air Force and UT from early in 1961, and step by step, the support of key people and organizations" was secured.

For a man who supposedly did not know how the system worked, Dr. Goethert had made remarkable progress with it. Strong behind-the-scenes support came from **Dr. Marcel K. Newman,** who had recently become Goethert's close friend.

"I arrived the first time at AEDC in June, 1960, as a consultant interested in the operation of the many unusual and large testing facilities," Dr. Newman said years afterward. "There I met Dr. Goethert, and, having my family with me, also his family. In many discussions during the summer I became convinced that, given the right conditions, the chances of strong growth of the UT Branch graduate program into an eventually self-supporting institute were excellent."

At this time, 1960, Newman was in his tenth year as professor and department head at Notre Dame. He received a small research contract dealing with modification of the J-2 Test Cell at AEDC from **Eino Latvala** at ARO. He completed the work during the winter of 1960/61 and submitted a report entitled, "Ejector Heat Transfer."

Dr. Young remembered that Newman, in the summer of 1960, had "so impressed Goethert that he suggested that the University try to hire him." He said Goethert got Newman appointed as professor of mechanical engineering at UT Knoxville, assigned to the AEDC graduate program, with a salary that was two thousand dollars higher than Young was drawing as director.

"This distressed Dean Granger that a professor was making more than the director," Young remembered. "The next year, I got the biggest percentage raise that I ever got, though Newman's salary didn't bother me—he was older and more experienced."

So **Dr. Marcel K. Newman** returned to Tullahoma in 1961 to accept a position as professor of mechanical engineering in the UT graduate school at Arnold Center.

"I divided my time between consulting activities in the Planning Group of the Office of the Director of Engineering (Goethert) at ARO, teaching various subjects in the UT graduate program, and **generally supporting the development of the not-yet formed Space Institute,"** Newman said. His activities in the Planning Group led to a technical memorandum in October, 1962, entitled "Space Environmental Testing of Nuclear Vehicles and Systems."

"In my activities related to the development of the UT Graduate School, AEDC Branch, into a self-supporting institute, I worked very closely with Dr. Goethert. Speaking of financial support, we knew very well that to find a benefactor was not in the cards." The next year, however, Newman would succeed in winning strong support for the project from a group of his fellow engineers.

Speaking of those early days of planning, **Newman** said, "We were confronted with a new concept of engineering education, the joining of industry and academics in the support and strengthening of engineering education. Having been for a period of ten years a member of the ad hoc Engineering Council for Professional Development's Accreditation Committee, I was well aware of the problems and opportunities that were to be confronted. The idea was already tried by AEDC when the proposal of the von Karman Institute was made. Unfortunately, that attempt failed because proper financing could not be obtained."

By late summer, 1961, Goethert's dream had picked up substantial momentum. He recalled that with "effective support" from **Harvey Cook,** chief of Goethert's scientific staff at ARO, Goethert arranged for a meeting at AEDC headquarters in August with **Colonel Charles Alexander** and **Colonel Jean Jack** of AEDC Headquarters. **Goethert, Cook** and **Dwight Goodman** of ARO also participated. A similar meeting, with **Dr. Young and Dr. Newman** joining the others, was held in September.

"After several more meetings of the above persons and others," Goethert wrote, he took the idea of the institute to "the main direct beneficiaries"—**Bob Williams, General Sverdrup, General William Rogers,** AEDC commander, and his staff. The Air Force and ARO were immediately in favor of the proposal, he said, so the next step was to present it to the UT president, **Dr. Andrew David (Andy) Holt.** A meeting was arranged. Bob Young thought this is where the dream would shatter.

"Goethert decided it was time he and General Rogers and Bob Williams proposed the institute to Andy Holt," Young said. "I figured this would be the end of everything, including the AEDC graduate program. I imagined Dr. Holt would throw Dr. Goethert out of his office for making such a wild proposal for the expenditure of University funds."

Andy Holt, a personable and powerful man himself, later would call Goethert "the most persistent man" he had ever met. He also would joke about the opposition that Goethert encountered on some levels of the UT administration.

Young was dumbfounded at the results of the meeting: "Much to my surprise, Goethert came away from the meeting with Holt with the understanding that he (Holt) would ask the legislature for a million dollars! What I didn't know was that Andy Holt was also very progressive and immediately saw the political and academic advantages of the Space Institute."

Goethert wrote that Holt "assured his strong interest, but stated that finding funds for both construction and operation of the proposed institute was the overriding condition."

Goethert later wrote that the president's concern was understandable, considering that the university was "already greatly strained to secure" necessary funds for the rapid expansion of graduate programs under Dr. Hilton Smith's leadership on the Knoxville campus. It is doubtful, however, that Goethert was as charitable at the time that he was battling to get his plan approved.

In a letter dated December 2, **1961,** Dr. Holt wrote Goethert that the preliminary outline "suggested by you and your staff appeals very strongly to me. With the refinements which broader study would make possible, this plan should appeal to the University as highly desirable." He warned, though, that the major problem "is finding the money needed for its establishment (and operations) . . ."

Money, indeed, was one of the major obstacles. Later, Goethert wrote: "In the early years, Dr. Holt, **Dr.** (Herman) **Spivey,** Dr. (Edward J.) **Boling** and Dr. Hilton **Smith** recognized the great advantage of an institute for space sciences in Tullahoma, but they also recognized the difficulties in securing the funds both for construction and operation of the new institute. On the other hand, **at the lower organizational levels** of the University, there was a **wide-spread pessimism** and even **resistance** to the idea of a new UT institute because the available funds were not even large enough to meet the financial requirements of the existing UT colleges and departments."

Dr. Young remembered that during early discussions when projections called for an annual budget of $700,000 for the proposed institute, Dean Granger of the UT College of Engineering "pointed out to me that $700,000 was about the total budget of the UT College of Engineering. That illustrates the desperate shape they were in, and our brave plans amazed him."

Despite concerns about money, Dr. Holt proposed going ahead with the idea contingent upon **two conditions.** First, he would make sure that any funds for the proposed institute would be granted by the state **only as additions** to the construction and operation funds of existing UT campuses. Under the second condition, Goethert assured the UT president that a major portion of the operational expense would be covered by research contracts and grants that the faculty would acquire from sources outside of the University.

Goethert had an ace up his sleeve in **General B.H. Schriever,** commander, Air Force Systems Command. Schriever was a strong supporter from the start. On December 16, 1961, he wrote Dr. Goethert that "an institution developed along the lines suggested in your proposal would enable us to take full advantage of the opportunities that exist at the Center for the development of cooperative educational and research programs. I believe we should move ahead without delay with the creation of the proposed Tennessee Space Institute. You have my whole hearted support in this endeavor, and I will be pleased to help you in any way that I can."

Schriever, born in Germany in 1910, came to the United States when he was six years old and settled with his family in New Braunfels, Texas. As a member of the 19th Bomb Group in World War II, he flew sixty-three missions against targets in the Solomons, New Guinea, the Philippines and the Ryukus. (During his service in the South Pacific, Schriever also became close friends with **General Leif J. Sverdrup.**) In the charismatic Schriever, Goethert had a powerful ally who would back him time and time again.

Schriever suggested that Goethert develop the institute's land requirements in cooperation with General Rogers and submit a proposal covering those. Schriever also appointed **Colonel Edgar Masters** as liaison between the planners of the institute and the Air Force Systems Command headquarters in Washington. Goethert remembered that Masters was "very successful in this arrangement until his transfer in 1967," and Goethert also acknowledged the "behind the scenes" support of **Colonel Barney Marschner** of the Systems Command. A few years later, Goethert would talk Masters into leaving his retirement home in Hawaii to work at the Space Institute.

As 1961 came to a close, Dr. Goethert had to be pretty pleased with himself. He had garnered prestigious support for his dream. He had sold the UT president on the idea, and not only were the AEDC commander and ARO's president behind him, but also the commander of the Air Force Systems Command. Still, there were many obstacles, not the least of which was the challenge of finding a large amount of money.

The new year would be a busy one with the Tennessee legislature wrestling with a political hot potato called "reapportionment" and **Frank G. Clement** winning election to his third and last term as governor. It also would be the year that things started snowballing for the Space Institute, and a major victory in overcoming fears and objections would open the door for success at last.

✦

Chapter Four

'OPPORTUNITY OF A LIFETIME'

OBJECTIONS to the idea of establishing a graduate center or institute in Middle Tennessee, far from the main UT campus in Knoxville, were strong. Key players, all friends of UT, later agreed that not everyone "on the hill" looked favorably upon efforts to open an institute. Some said it was a battle over turf. Some called it professional jealousy. Others believed that Goethert's proposal represented unwanted competition. Unquestioned is the fact that money, or the lack of it, prompted much of the opposition.

Dr. Frank P. Bowyer Jr., the man whose optimism and enthusiasm convinced his fellow UT trustees that a space institute not only was feasible, but that it was a great opportunity, and that UT should play a major role in it, said that money was a major cause of hostility toward the proposed institute. On February 4, **1989,** Bowyer recalled that there had indeed been opposition, but he said it was not toward the project or toward the **concept** of an institute.

"I don't really recall on the board, as there is on some issues, any violent opposition when it was mentioned by the administration," Bowyer said. "However, neither was there a lot of enthusiasm. It was more, I think, a lack of information, a lack of knowledge about it. Maybe some of them just weren't as excited about the thing as I was. I just felt it was the greatest thing that could happen at that point in time in the history of the United States and in the development of the aerospace program, that UT could have such a substantial role in it. That's what motivated me."

Not only was there opposition within the administration toward the project, Bowyer said, but "to a degree, there developed some antagonism and opposition to Dr. Goethert as a person. In all fairness, however, opposition to the project was not that it was not a worthy endeavor, but the university had so many demands and was on a limited budget, that some felt that unless some very special additional direct funding was made available, there was no way to absorb it into the university system."

No one questioned Goethert's scientific knowledge, ability and background, Bowyer said, but some were hesitant about "some of his concepts because he was so eager and there was apprehension about moving too rapidly." In short, Goethert's aggressiveness and impatience bothered a lot of people in Knoxville. Bowyer said the administration had a problem with Goethert's eagerness. "His requests became almost **demands.** They didn't like being pushed so hard by Dr. Goethert for him to acquire what he needed, to attain what he wanted to attain, but the squeaking wheel gets the grease, and he would squeak plenty loud!"

The man with the sky-blue eyes and independent shock of hair was indeed a driven man—aggressive, stubborn, impatient with delays and protocol. Perhaps these traits, which irritated even those who loved him, had helped him survive the trauma of losing his house, job, security and prestige in 1945 and having to beg for food while inching his way toward his family after the fall of Berlin. How had it affected the once-revered scientist to cower in darkness and hear screams of young German girls and pleadings from their mothers and to be powerless to protect them from the brutal Russian soldiers? Or to be thrown into a basement jail? He carried with him memories of that Christmas morning in 1943 when he and his beloved Mutti—she great with child—had walked with their three children past the flames and rubble of their native village after a night of heavy bombing and tried to explain to their frightened youngsters why the world was on fire. Whatever the cause, Goethert's impatience was never more evident than in his insistence that the path to strength for his adopted country lay in advanced education.

"I don't feel that there was at any time a personal dislike for him," Bowyer said. "He gave them problems. Underneath it all, they had a great respect and admiration for him. It was just that his methods were so entirely different. . . . He really **demanded** instead of making diplomatic requests and being willing to go through channels. He would say, 'Don't waste time on something you can leap frog.' He told me that even in scientific research it was not uncommon to leap frog a step . . . that if your projections were right, you could save money and time. He was so positive about level five that he didn't want to spend time on levels two, three and four. He was impatient, but properly so. He was eager to get the show on the road.'' Goethert, Bowyer believed, ''never understood protocol, America, or the university system.''

Dr. Frank Bowyer

One man who heard about Goethert's visions of a space institute ''straight from the doctor's mouth'' was **Dr. Ewing J. Threet,** Manchester dentist and a state senator from 1961 to 1963. He was born in Midland, Tennessee, when this community in Rutherford County had two country stores and one physician. He received his D.D.S. degree in March, 1952, from the University of Tennessee Medical School in Memphis, and his master's in pedodontics in March, 1953. After a year with the Memphis and Shelby County Health Department, Threet opened his practice in Manchester. As a resident of 1800 Sycamore Circle, he was a close neighbor of both **Goethert** and **Bob Kamm.** He was Goethert's dentist from 1955 until Goethert's death, and during those thirty-three years, Threet put forty crowns on the scientist's teeth.

''He was very hard on his teeth,'' Threet said. ''He wore them off.'' His ''powerful gritting'' was indicative of his strong drive and impatience. Goethert didn't like to make appointments. When he had a problem, he felt that he should be able to take care of it then and there, and usually Threet accommodated him.

''His visits usually lasted from one to one and a half hours. I enjoyed them. He always wanted to talk. I tried to be patient and listen.'' Threet said that other dentists, including **Dr. Manual Sir** of Nashville and formerly of Manchester, performed at least ten ''root canal'' procedures on Goethert.

Goethert lived two doors from one of Manchester's most controversial and colorful mayors, **Clyde Vernon Myers.** Threet lived between Goethert and Bob Kamm, **Alfred Windmueller** (who also came to the U.S. from Germany after the war), and **John (Red) Wiltshire,** a noted stone sculptor. Another neighbor, **James G. (Red) Jarrell,** was elected superintendent of Coffee County schools in 1964 and served in this post for several years. Sycamore Circle had one of the steepest hills in Manchester, and when it snowed, children sledded downhill from Threet's house.

Threet remembered early mornings when he would hear Goethert leaving for work in his Volkswagen, sounding ''like a freight train.''

''I always knew when he was leaving home. I would hear him fire up his V.W. and roar off. He figured everyone would get out of his way.''

Threet served as a Manchester alderman from 1955 to 1957 and began a two-year term as state senator in 1961. Because of a ''rotation'' agreement that existed prior to reapportionment, Threet did not seek re-election, and **Ernest Crouch** of McMinnville was elected to the Senate. Threet made an unsuccessful run for the Public Service Commission. He was elected as a delegate in 1965 to the Limited Constitutional Convention that raised the pay of legislators and changed the meeting of the lawmakers from every two years to an annual basis. (As a senator, Threet had been paid four dollars a day.) **Governor Buford Ellington** named Threet to the Tennessee Industrial and Agricultural Commission in 1960, and in 1964, Governor Clement reappointed him for four years. The group's function was to attract industry to Tennessee, and members made several trips out of state. The commission was succeeded by the Office of Economic Development.

"I was really interested in trying to do something for our county and our area," Threet said. "Hence, the **Space Institute** fit right into all our efforts as we were repeatedly told that education was a key, and I believed it."

It was Threet who, in December, 1961, arranged a meeting with Governor Ellington at which time Dr. Goethert, General Rogers, Mr. Williams and Dr. Holt presented the proposal for an institute. Threet's relationship with Ellington was a bit cool because the senator had supported reapportionment of the legislature, which Ellington, with strong ties to rural areas, opposed. (The next year, the U.S. Supreme Court ruled that Tennessee citizens could sue in federal courts to force reapportionment, and the General Assembly, in special session in May and June of 1962, passed a reapportionment bill—a measure later ruled inadequate by a federal court.)

Goethert noted that Ellington was "highly interested in the idea" of a space institute and suggested that the officials be invited to present the institute concept at the next meeting of the UT trustees.

Threet's friendship with Goethert became closer after he began pushing to get the space institute. "We, Crouch and Simon, met at his (Goethert's) house and laid the groundwork. I don't think Goethert understood the political system of our state. If so, he paid it no attention. In Germany, if the man in command said it would be done, it was done. I told him: 'We must keep everybody informed.' He would say, 'Ah! We got to move quickly!' I'd tell him how we had to work with the legislature, the UT board, the UT president . . . he was so impatient. He'd call. He thought I'd have information. I'd say, 'No, I'm laying the groundwork.' "

Occasionally, Threet would prevail over the persistent Goethert, such as the time when Goethert was scheduled to meet with a group from Germany. Threet argued that Goethert needed a crown, but his patient argued that he didn't have time. Threet insisted and "numbed him up . . . and he had to go to the meeting that way."

UT was "not enthusiastic about a space institute, if for it at all," Threet said. He credited **Frank Bowyer** with getting the concept approved. Bowyer "held the key to getting approval of the trustees," Threet believed. "He was the one who made the motion that the Space Institute be built and that UT accept the responsibility for it."

"I was first approached about the space institute by Dr. Threet in the mid-winter (January) meeting of the State Dental Association in Nashville in 1961," Bowyer said. "I had no factual knowledge of the institute before that. He stimulated my interest." **Jack Shea** was with Threet the day they talked with Bowyer in a suite of the Andrew Jackson Hotel. Bowyer remembered the encounter as "extremely exciting, fascinating and challenging." He was immediately interested in the project's success. The logical procedure, he thought, was for the university to take on the proposal for a space institute as a special project.

It was Threet, too, who first introduced Bowyer to Goethhert a short time later in Tullahoma. Bowyer was surprised. He had anticipated seeing a man of "more robust statue, physically, and a very positive, domineering type." He found that Goethert's physical appearance was "very small," but Bowyer was "more surprised by his manner. He was very mild mannered, very pleasant and down to earth. He was not the superior type (despite) all his vast knowledge and notoriety. I got no impression that he was trying to project a superiority type thing at all. He was warm, personable. His character and characteristics aided greatly in winning me to his side as a person and substantiated my interest in the entire project. My relationship with him was always extremely pleasant. He tried to explain things in words that I could understand."

Threet, who was present during Bowyer's rather glowing description of his "warm, personal relationship" with Goethert, explained that Goethert "had two types of personality. He could be very charming, but he also could be very demanding." Apparently the chemistry between Goethert and Bowyer was such as to bring out the charm. Of course, Goethert could afford to be charming with Bowyer because he was a good man to have on his side. Bowyer, a staunch Democrat and a personal friend of both Clement and Ellington, made good use of his Nashville contacts while campaigning for the institute. His greatest influence, though, was with other trustees.

"I was so enthusiastic about it," Bowyer remembered. "I would say, 'We'll find a way. This is something we've got to do!' Perhaps my responsibility as a trustee to the overall university was tempered a bit by my strong enthusiasm for the institute. I also did some individual lobbying in Nashville, in the legislature and around, to support an appropriation for the project."

Bowyer felt that the problem that some UT officials had in dealing with Goethert was "sound to a degree" and not based on a serious dislike of the scientist.

"They had a responsibility for the whole university, and he was interested in his project. His whole interest was in Tullahoma and the institute, and he could care less about medical school or college of education or anything else."

Threet firmly believed that Goethert never "fully understood" the political system. "On many occasions, he would say, 'Ah! Just go to the governor; he'll get everything done.' Dr. Holt said he never met a man who thrust himself like Goethert."

It was during a meeting of the UT Board of Trustees in Memphis on **February 5, 1962,** that Dr. Bowyer made his momentous motion on behalf of a space institute. President Holt had introduced General Rogers and Dr. Goethert and explained to the trustees that he and other UT representatives had been working with Rogers and Goethert relative to establishing a center for training scientists and for research in the "great and rapidly growing field of space technology." Goethert called attention to the size of Arnold Center and the amount of money being spent on it each year and its value as a great missile research agency. It was most important, he said, that a civilian organization be developed near and in cooperation with AEDC, and he presented a brochure showing the need for such a civilian organization. The brochure focused on advantages to the state, Air Force and United States. Goethert noted that the Air Force was considering donating land for locating the proposed institute. General Rogers told the trustees that the Air Force was "tremendously interested." Except for certain classes taught exclusively at the Wright-Patterson Air Force Base, he said the Air Force depended on civilian institutions for the scientific training of personnel. A graduate center such as the one proposed, he said, would be invaluable in training military personnel while also providing a source of scientists from which to draw.

UT Vice President **Herman Spivey** noted that the proposed graduate center would be a part of UT's graduate school, subject to its rules, regulations and standards, and it would meet the same requirements for Ph.D. and other graduate degrees required on the Knoxville campus. It would be unwise, he said, to try to make it a separate institution since this would involve problems and difficulties with accrediting agencies. Spivey submitted a statement in reference to the graduate school that UT had operated at Arnold Center since 1956, saying it would be simple to expand that branch to become the proposed Space Technology Graduate Center.

During the lengthy discussion that followed, Dr. Bowyer stressed advantages that would accrue to the state and to the entire southeastern area from industries locating because of the proposed institute. He had found that the production of materials and devices in connection with space science was a multimillion dollar industry in California. Goethert commented that such industries had grown up around the California Institute of Technology and the Massachusetts Institute of Technology. Governor Ellington, board chairman, and other members expressed interest, and President Holt discussed sources that might be approached for funds. General Rogers and Dr. Goethert then left the room, and Dr. Bowyer, chairman of the standing committee on engineering, asked for the privilege of making the motion regarding the proposal.

"Gentlemen," he said, "we are about to vote on a matter that could have a great influence on our survival and our future destiny. Being cognizant of the magnitude of this decision, with deep humility, I move, Mr. Chairman, that the Board of Trustees of the University of Tennessee enthusiastically support the plans formulated thus far and authorize our administration to proceed vigorously with further plans for the establishment of a Graduate Center of Space Technology at the Arnold Engineering Development Center."

The resolution offered by Bowyer authorized and directed the UT Administration to "proceed with the working out of a graduate and research program in Space Technology in cooperation with appropriate Arnold Center officials and in harmony with University of Tennessee Graduate School policies and regulations." It also authorized UT officials, in cooperation with Arnold Center leaders, to "seek funds for the development of a plant and the operation of the above mentioned program, and with the aid of the governor, to secure the necessary funds for the preparation of architectural drawings and other information and material which would be helpful in the solicitation of funds."

Twenty-six years later, Bowyer recalled that the groundwork had been "carefully laid." He said the late **Julian Harris,** for many years public affairs director for UT, "polished" the resolution and helped with its wording. (Harris began his writing career in the late thirties as a cub reporter on the Knoxville News-Sentinel, under the tutelage of **Morris L. Simon,** who was assistant city editor.)

Charles D. Lockett seconded Bowyer's motion, and it carried with all present voting "aye" except **Clyde M. York,** who asked that he be recorded as not voting. The board's enthusiastic endorsement marked a turning point in the long struggle. As **Colonel Harrie Richardson** later recorded, Dr. Goethert had laid the foundation by arousing the "interest of people all over Tennessee by literally stumping the state. He approached **Buford Ellington** . . . he spoke at civic luncheons, board of trustee meetings, Chamber of Commerce affairs . . . and he inspired his audience to generate an intense interest in the proposal."

Richardson also pointed out other vital supporters who took a special interest in the project in 1961: "**Harvey M. Cook Jr.**, at the time president of the Tullahoma section of the old American Rocket Society, was an active promoter of the institute. So were **Harold J. Black,** director of engineering and development for the Aerospace Structures Division of AVCO in Nashville, and **Edward M. Dougherty** of the Tennessee Society of Professional Engineers, which strongly endorsed the proposal."

In sanctioning the action, Dr. Holt had emphasized that the proposal should have special appeal to UT "since it allows the use of appropriate resources at Arnold Center, human and laboratory."

A groundswell of interest and support following the UT trustees' endorsement was reflected in stories in state and local newspapers in 1962, and on April 2, **Missiles & Rockets** magazine reported that the Air Force was backing the proposal. Of course, General Schriever had already strongly endorsed it, and in April, Goethert, Rogers and Sverdrup reported to him on the progress of the project. The magazine quoted Goethert as saying that one major goal was "to form together with the new research institute in Huntsville, Ala., the twin focus points of aerospace research and industry of the south, and possibly also, of the entire world."

On February 15, 1962, Dr. Goethert took Governor Ellington on a boat ride on Woods Reservoir to show him the site for the proposed Graduate Center for Space Technology. Along for the ride also were **Bob Williams, General Rogers,** and **Colonel James O. Cobb,** deputy AEDC commander. Ellington also toured testing facilities at Arnold Center. **Bob Young** remembered this visit by Ellington: "Although he had several million dollars in surplus (state) funds, he would not pledge any funds to start planning for an institute until the concept was approved by the legislature."

Ellington never showed quite the enthusiasm for the proposal that his leap-frog successor, **Frank Clement,** did, and it was Clement who would drive the project through to completion. He would do so with a lot of support on the grass-roots level.

Tullahoma's Chamber of Commerce directors in April, 1962, authorized a resolution expressing "hearty support" after first questioning whether the Chamber's support would conflict with its backing of

Dr. Goethert is showing Gov. Ellington exactly where he wants the Institute to be located. On the left are Robert M. Williams, General William Rogers, and Col James O. Cobb.

a community college. (**James C. Murray,** long-time editor of The Tullahoma News, said that experience in getting the Space Institute was a "real learning experience" for the community in going after Motlow State Community College, which opened in September, 1969, in Moore County.)

On May 12, **Rudy Abramson** reported in The Nashville Tennessean that one hundred and seventy-five members of the Tennessee Society of Professional Engineers from around the state had adopted a resolution, introduced by **Clark Mann,** past president, supporting the graduate center. **Dr. Marcel Newman** told engineers at a luncheon on May 11 that the institute idea had gained impetus after it became known that Russia was outstripping U.S. space capabilities in several areas, but especially in large booster engines. Then, Newman said, the fact became known that the Russians had been involved for the past three decades "in a quiet build-up of their educational system to produce high caliber specialists in the sciences and technology" while little such effort was being made in the United States.

"Being a member of the society," Newman said twenty-eight years later, "I was asked by Dr. Goethert to address the group in the afternoon session on May 11, 1962, explaining the benefits being derived from an academic institute for the local area, the state of Tennessee, and the country, and to urge them to support our plan for the formation of the Space Institute in all sections of the state of Tennessee."

The response of the TSPE members was "overwhelmingly gratifying," Newman said, and led to important activities all over the state, "thus paving the way for the success achieved by us in obtaining legislative approval and funding for the establishment" of the institute.

On June 19, 1962, Air Force Secretary **Eugene Zuckert** agreed to give "serious consideration" to a proposal to establish an academy for training of space scientists at AEDC after a meeting with **Senator Kefauver, Goethert, Holt,** and a representative of Senator (Albert **Sr.**) **Gore's** office at the Pentagon. Kefauver said Zuckert was "quite sympathetic."

On June 30, The Tennessean reported that Chattanooga Mayor **P.R. (Rudy) Olgiati,** running for governor, supported the institute, saying it would be a "financial bargain and would move the state's education system into the frontiers of the space age." State Representative **Ernest Crouch,** running for the Senate in the August primary, had made the space institute a central issue of his campaign. First elected to the state House in 1955, Crouch had served one term and then was elected to a two-year term in the Senate. He went back to the House for two, two-year terms, and then in 1962, after the senatorial districts had been changed, he ran against **Pete Haynes** of Winchester.

The space institute "was an issue in the campaign," Crouch confirmed in 1988. "Pete said, 'He's (Crouch) daydreaming; it will be twenty years away.' I said, 'No, it is here **now.'** " Crouch carried every county in his district—Warren, Coffee, Sequatchie, Marion and Franklin—and received forty-nine percent of the vote in Haynes' home county of Franklin. "He said I was pipedreaming," Crouch later said. "I replied that we can't go forward looking back." Crouch stayed in the legislature until 1985, establishing a record of thirty years of consecutive service in the legislature. Crouch and **Charles F. Hickerson** of Tullahoma, who had won the Democratic nomination for state representative, were briefed on institute plans in September, 1962, during a meeting of the Tennessee Society of Professional Engineers. While he had supported the institute during his campaign, Crouch said it was not until after he took office in 1963 that **Goethert, Threet** and **Simon** really convinced him that the dream could come true.

Re-enacting a meeting held in the early 1960s at Goethert's house are, from left, Sen. Ernest Crouch, Dr. Ewing J. Threet, Goethert, and Morris L. Simon. (This meeting was in August, 1974.)

In September, 1962, the Air Force agreed to make a four-hundred-acre site available for the institute, and Goethert headed a committee, composed of representatives of UT, Arnold Center, and ARO, Inc., to plan for it.

The UT trustees, in a meeting at Arnold Center on September 17, heard President Holt warn that UT must prepare for a huge increase in enrollment or "become mediocre," renewed their February endorsement of the institute, and made plans to start "definite action" at a meeting in November. Holt told the board that an investigation would be made into a possible source of federal funds for construction of the institute. **Governor Ellington,** as board chairman, said he was prepared to make state funds available for preliminary plans and studies. Goethert also addressed the trustees, outlining again AEDC's vital role in the nation's space and defense program. ARO officials were urging that classes start in the fall of 1963.

In October, 1962, **Dr. Frank E. Goddard, Jr.,** assistant director for research and advanced development of NASA's jet propulsion lab, told a reporter that Tennessee "has an enormous opportunity" to better its economy and educational system by linking AEDC to the state university system. (Dr. Newman had been cautioning publicly about a "great gap in education" in the United States.)

Frank Clement, campaigning in Tullahoma, had promised his friend, Morris L. Simon, that if elected governor, he would see that the space institute was built. On August 2, Clement won the Democratic nomination in a three-way race, and on November 6, in the General Election, he was elected as governor. Three new congressmen also elected were **James H. Quillen, William E. Brock,** and **Richard H. Fulton.** (As Fifth District congressman, **Fulton**—later mayor of Metro Nashville—would be a strong supporter of the Space Institute.) Shortly after Clement was elected, **Andy Holt** met with him to brief him on the institute plans.

At this time, an unusual display of unity was evident among cities and counties in the area, but the support was not limited to the smaller towns. A liaison committee of Manchester and Tullahoma Chambers of Commerce, headed by **Dr. Threet,** was making plans in November to confer with Governor-elect Clement. Meanwhile, a committee of the Nashville Area Chamber of Commerce recommended endorsement of the institute after **Harvey M. Cook,** assistant director of engineering for ARO, outlined plans to them. Cook said the proposed institute offered a chance to parallel the University of California's Institute of Technology and NASA's jet propulsion lab.

Addressing engineers on December 4, 1962, in Memphis, U.S. **Senator Estes Kefauver** said the institute was the best opportunity the state had had to increase its industrial base through its educational program. **D.J. Goodman,** a national director of the National Society of Professional Engineers and manager of the electrical branch of AEDC, also stressed to the (Tennessee Society of Professional Engineers) gathering that the institute was urgently needed.

Late in 1962, Threet called Goethert, suggesting that it would be a good idea if legislators could visit Arnold Center. Goethert responded: "Excellent! When?" Threet introduced a resolution authorizing the visit, and in early December, two bus loads of lawmakers (twenty-one senators and thirty-two House members) showed up at Arnold Center where they were greeted by Dr. Goethert, who took them on a tour of the facility and then to the Officers Club. Threet had gotten the AEDC commander to extend an invitation to the solons.

"This (the legislators' visit) was a good move," Threet said many years later. "It helped get us all familiar with the (institute) project. A lot of things were going in our favor. The governor was very strong for it. He had a good rapport with the legislature. This was at a time when space was very big. The timing couldn't have been better. **Morris Simon** helped coordinate our efforts."

Simon played a major role in getting cities and counties in the area to pledge five hundred thousand dollars toward getting the Institute established. Clement had told Simon that he would like to see a show of support from the area. Simon and **Attorney G. Nelson Forrester** spearheaded efforts to get the money committed, calling on governmental leaders to support it. The Tullahoma News, on December 14, in a front-page editorial entitled "We Can Help Get the Institute," urged surrounding cities and counties to raise the money. On December 23, 1962, Coffee County Court, with thirty-three "ayes" and two abstentions, pledged fifty thousand dollars.

Tullahoma Mayor **Jack T. Farrar** was elected chairman of an informal committee charged with raising two hundred and fifty thousand dollars, and the city of Tullahoma pledged to provide half of this. Winchester and Franklin County pledged fifty thousand dollars, Bedford County and Shelbyville pledged twenty-five thousand, and fifty thousand dollars was assured to come from either Manchester or Coffee County. Warren County indicated it also would help.

In Memphis, newly elected **Congressman Dick Fulton** was saying that the proposed institute could "stem the tide of young engineers leaving the state to find jobs."

On January 3, 1963, State Representative **Charles Hickerson** said the effort to raise five hundred thousand dollars was only about sixty-three thousand dollars short of its goal, and U.S. Representative **Joe L. Evins** said he had been assured that Tennessee would get federal funds for the institute. **George L. Benedict Jr.,** state industrial development director, enthusiastically supported the movement for an institute, calling it the "greatest talking point" in the recruitment of new industry.

During the first week of January, Clement publically pledged that the institute would be established and, after meeting with Air Force Secretary **Zuckert,** Clement promised to seek funds from the legislature. The 83rd General Assembly convened on January 7, and shortly afterwards the governor asked the lawmakers to appropriate $1,250,000 for the space institute.

The Tullahoma News, on January 25, 1963, reported that bonding attorneys had advised Mayor Farrar that cities and counties could **not** legally borrow funds to help build the institute. However, the "show of support" had not been in vain, and this support reached all the way to Nashville, where the Greater Nashville Chamber of Commerce hosted a **"Space Institute Breakfast"** at the Municipal Auditorium on January 30—an event that drew more than five hundred leaders from throughout Middle Tennessee. Goethert and all of those who had worked closely with him on the dream were jubilant at this almost unheard-of cooperation between rural and big city leaders.

At the breakfast, **Governor Clement** advised cities and counties to forget efforts to raise money, saying, "This is not an area vocational school," and the state, rather than local governments, should support the institute. Calling establishment of the institute the "opportunity of a lifetime" for Tennessee, the governor assured those at the breakfast that "We'll find the money with the cooperation of the General Assembly." His optimism was matched by a general positive attitude that swept the state as Tennessee Jaycees, the Oak Ridge Chamber of Commerce, Lions Clubs, and other civic groups joined engineering societies in endorsing the concept of a space institute near Arnold Center.

With his usual exuberance, **Dr. Goethert** addressed the breakfast crowd, showing slides as he described the facilities and the mission of AEDC and explained his vision of a close partnership between Arnold Center and the proposed institute, a "cooperative effort between the state of Tennessee through its university, and the United States Air Force through the Arnold Engineering Development Center." Goethert said the investment in AEDC's facilities totaled $340 million and added that "new facilities are continuously added at a rate of ten million dollars per year." He wondered how many people knew that "here at the gates of Nashville, in the center of our state, full-scale engine firings like those at the much publicized Cape Canaveral in Florida, take place on a routine basis." Showing slides of 1962 launchings of Saturn boosters, Goethert drove home the role that Arnold Center played in their successes. "Engines of the Saturn booster size can be checked out under simulated launch conditions in the large altitude rocket facilities which exist at no other place than at the Arnold Center," said Goethert as he zoomed in on the tremendous assets that he said also would be at the disposal of the institute. He emphasized the shortage of space engineers and scientists, noting that "at the present time, no Tennessee university offers education leading to a formal degree in Aeronautical Engineering or Space Technology."

While some in the audience may have had trouble with the little doctor's accent, they all clearly recognized his enthusiasm, and they grasped his message as he charged that "our young Tennesseans (now) must leave the state to get the formal education enabling them to participate in the greatest engineering endeavor of our time—the challenge of advancing into space." This situation, he promised, would be corrected by the space institute. "A close interrelationship between education, research and development work for practically all of our nation's airplanes, missiles, and space vehicles would put **life** into the abstract teaching of science," Goethert declared. He emphasized economic benefits that would accrue, cited the dual roles of professors who would spend half their time teaching and working with students and half conducting research on chosen subjects, both at the institute and in the AEDC laboratories, and calling for the "closest possible liaison" with other research centers.

Isolation of the institute was "impossible," Goethert warned, and must be avoided. "The times of splendid isolation for our scientists and professors is gone forever," he said. He visualized establishment of complementary programs with other centers such as Huntsville and Oak Ridge. While recognizing the importance of one university effectively managing the institute, Goethert said other Tennessee universities could pursue research and use the AEDC facilities on the "same free academic basis" as UT. Goethert told the Nashville audience that the institute itself would merely provide an opportunity. The quality of its professors would make the difference between "success, mediocrity or failure," he cautioned.

"Space science is a problem of the entire human race," he continued. **"It requires the best minds from all over the world."**

Noting that some European university people had already expressed interest in cooperating in an exchange program, Goethert spoke of a "natural inroad available" for such a program since "I, myself, hold a professorship at a well-known European university."

In his conclusion, Goethert spelled out a philosophy that would undergird his career at the institute—a vision of an institute that placed excellence in academics above all else. He gave the blueprint:

"Let us not make the mistake of believing that all that is necessary is to erect a minimum square building at the shore of the AEDC lake and to nail a shingle **'Space Institute'** over the doorstep. This will not attract the most essential element for success, scientists and engineers. It could only mean a waste of funds. The architecture of the campus must reflect the challenging goal of mankind to venture from the confines of earth into outer space. It must inspire scientists and express a vision of an institute equaling or exceeding any other space center on this continent or abroad.''

To thundering applause, Dr. Goethert closed with this challenge: ''We should grasp this one-time offer and pursue it with **urgency, vigor,** and **vision.''**

A lot of work was yet to be done, a lot of bridges yet to be crossed, but no one close to the project, or to the intense little man from Sycamore Circle, left that breakfast with any doubt that the picture he had painted would one day become a reality. They did not know that even FDR's son would be unable to nail down federal funds for the proposed institute (with three Tennessee congressmen voting against President Kennedy's ''distressed area'' bill.) Even Goethert did not suspect the battle royal that he soon would face with Middle Tennessee Girl Scout leaders who threatened to bring hordes of Scouts together in a massive protest against plans to locate a space institute on their territory. And no one that day would have dreamed that one town would seek to buy the very land on which Goethert planned to locate his precious institute—and to use it as a garbage dump.

Everything looked rosy that morning.

✦

Chapter Five

A Place on the Peninsula

BY FEBRUARY, 1963, **Governor Frank Clement** was pushing for approval of a tax bill designed to raise additional revenue to support his proposals for new services, including establishing the Space Institute. Clement promised there would be "no deficit financing in my administration," and he warned that there would be no industry without passage of the tax measures.

The Tullahoma News, in a front-page editorial, argued that more taxes were needed if more services were desired. "You can't eat cake without paying more than the price of bread," the editorial began. The News went on record as "favoring paying more money—more taxes—in order to obtain the increased services which we and other citizens need and have asked for. Tullahomans and Coffee Countians might do well to remember that the establishment of the UT Space Institute at Arnold Center is contingent upon increased revenue for the state."

Senator Ernest Crouch said, "Our area stands to gain ten dollars for every dollar we pay in taxes. In addition, some eighty percent of the proposed increase in spending will go to education, especially in the grades one through twelve area." He said the area would receive such benefits as the Space Institute, higher teacher salaries and resurfacing of the Viola-Winchester Highway. The Space Institute would bring Tennessee more revenue than the entire thirty-one million dollars sought in new taxes, Crouch said.

On February 15, The News reported that the Space Institute bill would be introduced the next Tuesday by **Charlie Hickerson** and **Ernest Crouch.** This edition of the newspaper contained a letter to the editor from Crouch concerning the institute in which he wrote: "It has been largely through your efforts in bringing to the attention of our area and to the people around the state the great import of this project. I would say that your newspaper has been the instrument that has really sold the Space Institute and followed the program through to become a reality."

It was indeed about to become a reality—at least on paper. On February 27, 1963, companion bills were introduced by **Senator Crouch** and in the House by Representatives **Charles F. Hickerson Jr.** of Tullahoma and **Morris Hayes** of Winchester, establishing the Institute. The bills provided for an initial expenditure of $1,250,000 to construct and equip the buildings on land to be made available at Woods Reservoir by the Air Force. A provision also was included for an additional $250,000 to match funds that supporters hoped would come from the National Science Foundation.

On March 7, the Senate unanimously approved the bill on final reading, and it cleared the House on March 12 by a vote of 85 to 0 and was sent to the governor. Clement signed the bill at 9:30 A.M. Friday, March 15, authorizing construction of the Institute

Dr. Goethert, Robert M. Williams and General William Rogers watch as Governor Clement signs a bill appropriating $1.2 million to establish UTSI. Just behind General Rogers is Senator Ernest Crouch. Others present included Lt. Gov. Jared Maddox and J. Howard Warf, state commissioner of education.

near AEDC with work to start in mid-summer. **General Rogers, Bob Williams,** and **Dr. Goethert** were present for the signing.

On March 14, The Nashville Banner had editorialized that by taking affirmative action, the legislature had "developed a new educational opportunity for young Tennessee scientists/engineers, creating a pool of trained technicians for prospective industry, and above all . . . furthered the nation's efforts in space."

Dr. Bowyer said that Clement's getting the legislature to appropriate $1,250,000 specifically for the institute was "a kind of breakover point. The resistance on the part of the (UT) administration was not antagonistic to the idea of the greatness of the program, but they were responsible for the whole university system. The feeling was that they did not want to become involved in it if they could not do it first class . . . That's where **Frank Clement** became a big factor in shepherding that assistance from the legislature. His enthusiasm and participation as governor was very outstanding."

Senator Crouch remembered it a little differently. He said the University of Tennessee "fought us all the way. They were there when we passed the bill, and then they wanted to put it at Knoxville, away from the supporting laboratories at Arnold Center. But we won."

Created as a branch of the UT Graduate School, the Institute was expected to be in operation by the fall of 1964. Backers expected that it would cost from five to six hundred thousand dollars a year to operate it. Plans at this point were for an initial faculty of twelve and perhaps fifty students. Supporters made a valiant effort to secure a five hundred thousand dollar grant from the National Science Foundation as well as other federal money, but it was not to be. The state of Tennessee shouldered the total burden.

On April 30, General **Schriever** told a group at a dinner at the AEDC Open Mess that he fully expected the Air Force to officially approve the Institute. According to The Tullahoma News of May 3, special guests at that dinner included Tullahoma's **Mayor Jack T. Farrar, Dr. Ewing Threet, Lt. Gov. James Bomar** of Shelbyville, **Dr. Andy Holt, Representative Charlie Hickerson, Senator Ernest Crouch, Generals William Rogers** and **Leif J. Sverdrup,** and **Bob Williams.** Schriever retraced the origins of AEDC, noting that the facility was established under his staff supervision in the Pentagon during planning phases in 1946–47 and that on the last day of its 1949 session, Congress had passed a bill authorizing the center.

"Relations between the Space Institute and AEDC present a unique opportunity," Schriever said. "There is no facility of this kind in the free world today, and I would doubt if there is a facility of this kind even in the Soviet Union." He could not think of a "potentially greater source for advancing education in this complex space age than the marriage of this Space Institute with our laboratories here at AEDC." He assured listeners that the Air Force would do "everything possible to make this the finest of any institute in the whole United States." He noted that Arnold Center was not exclusively Air Force. "We run it," Schriever said, "but we always consider it a national resource."

Governor Clement had flown to Washington in late March to speak with officials of the Area Redevelopment Administration and the National Science Foundation, trying in vain to raise seven hundred and fifty thousand dollars for the Institute.

On April 2, the UT architect, **Malcolm Rice,** gave the go-ahead on preliminary plans, and in May, UT recommended the Nashville firm of **Woolwine, Harwood and Clark** as architects for the Space Institute. The UT trustees, meeting in June, approved the Nashville architects while okaying spending up to two million dollars.

Architect John Harwood, in November, 1988, recalled a sleepless Saturday night in 1963 when he realized that he had passed up a chance to design the Institute: **"Ed Boling** made a rather unexpected and hurried trip to Nashville where he called on me at my home in Brentwood. It seemed some situation had developed requiring immediate selection of an architect for the project. We were unable to agree on some of the terms of my employment, and Ed Boling left with my negative reply. I can still remember awakening (and tossing) during the night and realizing what a great opportunity I had missed—beautiful new site, new concept, no existing buildings to conform nor connect to; the beautiful lake and the rise of Cumberland Plateau in the distance. . . "

Remembering that Boling was staying at the Capitol Inn, Harwood called and asked Boling if it was too late for the architect to change his mind. It wasn't.

"I had almost dealt myself out of the one job that was the most satisfying in my some fifty years of practice," Harwood said, "and clients and staff members who were the most understanding, appreciative and pleasant I had ever had the good fortune to work with."

Dr. Young said this was Harwood's first chance to design a building that did not have to fit in with some sort of existing architecture, and that he spent much more money on the design than he received for it. Harwood was instructed to design a first-class building, allowing plenty of room for expansion. Young said the architect fully subscribed to these instructions and came up with several small separate buildings occupying essentially the present site of the Institute. During a meeting, someone suggested that because of frequent rain, the buildings should be connected by corridors. Prior to this, Goethert had had another person design a small, attractive building to accommodate an estimated dozen faculty members and director plus classrooms. (Young said early planning always centered around twelve faculty members, which he whimsically suggested might have been "biblical.") **Malcolm Rice** had then designed a modest, brick, bank-like building, which was to sit near the existing access road, with no view of the lake. Rice and **Joel Bailey** showed this drawing to **Bob Young** and **John Dicks,** who were skeptical about it and showed it to **Goethert,** who was horrified.

"He pulled out the building they had (previously) designed that sat out on the peninsula," Young said.

Due to an apparent need for connecting covered walkways, the idea of a single building slowly developed. Rice mentioned some circular structure in Rome and drew a sketch on a blackboard, suggesting that Harwood might consider something of this nature.

"So the covered walkway concept became a circular corridor, and the wings extending out represented the individual small buildings that John had originally come up with," Young said. "As a very fine architect, it was quite remarkable that Harwood so rapidly adopted the suggestion that Rice had made." Harwood made a tentative design for a building, called a firm in California that did color drawings, and, acting on Harwood's telephoned directions, the firm did the drawing of the structure that became UTSI.

The Tullahoma News on June 5, 1963, reported that an advisory group, consisting of representatives of the Air Force, UT, and ARO Inc. had been formed to recommend a program for the institute. **Dr. Hilton A. Smith,** dean of the UT Graduate School, **Dr. Robert Young** and **Dr. Marcel K. Newman** from the AEDC Graduate School, and **Dr. Joel F. Bailey** from the Knoxville campus, represented UT. Representing the Air Force were **Lt. Col. Jean Jack,** deputy chief of staff for tests at AEDC, and **Lt. Col. Charles B. Alexander,** deputy chief of staff for civil engineering. ARO was represented by **Edward M. Dougherty,** deputy managing director, and by **Dr. Goethert.**

Colonel Jack was assigned to AEDC in the fall of 1962. Shortly after funding was approved for the Institute, General Rogers appointed Jack as the AEDC project officer to work with UT in getting the buildings up. Colonel Alexander was to be his assistant. Years later, Jack said he always would feel that "had it not been for efforts and the expertise of Colonel Alexander, the buildings which comprise the Space Institute would not be those beautiful structures which arose on the jungle-like edges of Woods Reservoir."

Jack discovered early that support was scarce in Knoxville, and in 1988 he would recall "not-so-fond" memories of the "struggle and frustrations of converting a legislative appropriation of funds for the Space Institute into the beautiful collection of steel, stone and mortar, which make it in its setting on the edge of Woods Reservoir the outstanding entity which it has become." His pleasant memories were of his association with the "outstanding staff and faculty of the institute" and he believed that "these people made the institute the outstanding organization which it has become." After his Air Force retirement, Jack returned to the Space Institute and under the tutelage of Dr. Young and Dr. Goethert became a teacher in the Aerospace Department of Middle Tennessee State University, Murfreesboro. He enrolled in UT's graduate school at AEDC in the fall of 1963 and later took numerous courses at UTSI.

"I have always felt that Dr. Young was one of the real pillars of the Space Institute," Jack said in November, 1988. "Certainly he has been the one who more than anyone else shaped my post Air Force retirement activities."

By early summer, 1963, UT had been authorized funding for two full-time professors for the Space Institute, and an ARO spokesman said funding might be provided for **four** professors as ARO would share services and pay half the salary **if** the faculty members met ARO's requirements.

A September, 1964, opening was now a definite goal, and the formal process for transferring land for the Institute was under way. On June 19, **Franklin D. Roosevelt Jr.,** under secretary of commerce, at a press conference in Nashville predicted the institute would do for Tennessee what MIT had done for the New England area and that it would pull in hundreds of scientific-based industries. **Nat Caldwell,** veteran writer for The Tennessean, said that Roosevelt "elevated" the institute and the lock in the Cordell Hull Dam "above all the others." But the Banner, on June 20, also quoted Roosevelt as saying that no federal money was available for the institute because Congress had defeated President Kennedy's distressed area bill by a close vote of 209 to 204 on June 12. Three Tennessee congressmen—**Bill Brock** of Chattanooga, **Tom Murray** of Jackson, and **James Quillen** of Kingsport had voted against it. Roosevelt hoped those congressmen would "reconsider their vote if it is brought up again," but he said the three representatives who opposed it had "made it impossible temporarily" to obtain federal funds for the Space Institute.

On July 3, The Tullahoma News reported that **Dr. John B. Dicks,** associate professor of physics at the University of the South, Sewanee, had the day before accepted appointment as professor of physics at the Institute, becoming the first additional faculty member hired specifically for the Space Institute. He was to join the UT graduate branch at Arnold Center effective September 1, **1963.** For two years, Dicks had lectured for the UT-AEDC graduate classes as well as being a consultant to ARO. A graduate of the University of the South, Dicks had received his Ph.D. in physics in 1959 from Vanderbilt University and had taught a year at Tennessee Tech. The newspaper reported that Dicks' duties would likely be split equally between research and academics and that he would teach as well as serve as director for research in space technology. A native of Natchez, Mississippi, Dicks was to become a pioneer in leading the Institute into research in magnetohydrodynamics (MHD) and would establish records for securing research funding. Other full-time UT faculty members at the graduate school who would become faculty members at the Space Institute were **Dr. Young,** branch director and professor of mechanical engineering, and **Dr. Marcel Newman, professor of mechanical** engineering. **Dr. C. C. Oehring,** assistant professor of math, taught full time in the graduate program at AEDC, but he left before the Space Institute classes began.

Dr. John B. Dicks, Jr.

After Dr. Holt accepted Goethert's idea for a Space Institute, Young said he was directed to serve under **Dr. Hilton Smith** and later was given thirty thousand dollars to hire faculty for the Institute. This money (from UT) was not spent the first year it was available because "we still weren't underway with the project," Young said. "Looking back, I see that this was a terrific mistake although finding good men in the space business was extremely difficult."

On July 31, 1963, the Air Force announced plans to give UT three hundred and eighty-two acres as a site for the Institute and advised **Senators Kefauver** and **Gore** and Representative **Joe L. Evins** that the Pentagon had authorized the House and Senate Armed Service Committees to declare the land surplus. Valued at $114,600 by the Army Corps of Engineers, the land had been bought for $47,000 in 1950 when the government was buying land for Arnold Center. The unimproved land included a portion that had formerly been used as a managed game reserve.

In August, Goethert led AEDC's third annual aerospace work-study program with thirty-four students from seventeen colleges participating. **Bob Young** directed the program. Later that month, Goethert said bids for the Institute would be taken in the fall, and construction should start by early 1964. Architect Harwood was still awaiting formal approval of schematic drawings. Goethert said the first complex would have an auditorium that would seat seven hundred, including two hundred in the balcony, an instructional building with twelve to fifteen classrooms, an administration building with space for fifteen to twenty faculty members, and a temporary library. It would be a semicircular building on a peninsula extending into Woods Reservoir, bounded on one side by Rollins Creek. Goethert estimated enrollment would be from two hundred to two hundred and fifty students, including those enrolled part time.

Goethert had resisted any talk that the Institute should be located near the existing access road, a short distance form the reservoir. He also had been adamant that the Institute would be far more than a block building as some had suggested. He knew from the start that the architecture of the Space Institute must express his vision of the essence of the facility.

"Neither the University nor I recognized the scope of Goethert's plans nor his ability to accomplish it," Young said. "I did know that UT was sorely pressed to accommodate a rapidly rising undergraduate population and to initiate a full graduate study program. It was for this reason that I expected little support for the Space Institute concept from the

UT administration. But **Andy Holt** and **Goethert** were of the same adventurous type; they genuinely admired each other, and we enjoyed the full support of Dr. Holt. Much less sympathy for the whole idea existed in the UTK colleges and departments. Governor **Clement** was also a Goethert-Holt type, and he supported the idea strongly. As Holt promised, the UTSI initial funding came about in a fashion which demonstrated (at least to Holt's satisfaction) that these monies would not have been available to UT otherwise. At the very end of the legislative session, the initial $1.25 million funding for Phase 1 of the UTSI building (no auditorium, no E wing) was passed. Before Phase 1 was finished, the additional .75 million for completion was voted in the next legislature. But many people at UTK did not buy the idea that monies unavailable to UTK were being used, and we started off under a cloud. Goethert generated a lot of political pressure. He did not understand U.S. politics, but he surely understood how to generate political pressure. **Hilton Smith** once told me at a crucial decision point: 'Bob, the political pressure is so great that we must do this! I hope it works.' ''

While funds had been voted for UTSI by the summer of 1964, Young said ''little had happened.'' He recalled having seen the **first** architectural sketch of the proposed institute: ''In the fall of 1963, the UT architect sent me a sketch of a square brick building located on the campus about where the ECP/MHD building (was later) located. He asked that I approve it. I almost did, but by some luck, I showed it to Goethert. He was not impressed—it did not take advantage of the natural beauty of the site, and the Air Force would never support such a 'small Tennessee school house.' ''

Young said proponents of the Institute had been very fortunate in having **Jack Shea** available to develop ''under the careful direction of Dr. Goethert'' brochures that were prepared in 1962 to assist in the campaign for support. Those brochures, for the first time, had used the name: ''Tennessee Space Institute.''

Colonel Charles B. Alexander, the Air Force's representative on the committee seeking land for the Institute, said some wanted it built beside the access road. Architect Harwood, laying out an entrance to the Institute, first sketched a straight road, Alexander said. But Goethert said, ''No, I want an S-curve'' so the Institute could not be seen until the very last moment as one approached, and it suddenly burst into view. Getting his S-curve proved far easier than standing against the angry Girl Scout leaders when they learned that Goethert's Institute was to be located in the area of the Girl Scouts' campground.

''Goethert stirred up a hornet's nest,'' Dr. Threet said. ''I tried to tell him that hell hath no fury like a woman scorned, and he would say, 'Ah! This is more important than Girl Scouts!' But when all those mothers confronted him, he realized he had a problem. He backed down.''

Charlie Alexander was caught in the middle of the battle. Late in 1988, a few months before his death, he remembered getting a call from **General Rogers,** AEDC commander.

''The Nashville people—Girl Scout executives—had contacted him and said how horrible it would be to have all those men there and the girls just over the ridge,'' Alexander said. ''Rogers and I thought it was funny. I don't recall that we asked for any land beyond that inlet. At one time, we thought about putting dorms over there. Maybe we talked about going to the ridge. I made a lot of trips to Washington about it.''

Mrs. Shirley Kamm, who was Shirley Rogers Male at the time, said UTSI originally was proposed ''on a hill in the Girl Scout area with dorms at the top of the hill—fifty to one hundred yards from the Scouts' main camp.'' Writing under her maiden name, Shirley ''applied a lot of pressure. I wrote the Secretary of the Air Force, the Secretary of the Navy, senators, congressmen, the President, the governor (sometimes weekly).'' In reply, she received carefully worded letters from **Senators Albert Gore Sr. and Estes Kefauver, Congressman Joe L. Evins, Dr. Threet, General Rogers,** and many others. On June 22, 1962, Rogers wrote to **Louis Farrell Jr.,** chairman of the Scouts' Camp Development Committee, that he found it necessary to ''adjust the boundary'' of the area leased by the Scouts to provide the Girl Scout Council with a total of about forty-eight acres rather than the thirty-four acres of the original lease. The AEDC commander also said he would amend the lease to permit the construction and operation of a dock for swimming and/or mooring or launching of small watercraft by the Scouts. The area immediately to the west of the Council's area would be made available to UT for locating the Institute, Rogers wrote.

On January 17, 1963, Shirley, in response to her letter to Air Force Secretary **Eugene Zuckert,** got a letter from **C.W. Harris,** deputy chief, Engineering Division, Directorate of Civil Engineering. He said planning for the Institute did not require using the Girl Scout property, but would require a ''minor adjustment in property lines.'' He assured her that the existing house and waterfront used by the Girl Scouts would not be disturbed so there ''will be no interference with the building or boating or other water activities.''

Shirley also experienced a lot of pressure. People told her she would cause the area to lose the Space Institute. Shirley knew Dr. Goethert. She was confident that he would not let things go that far.

What finally happened, she said, was that the Secretary of the Air Force told Governor Clement that he had better get things straightened out with the Girl Scouts. On February 14, 1963, Shirley Rogers Male met with Governor Clement in his office with **Mrs. D. G. Faulkner,** Girl Scout Council president, **Mrs. Polly Fessey,** executive director, and representatives from the Camp Council, and an Air Force representative (probably **James Emmons—**the man the Air Force had charged with keeping Shirley happy.) Emmons called one day and asked Shirley to ''please send him copies of the letters I sent to the politicians and military as they all came to his desk to answer,'' she said. ''He wanted to be prepared when the phone calls came in.''

Eventually the Scouts were given some acreage.

''To be a Girl Scout camp,'' Mrs. Kamm recalled, ''it must be a certain number of acres—I believe thirty-two. They cut back and then gave us some land across the road . . . They gave us permission to put in a dock. Before, we had no access to the water other than to look at it. We couldn't swim there; we had to go to the recreation area.''

Before the happy ending, however, Shirley had numerous confrontations with Goethert. She gave this account of one: ''A picture came out in the paper, showing the location of UTSI. I called Dr. Goethert and asked, 'What's going on? They are putting your institute right in the middle of my land.' He said he didn't know anything about it. But then he said, 'All we have to do is refund your dollar fee and take it back.' We had just spent three thousand dollars from cookie sales money to bring in electrical power and build a shelter. I told him: 'If you do, I'll bring in all the Girl Scouts from Franklin and Coffee counties and sit around the flag pole at the A&E Building (at Arnold Center).' Goethert always said that I was the only one who ever beat him out of anything.'' She said that the Institute building was re-designed after the location was shifted to the peninsula.

✧ ✧ ✧

In late October, 1963, the General Services Administration gave notice that two tracts on the AEDC reservation—364.62 acres lying west of the Officers Open Mess, bordering Woods Reservoir—would be declared surplus by the Air Force.

Congressman Joe L. Evins reported a bit of bad news: The Area Redevelopment Administration had no money for the hoped-for federal grant.

John W. Harton, whose name would be closely associated with Tullahoma for many years, died on October 12 that year.

On November 9, the Executive Committee of the UT Board of Trustees voted to name the first big building at the Institute in honor of **Frank Clement,** and **Dr. Edward J. Boling,** UT vice president for development, said bids would probably be taken early in 1964. The Tullahoma News, in an editorial, stuck to its original position that it would be appropriate to name the Institute for **General Leif J. Sverdrup** ''whose brain child the whole project is,'' but acknowledged that ''perhaps it is appropriate that the first group of buildings be named for Clement.'' Perhaps, the editorial continued, some future building might be named for Sverdrup.

In December, Governor Clement was notified that General Services Administration had transferred three hundred and sixty-five acres of surplus Air Force property to the U.S. Department of Health, Education and Welfare (HEW). Representative Evins and Senator Gore announced that HEW would convey the title at no cost to UT.

On Wednesday, January 29, 1964, in a Land Grant ceremony at the Officers Club at AEDC, **Captain Richard Lyle** of Atlanta, regional director of HEW, presented the deed to Governor Clement. Lyle, a native of Brownsville, Tennessee, had previously served as Tennessee's commissioner of transportation and as director of prison industries in Tennessee. Accepting the deed, Clement said, ''This marks a new age for the people of the great volunteer state of Tennessee, the nation, and the free world.'' He predicted that UTSI would bring ''growth and progress of a high type that this area has not experienced before.''

''This is as important a single thing as has happened in your lifetime or mine,'' Clement told those (state, UT, AEDC, area and civic officials) attending. Still optimistic about getting at least six hundred thousand dollars from the federal government, Clement said construction would start in the spring.

At 2:30 P.M. on Monday, February 3, 1964, the deed was formally filed in the Register of Deeds office in Franklin County by **Drs. Robert Young** and **John Dicks.** According to a story published February 5 in the Herald-Chronicle, the transaction was noted by the register, **Roy Tipps,** and **Miss Ruth Ann Henley,** deputy register, in the presence of County **Judge C. O. Prince.** The judge anticipated that ''with good access roads in the making,'' the Space Institute would ''play a major role in the growth of Franklin County and all of the Elk River Basin.''

Few people realized what a victory filing the deed represented. In fact, what wound up as an extraordinary site for the Institute might instead have been the location of a garbage dump for the city of Sewanee. Dr. Young explained it this way:

After the legislature voted funds for the Institute in 1963, the Air Force began proceedings to transfer three hundred and sixty-five acres to UT for educational purposes—a "long, torturous path." First, the land had to be declared surplus by the Air Force, then given to the General Services Administration, which was to try to sell it. Simultaneously, UT put in a request for the land free of charge to be used for educational purposes.

"The city of Sewanee, not knowing what was going on, offered to bid on the land for a potential garbage dump," Young said, "and much confusion existed. Negotiations took much longer than anticipated. I remember General Rogers once telling me that he believed the Air Force should take back the land because of all the confusion." This was hardly likely since the Air Force had declared the land surplus to its needs forever. Young also recalled that General Rogers had suggested that some other sites were suitable, including one near the golf course, so Young called **Malcolm Rice,** who drove down from Knoxville a couple of days later and went with Young to call on the general for directions as to which sites they might consider.

"By the time we got there," Young said, "Goethert had apparently talked with General Rogers and sold him again on the (present) site because General Rogers said as far as he was concerned, there was no other suitable land. Mr. Rice and I went to look at the present site; he climbed in his car and went back to Knoxville. Dr. Goethert was determined that the Institute would be in its present location. He was correct, but in the early days it was hard to envision such a remote location."

Goethert and his staff also were saying that the proposed site was about four miles from AEDC, but Young and **Dean Dougherty** discovered that it was almost twelve miles by the way one had to go to get there. The dean suggested that land near the Officers Club was closer and more suitable. As Young said, "The problem was . . . no one had the vision that Dr. Goethert had at that time for long-range development of the property. We are very fortunate that he did stick to his guns and insist that that was the land to be used."

✧ ✧ ✧

At the annual meeting of the AEDC Credit Union's board of directors during the second week of February, 1964, **Aldo J. Zazzi** of Tullahoma was re-elected president, and directors voted to extend Credit Union coverage to include the UT faculty and staff connected with the Space Institute and graduate program at Arnold Center.

On February 28, The News reported that another faculty member—and the second specifically chosen for the Space Institute—had been selected. **Dr. William T. Snyder,** formerly of Knoxville and associate professor of engineering at the State University of New York, was to assume duties as associate professor of aerospace engineering in the summer of 1964. He had obtained his bachelor's in mechanical engineering from UT in 1954 and his doctorate in 1961 from Northwestern University. He had taught at North Carolina State before going to the Long Island Center of State University of New York at Stony Brook. For two years, Snyder had taught in the work study program sponsored by UT, ARO, and the Air Force at Arnold Center. His main area of interest was in fluid dynamics and magnetohydrodynamics.

Dr. William T. Snyder

Early in March, the state announced a delay in starting construction of the Institute until possibly July 1 because of renewed hope of getting federal money. Congress had recently authorized federal aid to higher education at both the undergraduate and graduate levels. Governor Clement and **Harlan Matthews,** state finance commissioner, feared that to start construction might shoot their chances of getting the federal aid. However, in April, UT advertised for bids on the initial structure and set 3 P.M. May 29 for opening the proposals. Bids were asked for a main entrance lobby and three wings—one containing administrative offices, one classrooms, and a library. Bids also were sought for an auditorium, conference room, and labs.

On April 25, **Rudy Abramson** wrote in The Tennessean that **Dr. Goethert,** "internationally known expert in aerodynamics," would probably be named director of the Space Institute, and that **Dr. Holt** would recommend him. Four days later, The Tullahoma News reported that Goethert, at this time vice president and chief scientist of ARO at Arnold Center, and **Dr. Robert L. Young** would head up the Institute. **Goethert** would be the administrative head while Young would be director of academic affairs.

Meanwhile, the Institute faculty was taking shape, and in May, **Dr. Arthur A. Mason,** research assistant in the Physics Department at UT, accepted an offer to come to the Institute in September as assistant professor of physics. Mason had received his doctorate in physics the previous spring at UT. In 1964, he had married

Dr. Arthur A. Mason

Helen Burnette, the executive secretary of the Physics Department where he taught, and while on their honeymoon trip to Florida, Art and Helen swung through the Tullahoma area, looking over the prospects. Dr. Mason became the third faculty member picked specifically for the Space Institute. As it turned out, UTSI also acquired its first librarian in the deal, but even **Helen Mason** did not suspect this at the time her husband accepted the teaching position.

Melson Contractors Inc. of Shelbyville submitted the apparent low bid of $1,127,000 on May 28 for constructing initial Institute facilities. **Sam Melson's** firm was one of five contractors submitting proposals at the University of Tennessee Center at Nashville. The Space Institute Advisory Committee set June 11 for a meeting to review the bids and recommend one to the UT trustees, who would meet in Knoxville on June 18. Others bids were $1,137,000 from Rentenbach Engineering Company, Knoxville; $1,179,652 from Summer Construction Company, Nashville; $1,182,000 from Foster-Creighton Construction Company, Nashville, and $1,271,642 from L. L. Poe Construction Company, Tullahoma.

On June 11, on the Knoxville campus, the UT administration recommended to the advisory committee, whose members concurred, that the trustees be advised to accept the low bid. (Dr. Young was chairman of this advisory committee.) The trustees accepted the bid on June 18 and also officially named Goethert as director, to assume duties on September 1, and Dr. Young as deputy director in charge of the educational program. Young assumed his duties immediately. **Colonel Harrie Richardson,** who by then was with Chrysler Corporation in Detroit, was named as Dr. Goethert's administrative assistant, effective July 15.

Goethert was quoted as saying he thought the Institute was "far ahead of most similar installations now being constructed or planned because of the facilities which AEDC will make available to graduate students pursuing research projects." He said those facilities were "unmatched anywhere in the world, and we plan to make the most of this situation." The trustees also had appointed Goethert as professor of mechanical and aerospace engineering on the UT faculty.

The Tullahoma News editorialized on the appointments, saying they meant that the Institute would be "in capable hands when classes start in the fall."

Still hoping to get federal money, the trustees accepted Melson's base bid on condition that he would give options to accept one or more alternates by April 16, 1965. These alternates included an offer to add a laboratory for $197,000, an auditorium and/or conference center, $306,000, and an option to build both for another $595,000. Work was scheduled to start on the second Monday in July, and Clement and Holt promised a formal groundbreaking within a month. Excavation got under way later in July, but the ceremony was cancelled because of difficulty in scheduling of key officials. Work also was begun on an access road.

Young recalled a meeting with **Goethert, Harvey Cook, Newman** and **Dicks** that summer, and then about the first of August "they announced that we would open in the fall, so we had to attempt to recruit full-time students. I worked like a demon, took no vacation. Our ARO consulting contract was greatly increased, and we were able to hire five graduate research assistants, including **Maurice Hale** and **Dick Shanklin**."

On August 14, 1964, **General B. H. Schriever** announced at Andrews Air Force Base, Maryland, the appointment of **Dr. B. H. Goethert** as chief scientist for the Air Force Systems Command, succeeding **Arthur G. Wimes Jr.**, who had held the position for four years and would continue as assistant deputy chief of staff for science and technology. Goethert's primary responsibility in this new position would be to prepare technical and scientific bases for decisions that had to be made regarding the Air Force research and development programs. It was a good deal for Goethert, putting him into a close association with General Schriever, but it also meant he would go to Washington, D.C., and Goethert was supposed to head up the Space Institute, which everyone knew was opening in September.

"Harvey Cook and **Ed Dougherty** came to me," Bob Young remembered, "and asked me if I felt Goethert should be appointed as part-time director of the Institute for that (one-year) period. He had had so much to do with establishing it, he was the only one they saw who was able to make it fly. Much to my chagrin, I had to agree. I recommended that he be made part-time director while he was moving to Washington."

(**Dr. Newman** said he also discussed with Dean Dougherty and others the question of whether Dr. Goethert was "the right type of man" to head the Institute. "But it was concluded, yes, let us take the chance.") **Newman,** who may have considered himself a likely candidate for the director's position, would, in addition to his professorial duties, become director of research at the new institute.

As it turned out, General Schriever asked Goethert to stay on a second year as chief scientist for the Systems Command. During those two years, Goethert spent most of his time in Washington and kept in touch with Dr. Young and the Institute's business largely by telephone.

"At the end of the second year," Young said, **"Harvey Cook** came to me to ask if UT would be interested in Goethert being full-time director. I said I would investigate. Shortly afterwards, UT decided yes, with the provision that they would first clear it with ARO. **Hilton Smith** came to **Bob Williams** to see if ARO would release him (Goethert). Williams agreed, and Doc returned in the fall of 1966 as full-time director and professor of aerospace engineering."

Young said some thought Goethert's dual role as director of UTSI and chief scientist at the Systems Command was a conflict of interest. It very likely was a conflict, Young said, but "his boss, General Schriever, supported him strongly, and the conflict of interest charge got no place. General Schriever was very interested in technical education for Air Force personnel, and he particularly liked the idea of graduate study with the opportunity to utilize AEDC facilities. As a young officer, he had much benefitted from a similar program at Wright Air Field in Dayton. He spoke often of the benefits of his academic program there with work in the U.S. Air Force engine (piston) facility. Thus we enjoyed very good support from him."

Retention of young Air Force technical officers was proving to be a problem, and both Schriever and Goethert felt that a set-up like UTSI would assist in retention.

But in September, 1964, as the big day approached, **Bob Young** had his hands full, making certain that there would be students and, since construction was just getting started on the Institute, that they would have a place to meet.

It was a beginning.

✦

Chapter Six

LET THE CLASSES BEGIN

CLASSES officially started on **September 24, 1964,** in offices provided by the Air Force at Arnold Center with five full-time professors and eight part-time faculty members. Eight full-time students were enrolled with two hundred and sixty-six others signed on part time. Full-time students included three junior Air Force officers: **Lt. Johnny M. Rampy** from Edwards Air Force Base, **Lt. Elmer E. Goins** from Vandenberg AFB, and **Lt. Jimmie D. Young** from Holloman AFB. Fifteen other Air Force officers were enrolled part time.

The five civilians who were enrolled as full-time (nine quarter hours minimum) students were **Richard Vair Shanklin III, Samuel Ralph Pate, Maurice Grimes Hale, Oliver Boyd Lee Jr.**, and **Wolfgang Bergt,** a doctoral candidate from the University of Aachen in West Germany. While some published reports listed only three civilians, records in the UTSI registrar's office confirm that five civilians and three officers enrolled full time for the fall of 1964. The civilians, except for Bergt, were either employed by or on leaves of absence from ARO, Inc. and had been attending the graduate school at AEDC. All five were residing in Tullahoma.

Hale had received his bachelor's degree from UT in June, Shanklin's came in 1959 from Duke University, Pate's in 1960 from Auburn University and Lee's in 1958 from UT. Bergt had just graduated from Aachen.

Students, faculty and guests must have felt a special excitement as they filed into AEDC's main conference room on the afternoon of September 25, 1964, for the first academic year convocation of the University of Tennessee Space Institute. Some of the sharpest scientific minds in the country were assembled there along with the cream of the crop from the Air Force, private industry, and the University of Tennessee. At 3:30 P.M., **Brigidier General Lee V. Gossick,** AEDC commander, welcomed the visitors and then turned the program over to **Dr. Andrew David Holt,** president of the University of Tennessee.

Under "Resident Faculty and Staff," the program listed the following: **Dr. B. H. Goethert,** director; **Mr. F. H. Richardson,** assistant to the director; **Dr. Robert L. Young,** deputy director for educational programs; **Dr. Marcel K. Newman,** research coordinator; **Miss Virginia Richardson,** registrar; **Mrs. Joyce Lewis,** executive secretary, and **Dr. John B. Dicks Jr., Dr. Arthur Mason,** and **Dr. William T. Snyder.** (Young and Newman also were professors and with Dicks, Mason and Snyder constituted the full-time faculty.) Adjunct faculty members were **Mr. James Cunningham, Dr. Martin Grabau, Dr. Paul H. Hutcheson, Dr. Richard A. Kroeger, Dr. Wendell S. Norman, Mr. J. Leith Potter, Dr. Leon E. Ring,** and **Dr. Peter A. Schoeck.**

Listed as "Distinguished Guests" were **Congressman L. Mendel Rivers,** Armed Services Committee; **Major General Perry M. Hoisington,** director, Office of Legislative Liaison; **General L. J. Sverdrup,** chairman of the board, ARO, Inc.; **Mr. Robert M. Williams,** president, ARO, Inc., and **Dr. E. R. van Driest,** director, Space Sciences Laboratory, North American Aviation.

Seated at the speaker's table were **General Bernard A. Schriever,** commander, Air Force Systems Command; **Dr. Goethert; Mr. Williams; Dr. Hilton Smith,** dean of the UT Graduate School; **Dr. Herman E. Spivey,** academic vice president of UT; **Dr. N. W. Dougherty,** dean emeritus of the UT College of Engineering, and **Colonel Edgar Masters,** officer in charge of the systems command's educational programs. **Dr. Alvin H. Nielsen,** dean of the College of Liberal Arts and head of the Physics Department at UT, also was a guest.

Dr. Holt introduced **General Sverdrup** (who had to leave before the convocation ended) as "the man who had the idea that we ought to have a Space

Institute, the one who pushed it, and a ball of fire if I ever saw one."

Dr. Goethert, the first speaker, beaming and buoyant, spoke directly to the students and faculty, challenging them to "burst the plateau and move ahead" in aerospace science. Scoffing at the idea that scientists had reached a plateau in space knowledge, Goethert declared, "A plateau exists only as long as there are no people imaginative and vigorous enough to burst the plateau and move ahead. That this will not happen is our—and particularly your—solemn responsibility."

General Schriever, the second speaker, agreed with Goethert and added that "timid souls never accomplished anything." He said a recent Air Force survey had "identified more technical opportunities than we can ever take advantage of, particularly in the field of aerospace," and declared that "This nation has unlimited horizons." Citing the value of the multi-million-dollar test facilities at AEDC to students in a graduate institute, the general said he knew the value of such facilities because of his experience at an engineering school at Wright Field before World War II.

"We will make the labs here at Arnold available to the Space Institute as long as I'm in the Systems Command," he promised, "and I'm sure that will be the policy anyhow. It just makes good sense."

Goethert also emphasized the relationship of the Institute with AEDC activities. Describing the Institute as "unique," he said it should prove even more valuable to military, industrial, and academic institutions in years to come. One example of goals yet to be attained was the development of controllable landings from space excursions instead of their dropping into the ocean by parachutes. Goethert also outlined his concept of the Institute's mission: "The mission of the new Institute, as we see it, is not that of a conventional university branch. It reflects a novel approach in the engineering—natural sciences field to have **education, research** and **development** closely aligned to the Institute's activities in these years. Progress of science and technology is so rapid that engineering would be hopelessly lagging behind new findings of research unless an extremely close association is maintained. On the other hand, also the teaching at the university is in grave danger to limp behind the rapidly advancing status of research and engineering systems." It was hard to say whether there was a greater shortage of students or of professors in the "advanced fringe areas of science and technology where the decisive progress of our time occurs. This new Institute is clearly aiming at overcoming these difficulties by linking it closely to the development work at the Arnold Center. . . "

Goethert also told the gathering that because of the opening of the Space Institute, the Advisory Group for Aeronautics of the North Atlantic Treaty Organization (AGARD) had scheduled a meeting at Arnold Center for October of the next year. He paid tribute to Governor Clement, Generals Schriever, Gossick and Sverdrup, and to Mr. Williams for their support, saying that without their continued backing, the "idea of the Institute would lose its life's blood."

Heavy rain and wind put a crimp in plans to lay a cornerstone at the construction site on October 28, but **Andy Holt's** sparkling wit, music by three high school bands, and predictions by Goethert and Schriever of things to come made the ceremonies at the National Guard Armory in Tullahoma a lively affair on that Wednesday afternoon. Another ray of sunshine came when Dr. Holt announced that UT would ask the 1965 legislature for another million dollars to finance additional construction at the Institute.

About two hundred people braved the weather for the 3:30 P.M. ceremony. The **Rev. A. Richard Smith,** pastor of Trinity Lutheran Church, Tullahoma, gave the invocation, and **Jerome Taylor** of Knoxville, chairman of the building committee of UT trustees, presided. Bands from Tullahoma, Coffee and Franklin County high schools performed individually beforehand and then joined in a stirring rendition of the National Anthem at the conclusion of the ceremony. Unfavorable flying weather prevented Governor Clement from attending, but **Lieutenant Governor James Bomar** of Shelbyville was on hand to unveil a simulated cornerstone. (Displayed next to the podium was the copper box, containing documents and photographs pertaining to the Institute, that later was placed in a corner of the building that would bear the governor's name.) Bomar emphasized that it was the **people,** through their lawmakers, who had provided funds to establish the Space Institute.

As Dr. Holt was speaking, General Schriever, his flight from Washington delayed by the weather, entered the armory. The UT president stopped speaking and stepped down from the podium to greet the general.

"You know how a master sergeant feels when the general walks in," Holt quipped. "That's just what happened to me."

Moments later, after introducing **Dr. Nathan W. Dougherty,** dean emeritus of the UT College of Engineering, Holt also introduced **Edward W. Dougherty** of Tullahoma, deputy managing director of ARO, adding, "Ed's greatest distinction is that he is the son of Dean Dougherty."

Dr. Young talks with Eldridge Goins, Jim Young and John Rampy, three of the first full-time students.

When Holt introduced cigar-smoking **Sam T. Wallace,** head of ARO's plant protection division and former National Guard station commander in Tullahoma, he said, "Sam's the one who got us in here out of the rain." Glancing around and not seeing Wallace, Holt added, "He's probably in his office talking to the weatherman right now!" He recalled that the first time he met **Bob Williams,** "I didn't know what to think because he is kind of quiet, and that is strange to us school teachers. But I found out that when he does speak, he says something." Of the deputy institute director for educational programs, Holt punned: "Bob's been with our program down here since 1957, and he is still a **Young** man."

Others introduced by Holt included **Dr. Herman Spivey,** academic vice president of UT; **Keith Hampton,** state personnel commissioner; Purchasing Commissioner **Grundy Quarles; Sam Melson,** contractor; Architect **John Harwood; Boyd Garrett** of Nashville, a UT trustee; **Kirby Primm,** a member of the UT Development Council; **Donald Sahli,** executive director of the Tennessee Educational Association; Dean **Hilton A. Smith,** and Dean **A. T. Granger,** both of UT.

Goethert and Schriever predicted that the Institute would be greatly expanded in enrollment and physical size within a decade. Increased enrollment and the attraction of the Institute in the U.S. and abroad would make a second campus area necessary by 1969, Goethert said. Schriever said enrollment would double by 1970, and the size of the faculty would grow at an even faster rate.

Praising the governor and the 1963 legislature for providing the initial funds, Holt said, "Governor Clement embraced the idea wholeheartedly when I first told him about the proposal for the Institute. Before I got through, he was saying my speech to me. The legislators also were sold completely on this project before I got to them, but we've found out that we should have asked for $2,250,000," and he added that

Founding faculty, from left, Dr. Grabau, part-time, Professor Snyder, Dr. J. Leith Potter and Dr. Leon Ring, part-time, Deputy Director Young, Director Goethert, Dr. James Cunningham, part-time, Professors Newman, Dicks and Mason, and Dr. Richard Kroeger, part-time. Dr. Newman also was Research Coordinator.

another million dollars would be requested "to get this thing done properly."

"We have a pretty good building that's going to get us in out of the rain," Holt said, "but we don't have much room in which to operate."

The UT president reviewed the state's unsuccessful efforts to obtain an additional seven hundred and fifty thousand dollars in federal money. At this time, the state still held an option in its contract with Melson to contract for additional units if funds could be obtained by the following April.

Dr. Goethert, after talking about the AGARD conference that was planned for the next October, predicted that **"Governor Clement** will welcome the visitors **in our new auditorium."**

"Two years from now," Goethert said, "I can envision that an additional building, already designed, will be there, and in 1969 the influx of students and others will be so great that another campus area, also designed already and for which there is ample room, will be opened." He said plans called for the instructional and conference buildings to be clustered around a central hall with dining facilities and that in the 1970's there would be student dormitories and other dwelling facilities for faculty and students. (He was correct, but those things would come about only after a long hard struggle.) "Let's be sure that we provide the facilities and opportunities for our young people of today," Goethert urged. "We must strive to make the (Space Institute) one of the best in the country and throughout the world."

Schriever assured his listeners that the existing curriculum and buildings were "just the beginning" because technological development was only starting. "The Tennessee Space Institute is uniquely situated to make a major contribution," the general declared. "It will have a quality unmatched in the world for research into propulsion, design and operation, advanced computer theory and research and nuclear power, with our location near Oak Ridge."

While first classes were being held in AEDC offices, the peninsula was being prepared as a site of the Institute.

Schriever noted that up-to-date aerospace research and instruction would be especially vital for the development of such things as supersonic transport planes, aircraft flying six to twelve times the speed of sound, and large-sized planes able to take off and land vertically. He predicted that all of these developments would come about by 1980. He also said that the Institute's budget should rise from about three hundred and thirty thousand dollars a year to the point that from one million to a million and a half dollars would be spent annually for new facilities, and he said the operating budget should be approaching five million dollars by 1975.

After a great majority of those in the audience rose in response to his request for all UT alumni to stand, President Holt drew heavy applause when he quipped, "May your tribe increase."

One person missing from the group of dignitaries at the ceremony was **Dr. Frank L. Wattendorf,** but his absence was not due to lack of interest in the Space Institute. In a letter dated November 6, 1964, Dr. Wattendorf, serving then as vice chairman for the Advisory Group for Aeronautical Research and Development, wrote that he had just returned from Paris and, too late, had received his invitation to the cornerstone ceremonies. While expressing his regrets at being unable to attend, he added, "As you know, for the past fifteen years I have been deeply interested in the creation of an academic center at Tullahoma complementary to AEDC."

General Lee Gossick and Dr. Goethert accompanied by Air Force officers inspect construction of A-wing. Gossick helped students get through that first year. (John Rampy is on the left of Goethert.)

Indeed, Dr. Wattendorf's interest in such an institution had not been overlooked. Among documents placed in the box for the cornerstone was a copy of a memorandum, dated June 1, 1950, and written by Dr. Wattendorf, at that time scientific advisor of the Air Engineering Development Division. This memorandum was generally regarded as the **initial suggestion** that a facility such as the Space Institute be established near Arnold Engineering Development Center.

Wattendorf's suggestion had, after all those years, been taken seriously, but for those early students, the Space Institute was anything but glamorous, and before contractors would get the mud cleaned up around the new building, Dr. Goethert himself would almost get thrown out of the place.

✦

Col. Harrie Richardson, left, leads National Advisory Board members on a tour of the construction. Professors included in the group are Dr. Marcel K. Newman, third from left, Dr. Bob Young, fifth from left, and Dr. Art Mason, seventh from right.

General William Rogers was a strong supporter of early efforts to establish the Institute. Here is a meeting between, from left, Trustee Gerome Taylor, Dr. Andy Holt, Dean Nathan Dougherty, General Rogers, Dr. Goethert, and Dr. Hilton Smith.

Bob Williams, left, ARO president, Dr. John Wild and Dr. Goethert had many meetings such as this.

Generals Shriever and Sverdrup were boosters of UTSI. In 1968 both appeared in a public lecture at the Institute. From left are Bob Williams, Dr. Goethert, General Sverdrup, General Shriever, and AEDC Commander General Lundquist.

John Stack, second from left, supported UTSI and helped get its library started. With him in the library are Art and Helen Mason and Dr. Goethert.

Celebrating Colonel Harrie Richardson's 70th birthday on June 27, 1968, are from left Janice Wheeler, Dianne Hall, Doris Billingsley, Virginia Richardson, Mary Ann Marchand, Richardson, Barbara Perry, Helen Mason, Martha Smith and Janet Jones. Behind Richardson are Rachael Stewart and Joyce Moore.

Chapter Seven

'A BRIDGE BETWEEN COUNTRIES'

WHEN Johnny Rampy got a call from the Air Force Systems Command in July, 1964, asking if he would like to attend the University of Tennessee Space Institute, his first response was "No, thanks." He had never heard of the place, which he assumed was in East Tennessee. A lieutenant at Edwards Air Force Base in California, Johnny figured he would continue working toward a master's degree at the University of Southern California. Then he mentioned the offer to his wife, **Marie.** Johnny and Marie had been sweethearts at Woodland High School in Alabama in the fifties when young Rampy, an outstanding quarterback, dreamed of getting a scholarship and a chance to play for **Bobby Dodd's Georgia Bulldogs.** But in the fall of 1954—his senior year—a dislocated shoulder spoiled those plans. He still wanted to go to college, but money was short, so he worked a couple of years for Lockheed Aircraft Company before being laid off. Prior to joining the Air Force, Rampy attended the University of Georgia and later earned his bachelor's degree from Auburn University.

As soon as Marie learned that the Institute was in Tennessee, she said, "Let's go!" After two years in the desert, the idea of getting closer to their families in Alabama was appealing. So they moved to Tennessee, arriving in August.

It was a shock, not only to Johnny Rampy, but also to two other Air Force officers—**Elmer Goins** and **Jim Young**—when they arrived to find classes meeting in the A & E Building, in Propulsion Wind Tunnel offices, and at other places at AEDC.

"We got here expecting to find a school," Rampy recalled years later, "and they didn't even have a building! Wherever they could find a conference room or some place to teach classes, they did. So we were a little disappointed. We caused Dr. Young some headaches because we had expected to find a school, and of course they were at that time trying to establish one. He (Young) was the vanguard . . . working very hard to get the Space Institute idea going. He was trying to get through all the publicity matters of establishing a university as well as conducting class schedules. We were disappointed because of no facilities." He remembered that it was "hectic, always, trying to get established classes while we were still at AEDC. I salute Dr. Young. He always had to arrange for classes, and he was always trying to make people feel at ease in some not-too-good situations. He was able to make things enjoyable for most people; his sense of humor made those first days more bearable."

Rampy knew that UTSI officials were having administrative problems with UT in those early days: "The University wasn't really all that enthralled with having this Space Institute located down here in Tullahoma, so there were things that had to be worked out, like how to **process a thesis,** how to get it approved. One of the most difficult things that I encountered was the fact that it was so hard to process a thesis through the system because we were remotely located from Knoxville."

Despite hardships, Rampy remembered that "Overall, it was a very good experience. We moved into base housing. The Air Force didn't know how to treat us. Here we were, students assigned to the Space Institute. There was some doubt in their minds as to what our role was in the Air Force here. **General (Lee) Gossick,** certainly one of the best commanders the Air Force could have, looked after us. He talked to us about the Space Institute, about the role the Air Force was playing in it, and about the importance of it . . . He made us feel welcome. A lot of the bureaucracy here within the Air Force didn't know how to treat us, but General Gossick told them he wanted it done this way, and they kinda did that. We didn't have on uniforms for one thing, yet we were coming in the gate of the center because all the classes were here."

While physical accommodations were poor, the level of instruction that those first students received was "extremely high," Rampy said. "We had good people who worked at the center as (part-time) instructors. They were the best that I've seen . . . the level of instruction here was the best." Young had a way of teaching "so that you could understand. He was extremely bright. He is the **kingpin** behind the Space Institute. Doc (Goethert) was the visionary, but if you look at any movement, you've got to have a man of vision and someone who implements it, and Dr. Young was the kingpin behind making the Space Institute happen."

While Goethert was away, Young handled diplomacy, convincing companies to send their best engineers to UTSI for an education. "Doc was heavily influential in that because he knew all the people—vice presidents and presidents of these aerospace companies—but Dr. Young was very good at implementing the whole idea. The Space Institute was blessed with having Dr. Goethert and Dr. Young as two main players."

The officers found the math program to be weak at the start even though **"Dr. Grabau** was an excellent mathematician. Our complaint was not on the quality of instructors, but the fact that we had such a limited number of math courses."

Rampy remembered having a math class from **Dr. Marcel Newman,** but most of his classes were under **Leon Ring, Dr. William Snyder, Dr. Walter Frost, Dr. Wendell Norman,** and **Dr. Young.**

While Goethert and Schriever both visited during the Institute's first year, Rampy said the student-officers dealt mostly with **Dr. Richard Kroeger,** who worked for Goethert and was thesis advisor for the officers. Goethert's office (as chief scientist for ARO) was still in the A&E Building, and Kroeger made space next to that office available for the three officers to use as a study room.

"Coming to a school **that wasn't a school,"** Rampy said, "one of the things that you didn't have was a place to work." Once, he remembered that an effort was made to give **Wolfgang Bergt** the office space, but the officers resisted, and other arrangements were made for the Ph.D. candidate from Aachen.

While records of that first quarter show that five civilians also were carrying full-time loads, Rampy only recalled two: **Richard Shanklin** and **Maurice Hale,** both on leave from ARO. "Dr. Young was very specific in telling me that Shanklin was the first civilian student they had," Rampy said, "and we were the first military people." (Actually, Hale, a native of McMinnville, who registered July 25, 1964, believed he was the first full-time civilian student. Records show that Shanklin processed a "Permit to Enter" UTSI on May 28, 1964.)

Rampy, Goins and (Jim) Young attended classes in the new UTSI building during their last quarter before completing their work for master's degrees in aerospace engineering in December, 1965.

Rampy spent seven years in the Air Force and worked five years with the Northrup Corporation in Huntsville before returning to Arnold Center in May, 1974. At the time of an interview on March 13, 1989, he was technical director, Deputy for Operations, at AEDC. He had kept in touch with **Elmer Goins,** who was practicing law in Dallas. Northrup was about to send Rampy back to California in 1974, and he was reluctant to leave the Middle Tennessee area, so he "wound up here (at Arnold Center) by default. I had always respected the professional people and their mission at Arnold Center."

Rampy concluded that the Space Institute was "a good idea, even with all the minor things that went on. Certainly, it is a major accomplishment to start something as significant as the Space Institute away from the University. This school, with time, will become a lot stronger (and with renewed scientific interest) the Space Institute will . . . become finally what Doc really wanted it to become. There are very few schools that have technical programs that are directed toward space exploration. Fifteen years ago, every university in the country was trying to gear up to that. Now there's only a few."

As for the civilians who first enrolled full time at UTSI, **Hale** received his master's in engineering science in the winter of 1966; **Lee** received a master's in electrical engineering in the spring of 1965; **Shanklin** got his master's in mechanical engineering in the spring of 1965 and his Ph.D. in aerospace engineering in December, 1971. **Bergt** received his Ph.D. in aerospace engineering in August, 1969. **Samuel R. Pate** received his master's in mechanical engineering in the winter of 1965 and in 1977, he earned his Ph.D. in aerospace engineering from UT, Knoxville. Like Rampy, Pate had received his bachelor's degree from Auburn in aerospace engineering. He first joined Arnold Center forces as an employee of ARO assigned to the von Karman Gas Dynamics Facility. By 1989, Pate was serving as senior vice president and general manager of the Southern Group, Sverdrup Corporation, and he became president in 1990. Years later, **Maurice Hale** came back and in March, 1987, got another master's degree, this time in Industrial Engineering with emphasis in Engineering Management. In earlier years he also did some flying (without pay) for Goethert and for **Colonel Henry Sherborne** in the Aviation Systems program.

✧ ✧ ✧

It was a rocky start, but 1965 moved the Space Institute farther toward becoming what Dr. Goethert had envisioned. Late in January, Governor Clement presented a budget to the legislature calling for $13,565,000 in UT construction funds. Senator Crouch, who headed the state senate's finance, ways and means committee, said that between seven hundred and fifty thousand and a million dollars of this budget would go toward completing construction at UTSI. On February 26, The Tullahoma News reported that the legislature had approved money for the work at the Institute, and Dr. **Edward J. Boling,** UT vice president for development, had "little doubt" that UT trustees would act well ahead of their scheduled June meeting to see that the work started. The University had asked for a thirty to forty-five day extension of its option to contract with Melson for completion of the building. Work was still under way on the first phase, and now funding had been assured for completing the planned auditorium-conference center, a second classroom and a laboratory. Melson's bid to complete the additional facilities had totaled six hundred and ninety-five thousand dollars.

In March, **Harrie Richardson,** assistant to the UTSI director, said the Institute would have a budget of six hundred and five thousand dollars for its first full fiscal year of operation. It was expected that the Institute would have fifteen faculty members and forty full-time students by the end of 1965, but, he said, "At the present, we have five full-time faculty members and eight part-time, if you count Dr. Goethert and myself among the latter."

As expected, UT in April gave approval for completion of the Institute, exercising an option to contract for it at the bid price of six hundred and ninety-five thousand dollars. The News reported on April 23 that Melson workers had started staking out the site of the auditorium the day before after getting the go-ahead from UT officials. A UT spokesman said that a two-point-four-mile access road was expected to be included in a June 25 contract letting by the state highway department.

Institute officials were hoping that the auditorium-conference room would be completed by October in time for a meeting of the Advisory Group for Aeronautical Research and Development (AGARD) of the North Atlantic Treaty Organization—an event that promised to bring about five hundred scientists from twelve countries to the new Institute. It was part of Goethert's plan to open the Space Institute with a bang. Meanwhile, UTSI's staff was growing. **Freeman D. Binkley** of Tullahoma, an employee of ARO for nearly eleven years, had become business manager for the Institute. In April, **Colonel Harold A. Ellison,** a retired Army officer, had joined the staff as administrative assistant. Ellison's military career had included five years on the staff of **Dr. von Karman** when von Karman was chairman of the NATO Advisory Group for Aeronautical Research and Development. Ellison had met Dr. Goethert while serving as an executive of the fluid dynamics panel in Paris.

On April 28, The Tullahoma News ran an editorial entitled "Predictions Coming True," citing the announced completion plans as evidence that predictions by Goethert and Schriever about the Institute's growth had been well founded. "All the signs point to an exciting future for the Space Institute," The News concluded. "We are fortunate to have it in our area."

In May, 1965, two other Air Force lieutenants—**First Lt. Thaddeus M. Sanford** and **Second Lt. Robert E. Wilson**— had joined Rampy, Goins and Young as the first full-time students sponsored by the Air Force at UTSI. Three other officers came the next month as full-time students. These were **Major Alfred R. Deptula, Lt. Kenneth L. Anderson,** and **Lt. John V. Kitowski.** Kitowski earned his Ph.D in aerospace engineering in the winter of 1968, the **first Air Force officer to earn a doctorate at UTSI.** His Ph.D. also was the first in aerospace engineering at UTSI. His professor was **Dr. William T. Snyder.** Sanford, who earned his master's in the winter of 1966, retired in 1989 as commander of the Flight Dynamics Lab at Wright-Patterson.

In June, **Dr. Frank Sevcik,** formerly a scientist for Chrysler Corporation's missile division in Detroit, was appointed assistant professor of physics at UTSI. Appointment of the native of Czechoslovakia and of **Dr. C. Lamar Wiginton** as assistant math professor, gave the Institute a total of seven full-time faculty members including Dr. Young, deputy director, Dr. Newman, research coordinator, Drs. Dicks, Snyder and Mason. Formerly a mathematics teacher at UT, Wiginton joined the faculty on June 1, 1965, having worked since January as a consultant. A native of Burnsville, Mississippi, he had entered UT in 1957 where he subsequently earned bachelor's and master's degrees in mechanical engineering and his Ph.D. in mathematics. He also had been a consultant in plasma physics with the Oak Ridge National Laboratory for three years.

Four more full-time professors would come on board by September 1. These included a husband and wife team, **Doctors Jimmy** and **Susan Wu,** who at that time were with Guggenheim Aeronautical Laboratory and Electro Optical System in Pasadena, California, **Dr. Walter George Frost,** who was completing doctoral work at the University of Washington, and **Dr. Gerhard W. Braun,** a former student of **Dr. Ludwig Prandtl,** coming from the naval laboratory at Point Magu, California.

Frost, a naturalized U.S. citizen, was born in Edmonton, Alberta, Canada, and had attended Mount Royal (Junior) College at Calgary for two years. His bachelor's, master's and doctoral degrees all came from the University of Washington at Seattle, and he had been a teaching assistant there. He moved to Tullahoma in August, 1965. He met his first classes in the A&E Building at Arnold Center. Years later, he remembered a day when the distinguished **Dr. Marcel Newman,** running late to a class of his own, decided to save time by climbing over a fence. In doing so, he set off an alarm, bringing security guards on the run.

Braun, born on December 23, 1907, in Danzig, Germany, came to the United States in 1945 as a consultant at Wright Air Force Base, Dayton. He later served as scientific adviser to the Directorate of Advanced Technology and in 1962 became chief scientist of the Air Force's Western Test Range in California, a position he gave up to become professor of aerospace engineering at UTSI for the term beginning in September, 1965. He also became coordinator of UTSI's exchange agreement with the University of Aachen, signed that fall. Braun had received a diploma of engineering at the Institute of Technology in Danzig in 1928, his master's in math also from the institute at Danzig and his doctorate in applied mechanics, math, from the University of Gottingen. Long after he retired on August 31, 1973, Dr. Braun remembered that "For his Institute, Dr. Goethert aimed right from the beginning at the highest goals," even though the Institute lacked manpower.

At the time UTSI was getting started, Braun said, "the re-entry of warheads and spacecrafts into the earth atmosphere still presented hard problems. Dr. Goethert had worked as Air Force chief scientist directly for the four-star General Schriever. The general thought so highly of Dr. Goethert's scientific abilities that he was willing to let a major contract on atmospheric re-entry to the young UTSI. Here was for our institute the possibility to become the center of atmospheric re-entry technology. Obviously, Dr. Goethert could not do this research all by himself. He asked **Dr. William T. Snyder** and myself to work with him on the expected contract. Unfortunately, neither Dr. Snyder nor I had any experience in the field of hypersonics as it was needed for this contract, so UTSI had to pass its first chance for becoming a center of excellence and reach its goal later."

Braun also would remember that Bob Young's "way of teaching and acting as associate dean has always appealed to my old-fashioned European thinking." Braun liked to call himself "old-fashioned" because, he said, of the influence that **Ludwig Prandtl,** his teacher and thesis supervisor, had had on him. He felt that Young had given him the freedom to "try following my old teacher."

Of those early faculty members, **Young, Mason, Dicks, Jimmy Wu,** and **Frost** would still be going strong when the Space Institute marked its Silver Anniversary. **Newman** and **Braun** had retired, **Snyder** had gone on to Knoxville to become engineering dean, and **Susan Wu** had opened a private consulting firm.

Mid-South Pavers of Nashville had submitted the low bid of two hundred thirty-seven thousand, nine hundred and eighty dollars to build an access road from the Institute to the AEDC access highway.

In July, about sixty-five military and civilian scientists of the Air Force Systems Command attended a two-week course in supersonic combustion ramjet engine technology at the Space Institute (in AEDC facilities). General Schriever spoke at a dinner at the Officers Open Mess for UTSI students, top executives from eighteen leading aerospace firms, and state and area officials and civic leaders on the first evening of the session.

By July 1965, the **Wu couple** had arrived, making a total of nine full-time faculty members on board within less than ten months after the opening of the Institute. Born in China, the Wu's came to the United States in 1957. She held a bachelor's degree from the National Taiwan University, received her master's in aerospace engineering from Ohio State University and her doctorate from California Institute of Technology. Her husband received his bachelor's in mechanical engineering from the National Taiwan University and after moving to the U.S., he received both master's and doctoral degrees in aerospace engineering at the California Institute of Technology. Many universities at that time would not employ a husband and wife on their faculty, **Susan Wu** later said, but they heard about a space institute in Tullahoma, Tennessee, and Goethert interviewed them (at the Holiday Inn) during one of his visits to Los Angeles.

"We tried to call the Institute," Dr. Wu said, "and the operator said she had no number for a space institute. We thought this was very strange that the operator had not even heard of the place. We wondered if there really was a space institute, but we knew that Dr. Goethert was not a fake." They soon realized why the number was not listed at that time—the Space Institute had no telephone because at that time it was still under construction.

The Air Force awarded a research contract to the Space Institute in mid-summer of 1965 to investigate the possibility of converting rocket exhaust energy into electrical power: magnetohydrodynamic (MHD) power conversion. **Dr. John B. Dicks** headed the project under the largest contract the Space Institute had yet received. Other contracts had included a grant for research study of MHD channel flows, headed by **Dr. Bill Snyder.** Dicks, who would become increasingly involved in the MHD program, at that time pointed out that the concept behind the research project was not new, noting that a testing facility under construction at Arnold Center would use an MHD generator.

On August 13, The Tullahoma News reported that Colonel Richardson had revealed plans to occupy the new building before October, saying that some classes would be held there by the latter part of September and all classes would be in the new structure by October 15. A separate story announced that a sixteen-member industrial advisory group **(National Advisory Group)** had been named, consisting of top engineering executives of leading aerospace companies in the nation. This group had held its first meeting on August 11 at AEDC. Its membership sounded like a Who's Who in aerospace research and represented an important plank in Goethert's drive to establish a close association between the Institute and industry. Charter members were **R. A. Herzmank,** chief propulsion engineer of the McDonnell Aircraft Corporation, St. Louis; **V. Tizio,** aerojet project chief of Republic Aviation Corp., Farmingdale, Long Island, New York; **R. F. Sweek** of Curtiss-Wright Corp., West Lynn, Massachusetts; **E. T. Williams,** chief engineer for advanced aerothermodynamics of Douglas Aircraft Co., Santa Monica, California; **D. R. Kent,** program manager for manned hypersonic vehicle studies of General Dynamics Corp., Fort Worth, Texas; **R. E. Wilshusen,** The Marquardt Corp., Van Nuys, California; **Dr. J. Frank Sutton,** vice president and research director for Lockheed Georgia Co., Marietta, Georgia; **H. M. Graham,** engineering director for the LTV Astronautics Division of Ling-Temco-Vought Inc., Dallas, Texas; **Herman Pusin** of Martin Marietta Corp., Marietta, Georgia; **William R. Collier** of General Electric Co., West Lynn, Massachusetts; **Reeves Morrison,** propulsion specialist for United Aircraft Corp., Hartford, Connecticut; **Arthur Gilmore** of Grumman Aircraft Corp., Bethpage, New York; **W. A. Martin,** industrial director of research and engineering of North American Aviation Inc., El Segunda, California; **George C. Grogan,** vice president and technical operations manager of Northrup Aircraft Crop.; **George S. Schairer,** vice president for research and development of The Boeing Co., Seattle, Washington, and **Edward M. Dougherty,** deputy managing director of ARO, Inc., Tullahoma.

"It was amazing to see the level of people who agreed to serve," Dr. Young later said, "but it was not surprising. A cover letter from General Schriever and a descriptive letter from Dr. Goethert engendered great enthusiasm among the recipients." This was understandable, for Schriever's ballistic missile budget was in the billions.

(In 1967, the first NASA representative—**Bob Kamm**—was named to the advisory group. At that time, Kamm was director of NASA's western operations offices, headquartered in Santa Monica, California. Soon afterward, he would join his friend Dr. Goethert at the Institute.)

In late August, the Upper Duck River Development Association proposed building a dam on Duck River, which subsequently materialized, creating Normandy Lake that became the water supply for both Manchester and Tullahoma.

With pre-registration coming up in mid-September, Dr. Young was looking for thirty full-time students, including for the first time four engineers assigned to the Institute by leading industries for graduate work in aerospace sciences. Young said two would come from North American Aviation Corporation and one each from Northrup Corporation and General Dynamics Corporation. By the time all classes were under way in the new building in November, enrollment totaled two hundred and forty-two students, with twenty-seven of these full time. Of the latter, eight were Air Force officers and fourteen were research assistants.

On Friday morning, September 17, The News announced that UTSI officials would sign an agreement on the following Monday providing for exchange of students and faculty members with the Technical University of Aachen in West Germany. An accompanying photograph showed three students—**Peter Wolf, Wolfgang Bergt** and **Peter Muller,** research assistants and doctoral candidates at UTSI—with **Dr. Gerhard Braun,** coordinator of the Aachen exchange program, and **Dr. Wilhelm Dettmering,** visiting research professor from Aachen. A fourth Aachen exchange student was **Hrishikesh Saha,** who would later become the first president of the Space Institute's Student Government Association (SISGA). Dettmering remained at UTSI for a couple of months, teaching courses in turbojet aircraft engines. Another visiting professor from Germany at this time was **Dr. Peter Schoeck** with the German Aeronautics Test Center Branch at Stuttgart. **Dr. Andy Holt** noted that the exchange program, in effect, had already started since four students from the University of Aachen were already enrolled, including Bergt, who had enrolled the previous year, and three who were entering in the fall of 1965. The agreement was formally signed during a convocation in the main auditorium at AEDC at 4 P.M., September 20, marking the start of the Institute's second fall quarter. Joining Dr. Holt at the ceremony were **Dr. Herman Spivey,** academic vice president, **Dean Hilton Smith** of the UT

Graduate School, and Dr. Goethert. Representing the University of Aachen were its rector, **Dr. Volker Aschoff** and **Dr. A.W. Quick,** director of Aachen's Institute for Aeronautical and Space Technology. About two hundred persons gathered in the main auditorium at Arnold Center for this historic ceremony, and the exchange program officially went into effect the next day. Presiding, Dr. Holt praised the contract for its importance to international science, saying: "We look forward to a long and mutually profitable association. We believe this agreement has international significance. This contract being in the new aerospace sciences is somewhat unique and promises a resulting program having an international importance."

Dr. Aschoff also praised it, noting that the contract was in the spirit of work accomplished by the late **Dr. Theodore von Karman.** He said the agreement joined forces with leading universities in the two-thousand-year-old city of Aachen. The University of Aachen, he said, had about two thousand foreign students from seventeen countries.

Dr. Goethert stressed: "We cannot just copy a program of another university. We must develop a unique program specifically tailored to the resources here and to the needs of the aerospace engineering industrial firms, research laboratories and governmental groups." He said the Space Institute would operate on three principles: That teaching, combined with research, presents the optimum form of operation, that teachers must have a successful background in engineering or must be closely connected with engineering work while teaching at the Institute, and that professors must stay in "living contact" with international aerospace programs by carrying out research contracts.

"We have to become a part of the worldwide scientific community," Goethert declared, "and the Space Institute will begin to form a **bridge** between countries and continents."

Meanwhile, the twelfth full-time faculty member—**William L. Morris**—had been named assistant professor of mathematics at UTSI. The Ohio native had received his master's degree in mathematics from the University of Cincinnati and started work on his doctorate in Knoxville in 1962, where he also had taught mathematics. He was senior research engineer for General Dynamics Corporation and had been a consultant to the Aerospace Corporation.

As last-minute preparations were being made to move into the new building, tragedy struck one of the men who had helped to get legislation passed to establish the Space Institute. **Charlie Hickerson** died in a Nashville hospital on October 7, 1965, two days after the car in which he was riding alone crashed near Eaglesville. He had served as state representative in 1963 and 1964 and had co-sponsored the legislation that provided the first state money to build the Space Institute. At the time of his death, Hickerson was president of the Coffee County Bar Association and city attorney for Tullahoma. It would be slightly more than four years later that another automobile accident would claim the life of still a greater booster of the Institute, **Frank G. Clement.**

In September of 1965, **Dewey Vincent** joined the UTSI staff as its first manager of the physical plant, and he still held this position on October 13, 1988, when he remembered that "We got partial occupancy (of the building) in October, 1965." Vincent had worked with the contractor constructing the Space Institute before joining its staff. It was November 19, 1965, however, before The News reported that the Institute would start holding all classes in the Frank G. Clement Building on the following Monday (November 22). According to this story, the "fifteen full-time faculty members, administrative staff and employees" had moved their offices to the new building the previous week. The administrative wing, the first wing for classrooms and faculty offices, and a library wing were complete. **Harold Ellison** said that a laboratory wing and a second classroom-faculty office wing were about eighty-five percent complete and should be ready for occupancy by the following February 1. Construction of the auditorium-conference center wing was expected to be complete by the next spring. Because of a shortage of materials, work had been halted temporarily on the lobby, which would have walls of tinted glass. Indeed, the lobby at the time the building opened was anything but finished. Plywood walls were thrown up on each side, and a latch-string lock was put on each door leading into the lobby space.

It had rained a lot that fall, and red Tennessee mud abounded. The auditorium was a long way from being finished, and **Dr. B. H. Goethert** was dead set on having his AGARD meeting in late October, bringing aerospace scientists from across the United States and Europe to the Space Institute. **General Lee Gossick** daily checked the site, walking out front, shaking his head, trying to decide whether the visiting scientists could indeed meet in the new institute as Goethert wanted or whether Gossick would have to make arrangements to accommodate the AGARD meeting at Arnold Center.

Leaves were turning brilliant yellow, crimson, and purple. Wind blowing in off Woods Reservoir had taken on a chilling edge. A pinch-hitting switchboard operator was shivering in the make-shift reception area. It was going to be an interesting winter.

✦

Chapter Eight

THE WIND BLEW SHARP AND COLD

"The connecting lobby was still open (no windows), permitting the free flow of wind and weather, rain and snow. One donned coats and scarves for the trip and made a mad dash across, praying not to slip on the terrazzo. It was necessary to pause at either end long enough to pull the latch strings on the temporary wooden doors before proceeding breathlessly into the opposite hall, apologizing to the receptionist for the accompanying wintry blasts . . ."

THUS Helen Mason remembered the winter of 1965–66, a time when trips between the administrative offices and the academic wings were major expeditions.

"When we moved in," **Dewey Vincent** agreed, "the lobby was unfinished, and there was no auditorium . . . so there was a temporary plywood wall across that corridor there where you turn to go down to A Wing. We moved into A, B and C Wings. We took occupancy of that building down as far as C-102—the one room beyond C Wing. C Wing is where the library was originally located. We didn't have any library furniture when we moved . . . that was a great big wide open space. The space up on the second floor above it was an unfinished area. The contract didn't call for it to be finished. We moved from the A&E Building in October." Moving was not a major job, he admitted, since there was little, other than people, to be moved.

In late October, a new telephone system was put into operation at the Institute: a dial private branch exchange with twenty-five telephone stations and six outside line circuits as well as two direct lines connecting the Institute with the AEDC switchboard. **W. E. (Bill) Stephens,** group manager for Southern Bell at the time, speculated that the Institute's telephone needs would double when dormitories were built during the next four years. (It would actually be more than six years before any dorms were built.)

Ruth Binkley, wife of the Institute's first business manager, **Freeman,** served as the first part-time switchboard operator at UTSI. It was a cold job that first winter, and workmen chopped a hole in the plywood wall and set a kerosene Salamander heater in the lobby area so that it would blow warm air through the hole to keep Ruth halfway warm. The first full-time operator was **Glenna Putnam Booker,** who had been Arnold Center's first chief operator. She began work at UTSI on January 2, 1966, and continued through November, 1969. In addition to serving as receptionist and operating the switchboard, Glenna also sold textbooks. In December, 1989, living in Nashville, she would remember that it was "so cold" when she started at UTSI, and "there was no roof on the auditorium, and it snowed." She also remembered that her job was fun: "I got to talk to aerospace presidents from all over the world." She remembered when **Martha Smith** (Goethert's secretary in the late 1960's) called her at home one Sunday, needing help in contacting someone in Washington with an unlisted number. "I went out to the Institute and got to work on it," Glenna said, "and after awhile, I got the number for her."

Before the move was complete, Dr. Goethert's long-awaited AGARD meeting drew near. General Gossick paced in front of the unfinished building, wondering about the weather, wondering about the muddy driveway, wondering about the barren insides of the new Institute. On Sunday night, October 22, 1965, before the AGARD meetings would start the next day, Manchester and Tullahoma Chambers of Commerce sponsored a dinner at the Arnold Center Officers Open Mess. Governor Clement addressed the visiting dignitaries, and **Dr. Ewing J. Threet** served as master of ceremonies. All of the visiting scientists—about one hundred and fifty aerospace scientists from the United States and Europe who had come to discuss the aerodynamics of power plant installation—area civic leaders and other officials were guests at the dinner. **Goethert was excited.** It was, afterall, an auspicious gathering that appropriately coincided with opening of the Institute, and he was determined that the meeting would be held in the new facility.

"It was held in Lower C because there was no furniture for the library in there at that time," Vincent said. "Our library—what few volumes we had—was being maintained over in Manchester in the home of the Masons. So they had this AGARD meeting in Lower C, and we put classroom chairs in there, and the Air Force shipped in a couple of sound-proof booths for interpreting. Lecturers would be talking in English, German, French . . . and be interpreted simultaneously back to the various attendees, who had headsets. We used that one room down beyond C Wing (C-102) where we all ate our lunches, where the vending machines were. A meal was catered in the hallways. Doc was gung ho on Internationals."

Dr. Young remembered that the moving was done in stages during 1965 and that "We were terribly crowded at AEDC. The Wu's were there, Frost, Newman . . . it was a trying time."

Dewey Vincent, who worked directly for **Colonel Harold Ellison,** said they started out with three hourly employees. The first man he hired was **Norman Adams** of Manchester, who worked for the telephone company during the day and as custodian at the Institute from 4 P.M. until midnight.

"We didn't have any maintenance people the first year," Vincent said. "We didn't need them because the building was still in warranty. If a light bulb went out, we changed it ourselves."

The second man Vincent hired as a custodian was **John C. (Charlie) Burton,** who had worked at Camp Forrest before becoming a barber at the old King Hotel in Tullahoma. Both Adams and Burton retired from their jobs as did **Charlie Childress,** another of the earlier employees hired by Vincent. Childress, who retired in the spring of 1988, "was never late a day in his life," Vincent said. "He's got a son who graduated from the Air Force Academy, a bird colonel."

Charlie Burton occasionally brought his barber tools with him. "He cut my hair," his former boss said, "cut Ellison's, too, sitting in the back of B Wing." There was talk for awhile of Charlie's opening a shop on the second floor.

A xerox machine, office supplies, and an old Model 85 offset press were in the first office (B-100) of B Wing when the Institute opened, and when

Dewey Vincent

secretaries wanted something printed, they would type it on a direct image master and run it on the press. **Rachael Stewart,** who later taught school in Winchester, was one of the first secretaries at UTSI. She may well have been the first Institute employee to suffer a football injury, too. That was the same Saturday that **Ruth Binkley** showed up wearing helmet and shoulder pads for a staff football game. It was an example of the "good times" that **Dewey Vincent** remembered from those early days.

"We had a football game out here one Saturday afternoon. Secretaries played the faculty and staff out in the front circle. We had officials and a chain crew. Ruth showed up with shoulder pads and a helmet. It was touch football, but during the course of the afternoon, one secretary (Rachel) got her leg broke, and another got her finger broke. We were just having a good time."

Ruth Porter Binkley moved with her husband to Tullahoma in 1951, and they both went to work for ARO, Inc. at AEDC. Her job was in the personnel office, located in an old barracks at Northern Field. In 1954, Ruth was selected as AEDC's first "Water Carnival Queen." Two years later, she left her job to devote time to her growing family, but when Freeman became the Space Institute's first business manager, she went to work as a clerk-typist for **Dr. Marcel Newman,** who was deeply involved in a "Scramjet" short course. In the years that followed, Ruth often worked with her husband in the business office, including a lot of Saturdays. She also worked for Dr. Goethert and, after his retirement, she was an executive secretary for his successors. At the time of her retirement on May 31, 1989, she was an administrative assistant in the dean's office.

Vincent remembered that while the staff was small, they got along well and "coming to work was a real pleasure" in those early days. "On Friday, somebody would come down through the building, stick their head in the door, and say, 'Let's all get together and go over to the Officers Club tonight for dinner.' We would—the entire staff and a lot of the students. We'd put one big table together, just like one big family. Doc (Goethert) worked hard, but he played just as hard as he worked. He could run right up to a point and quit and go to playing. Loved to play tennis."

Doc also liked parties, and Vincent remembered the first Christmas party held at UTSI—one that Goethert did not attend since he was in Washington. Vincent was in Ellison's office when the colonel was talking by phone with Goethert, who was at Andrews Air Force Base.

"Ellison was very precise," Vincent said. "Should have been teaching English. He called Goethert 'Bern**hard.**' I heard him say, 'Oh, by the way, Bern**hard,** we're having our Christmas party tonight. I wish you could be with us.' A blank look came on his face. He paused, then, and looked at me and said, 'He hung up on me!' Doc loved to go to parties. It was too late for him to come, otherwise, he'd been here." That first party was held in A Wing.

Indeed, the Institute has had its lighter moments, ranging from birthday observances in the wing offices to Christmas dinner dances, from paper airplane contests and major anniversary celebrations—pot-luck dinners, balls, picnics, watermelon cuttings, and dinners. One of the early anniversary celebrations featured a pot-luck dinner served and eaten in the Administration Wing hall and a dance in the foyer. Later ones have featured sit-down dinners in the cafeteria, followed by formal balls with live bands. Christmas was usually celebrated twice—the Space Institute Student Government Association traditionally sponsored a dinner dance, usually at the Officers Club, and a pot-luck luncheon was always featured on the last working day before the Christmas holidays.

Vincent, who chauffeured Goethert on many occasions, said "Doc never was one to get to something on time. He always liked to be just a little late where everybody could see him when he came in. He liked to be seen." One Sunday, Goethert asked Vincent to drive him to a meeting in Oak Ridge. One of Goethert's sons was visiting, and when Vincent got to Manchester, he found that Goethert was still having breakfast. Goethert apologized for being late and suggested that he sit in the back seat and prepare a few remarks in case he was called upon to speak. Vincent vowed that he would get his boss to the meeting on time despite their late start. There were no interstates between Manchester and Oak Ridge at that time. Dewey pushed the state car hard and arrived in Oak Ridge with time to spare. But Goethert wasn't ready to go to the meeting.

"He wanted to go shopping," Vincent remembered. "He carried me to a bowling alley, and we sat and watched women bowl. He was determined to be late . . . I had got him there thirty minutes early, and he watched the ladies bowl just so he could be late. He was going to have it his way."

Vincent knew Goethert as a great believer in interdisciplinary approaches: "He was always opposed to different departments. He liked to interface. He was opposed to different little groups sitting down at the cafeteria, eating lunch together, like the physics group or the metalurgists. He liked for them all to sit and eat together. He said if you could get engineers and (other) people talking to each other, you could learn a whole lot and get a lot done. The MHD group used to all get together; he didn't like that. He wanted everybody to mingle . . . all the different disciplines. . . . You go to Knoxville, and you're structured in departments. He was much opposed to this." If one group was doing one type research and another a different type, Goethert believed they needed to talk to each other. "That was one of Doc's philosophies," Vincent said. "He used to stress this. And he was great on ceremony."

Access to the Institute was not convenient that first winter. To get to work, employees would drive down the road in front of AEDC's Gate 2 to the pumping station and to the recreation area, where the paved road stopped. Later, the state built a link to the access road, and it was paved and ready except it did not have a center line, which the contract required before opening the road for traffic. The contract specified that the state had to paint the line. **Sam Wallace** was in charge of security at AEDC and, Vincent knew, was "quite a politician. So I called and asked if he could do us a favor. I said, 'We'd like to use this new road, but it has no line, and the state is supposed to do it. If we could get it painted, we could have a straight shot into work and save us a lot of miles.' Wallace said he would take care of it. Next day, state crews showed up, painted the line, and opened the road that same day."

The contractor worked all that first year, finishing the building. The entire complex had been designed, but the first contract only covered part of the work, including a water treatment plant. Before the first phase was finished, Melson was awarded a contract to finish the rest, including lobby, auditorium and D and E Wings.

The Institute's water came from Woods Reservoir, and with its water treatment plant, UTSI ran its own purification system for years, but it "got to where it was not economically feasible to operate our small plant because the cost per gallon was so high," Vincent said, "and there were so many regulations. We saw an opportunity to tie into the city of Estill Springs' water supply, and we shut this one down. This saved lots of money."

Charlie Burton gave up barbering at the King Hotel and started to work for Dewey Vincent on October 23, 1965—just in time to help with the move into the Institute.

"I moved 'em over from the A&E Building," Burton recalled on November 30, 1988, after his semi-retirement the previous January 28. "Moved 'em in my station wagon. I unloaded the first typewriter—it was red or pink. **Rachel Stewart** got it."

Only one room—the one that would later become **Dr. Susan Wu's** office—was complete when Charlie hired on; it was occupied by Ellison and Vincent. According to Burton, this was the only room with tile; the others had concrete floors, and no doors had been hung.

"Whenever I came out here, there was nothing but a pile of gravel and a buncha red mud. The library wasn't finished. They had a AGARD meeting back there, the biggest thing I ever saw. I never saw so many flags! All they had in there was lights and sound booths."

Burton worked from 4 P.M. to midnight as janitor. One evening, **Harold Ellison** and **Dewey Vincent** told him: "We want to get everything looking good. **Old Doctor Goethert** will be here tomorrow. We want everything spic and span for him."

"Every night we had to get the clay mud out that the carpenters had carried in during the day," Charlie said. "That night, I looked over there and somebody had went in that door. I went over there. He was throwing papers out there at that switchboard just here and yonder. And, Buddy, I reached up there, and I said, 'What are you doing in here?' I caught hold of his shoulder. He looked around and started talking that German talk, and I couldn't tell what he said. I said, 'You'd better talk right, so's I can understand you!' So he started talking English then, where I could hear him." The stranger talked clearly enough for Charlie Burton to soon realize that he was challenging "the boss," **Dr. B. H. Goethert.**

"You're not supposed to be here until tomorrow," Charlie stammered, and Goethert answered: "You can't ever tell when I'm supposed to be here."

"That's the first time I had seen the man," Charlie said. "That's how I met him, right there (at the receptionist's desk). He had come in from Washington. Here one day, and he went back."

Burton also remembered UTSI's first mascot and how **Dr. Bob Young** "boiled for days" after somebody stole Snowball. "Prettiest white dog, but he was wild. Nobody could get close to him. They had put in this eating place (where the short courses later were held) with vending machines . . . hash and things like that. I'd get a can every day and set it out for Snowball. I got him tamed. He'd lie by the road, waiting for me. Knew my car. By the time we got E Wing opened downstairs, we decided we'd keep him as a mascot. We made up money, and I took him to town to get him vaccinated. They said they couldn't tell which one was driving, the dog or me. He was sitting up there like he owned the car. Everybody fed him, and he was pretty and fat. He never got dirty in all that mud; he looked snow white. A student stole him and carried him up into Kentucky. Dr. Young got so mad, he boiled for a week or two, and he was going to go get that dog. We had finally found who had carried it off. He was supposed to bring him back, but he never did."

Born in Moore County, **Charlie Burton** had grown up farming. On December 30, 1940, he started work in the hospital at **Camp Forrest.** He was promoted to junior cook, a job he held until the camp closed. He turned down a chance to transfer to Pearl Harbor and instead went to work in Oak Ridge. Later, while driving a truck out of Nashville, he became ill and was unable to work for several years. This led to his decision to enter a rehabilitation program to learn barbering.

"They all but throwed me out," he said. "I hadn't finished high school, but I had taken an exam in Murfreesboro, and my grades were too high. I went to **Frank Clement's** office. He was a good friend. He said, 'Go back to the rehab office; they won't throw you out again.' They rolled the carpet out, and I was in barber college the next day."

One of Charlie's best customers at the King was **Dr. Jimmy Wu.** "Whenever he first came to Tullahoma, I cut his and his kids' hair, and he said nobody else in Tullahoma could please him." Dr. Wu thought it would be a good idea if Charlie would set up shop at the Institute. "He got permission through Knoxville and an act of Congress, I reckon, to get me to put a chair up where the drafting room is (C-Wing). It was empty then. I kept telling him the cost was too much, that we'd have to have it inspected and hot water put in. It was a big space; we'd have had to put in a wall. There was no heat up there. Then it turned out that we could only do it on Saturday and Sunday, and I was not fixing to do that. The reason I had quit was to get out of cutting on Saturday."

Through the years, before retiring as senior materials control clerk, Burton worked as janitor, yardman, mail clerk, printer, projectionist; he kept vehicles gassed, made trips to Nashville and Knoxville for supplies, and operated the water plant. He remembered that Goethert's first secretary was "stubborn." Once while he was dusting her desk, she complained that a pencil was not lying in the right place, so Charles didn't dust her desk anymore. He also remembered a gun-toting preacher who worked nights. Once, Charlie attempted to deliver a message from Dewey relative to the man's duties, but the preacher interrupted, flopped a huge pistol down on the table, and said if Dewey had anything to tell him, he could tell it himself. This sufficed; Charlie didn't deliver any further messages.

"I guess he (the preacher) didn't have any heat where he lived. Some nights when it got real cold, he would bring his wife and kids out here, and they'd sleep on the tables upstairs."

But Dewey Vincent was "the best boss I ever had. He has been a daddy to me. He's the only reason I stayed all those years. He'd say, 'Aw, Charlie, let's try it another day . . . see what happens . . .' Doc wanted things done **his** way. I showed a film of a plane coming across the screen the wrong way . . . I hadn't rewound the film. Doc didn't catch it, but somebody told him, and I got chewed out!"

A bit of ingenuity by Charlie and Dewey one summer day brought a stream of curious workers from Arnold Center. Charlie told it this way: "We only had a push mower—eighteen-inch, self-propelled—to mow all this grass. Buddy, it was hot! We figured out how to make it cut itself, and all I had to do was put in the gas. We got us a six-inch post, put it up in the middle of the circle (in front of the Institute), took a rope, wrapped it around and around all the way to the top and tied it to the mower. Then we cranked her up and turned her loose! I think ARO closed up so everybody could come by to see it."

When the newspaper reported that all of UTSI's classes would be moved to the new building, the story also noted that **Mrs. Arthur (Helen) Mason,** Institute librarian, would "begin cataloguing 1,000 books, which have been ordered by the Institute faculty," and that the library would go into operation after this job was completed and furnishings had arrived. The library office actually became available in January, 1966.

Becoming UTSI's first librarian, a position she held for twenty-four years, came as a surprise for **Helen Mason** who, until she moved to Manchester soon after she had married Art, had been administrative secretary of the UTK Physics Department. It was Doc Goethert's idea for her to be the librarian. In March, 1965, The Republic Aviation Corporation, acting on a suggestion by Vice President **John Stack,** gave a thousand dollars to the Space Institute for the purchase of books in the areas of hypersonic and supersonic combustion. Stack made a personal contribution of two hundred dollars. **Art Mason** suggested to Goethert that the Space Institute put up money to start a library. Goethert put ten thousand dollars into the budget, then asked: "Who will order the books?" The faculty was polled as to which books should be purchased, but Goethert persisted: "Who will order them?" Then he answered his own question: "I know! Mrs. Mason will." He instructed her to proceed.

"She started the library in a back bedroom of our house for three to six months," her husband said. "Then they opened the library in the C Wing where the business office is now."

Goethert, still in Washington with the Air Force Systems Command, assigned Dr. Mason the task of obtaining a list of desirable acquisitions with which to start building a collection for the library. Mason circulated a memo to the faculty and to many AEDC scientists, requesting suggestions and getting a good response. Mrs. Mason was put on the payroll as a temporary, part-time secretary, and assigned the duty of ordering books and checking invoices as books and periodicals arrived. Once the library office opened, a year's accumulation of journals (about one hundred and sixty subscriptions) and perhaps a thousand books were transferred to the prison-made shelves in the library office, arranged alphabetically by author until Library of Congress cards could be obtained and the books catalogued and processed. Mrs. Mason remembered that the main library space "was still a construction site—dirty concrete floors, five-gallon drums, scraps of building material, unpainted walls. . . ."

Joyce Moore, a newly employed secretary, spent the early part of 1966 typing orders and cards for the library. Later, she was assigned as secretary to the faculty and moved into Upper B Wing and later to Upper E Wing. In 1975, after being away for about three years, she returned to the Space Institute as a secretary for Mr. Kamm. At the time of the Silver Anniversary, she was still in A Wing, serving as coordinator of contracts. Her father-in-law, **Athol F. (Jabo) Moore** of Winchester, was superintendent during the initial construction of the Space Institute. **Mary Ann Marchand** was hired in early 1966 as an assistant to Mrs. Mason, who said she "filled that position very capably until mid-1968. She became expert at lettering call numbers on book spines and had fifteen hundred ready for use when the library was opened for use in May, 1966."

Librarians were in short supply in 1965, but since Goethert apparently saw no reason why she could not satisfactorily perform as the UTSI librarian, Helen Mason settled into the job with enthusiasm and dedication.

"We bought and diligently studied two or three books on library practices," she said, "consulted with **Miss Olive Branch,** one of the librarians at UTK, and spent many hours on the telephone getting help from the AEDC librarians. Much credit is due **Mary Colvin, Sally Matlack, Gerald Beckham, Joe Ashley, Della Burch,** and many others who helped us over the years. They taught us (Helen and Mary Ann) nearly everything we needed to know to start and staff the library and provided us with a stream of loans to supplement UTSI's holdings. **Gaye Goethert** (Doc's daughter-in-law) and the present staff at AEDC have continued the generous support begun in 1965." (In the late 1970's, **Suzy Qualls** of Manchester worked part-time in the library.)

In 1989, Helen recalled that **Marjorie Joseph** of Manchester was responsible for the growth of the Technical Files (reports) in the library. She joined the staff in the 1970's.

"We had interviewed two persons when **Dr. Roy Joseph,** a recent addition to the Electrical Engineering faculty, showed up one day to announce that his wife was a librarian and might come for an interview if we were interested. We were, she did, and we found her to be a delightful person, willing to try to switch from being a children's librarian to being a documents specialist librarian on a graduate school level." Goethert, still director of the Institute even though miles away, agreed, and Mrs. Joseph became the third person on the library staff. Two had Master of Library Science degrees, and each person worked thirty hours a week. This level of staffing continued for several years. (Dr. and Mrs. Joseph of Manchester were still at UTSI when the Institute marked its twenty-fifth year. Also at this time, **Carol Dace,** wife of **John Dace,** an engineer with ECP, was library assistant, and **Brenda Brooks** was library clerk.)

While the library was initially planned to have both floors of the C Wing, the upstairs was not finished for quite a while. Collections grew, so the reports were moved upstairs and for awhile, Mrs. Joseph's closest contacts were with a few faculty and students who shared that area.

"This separation of staff and resources was not good," Helen Mason said, "but it lasted until the space was reassigned due to the growth of the Energy Conversion Programs. Meanwhile, Marge diligently set about learning the special problems of special documents librarians, and there were many problems (including) relationships with agencies such as the Defense Documents Center where one ordered Department of Defense publications and others within their purview, NASA, AGARD, ERA, EPA, and their predecessors. . . ." Other duties included finding sources for materials, identifying poorly cited documents, devising classification and cataloging systems. As the Institute's interests expanded, so did the library's problems.

"We were pleased to have a class from Peabody come down to visit us to see our different special documents system," Mrs. Mason said. "Except for AEDC, which was closed to them, ours was the only such library within many miles."

In July, 1967, **Mary Ann Marchand,** assistant librarian, obtained permission from Dr. Goethert to organize an art show at UTSI. An art student herself, she contacted artists from as far away as Huntsville and Chattanooga, and the display, held in the library, was well received. Later, Mary Ann became a professional artist and moved to Atlanta, painting under the name **Mary Booth-Owen.**

Over the years, the library received funds totaling sixty thousand dollars from Health, Education and Welfare grants. Some funds were used for cultural materials, some for minorities and disadvantaged students, but most went for standard library acquisitions. The library also received about two thousand, five hundred dollars in the early years from the American Institute of Physics for journals. After **Dr. Martin Grabau's** death in 1965, his wife invited UTSI to choose from his library, and ninety-one volumes were chosen for the UTSI library. Grabau, a Harvard Ph.D. and professor of Electrical Engineering, had taught for all eight years in UT's graduate program at AEDC (1956–1964) and was also on the Space Institute's faculty in 1964 and 1965. Other individuals who have contributed collections or money have included **Wallace McBride** of General Electric, **Herman Pusin** of Martin-Marietta, **Dr. Kennedy Rupert, H. K. Matt, T. C. Helvey, R. S. Sleeper, J. R. Connell** and **Colonel Wilson J. Boaz,** associate Aviations Systems professor from March 15, 1977, until early 1980. Long-time contributors also included **Drs. Susan and Jimmy Wu, Newman and Young.**

"**Colonel Masters** made annual gifts to the Alumni Fund," Mrs. Mason said, "as does **Dr. Mason,** and **Dr. Trevor Moulden** has made several valuable contributions." The AEDC library has made many surplus volumes available to the Institute's library, including bound journals when much of their collection was converted to microfiche. When NASA closed its western operations office in Santa Monica in 1968, **Bob Kamm,** who was director of that office, gave UTSI most of the holdings and some equipment from that office. Since 1967, when members donated subscriptions to two journals, the AIAA Student Chapter has supported the library with gifts. In 1969, they used prize money from the Bendix Competition award to establish a Visual Aids Library Collection, which was enlarged by a gift of slides, movies and photographs from Boeing's Seattle plant. Student Government Association members also have provided other video materials, journals and general reading materials. In 1973, **Dr. Kennedy Rubert,** then a member of the UTSI faculty, donated a thousand dollars for the **Diederich Memorial Collection** of propulsion publications. (**Herman Diederich** was a professor at Cornell University.) UTSI matched the donation, and books and reports were bought to complete a modest collection.

"We were particularly fortunate to obtain some reports from **Johns Hopkins,** and we had them bound into two volumes," Mrs. Mason said. "These were difficult to track down and obtain, having gone quickly out of print."

UTSI officials found themselves with a fine new building and no furniture and no money to spend on furniture. **Dr. Arthur Mason** said he and **Freeman Binkley** looked at "some very nice furniture," but there was no money, so they turned to UT officials in Knoxville, but couldn't convince them "that we needed fancy furniture such as we had looked at. Eventually, UT agreed to loan money for furniture if the Institute would buy where UT bought it, in North Carolina."

Bob Young said that **Colonel Ellison** was put in charge of getting furniture and told to "lay out the finest. We got in touch with a very exotic furniture supplier in North Carolina, and they sent out a representative to lay it out for us. By the time the building was completed, it turned out that we didn't have any money for furniture. The only furniture we have that came through the consultant is the airport-type furniture and round tables in the lobby. The University essentially just abandoned us on the furniture business, saying we had no money left, and some very, very bitter discussions occurred between Goethert and Dr. Boling, who at that time was (UT) vice president of development. Boling kept recommending that we use quite a bit of prison furniture, which was cheap and sturdy, and of course this didn't at all correspond to Dr. Goethert's idea of furniture to complement what was a really beautiful building. Our book cases were made in prison. Then we reached a compromise. Some money was found for a modest amount of furniture, and it was put out for bids. Very soon we received a very harsh call from **Morris Simon,** who had an interest in Webber's Inc. (an office supply store in Tullahoma), saying that they had not been invited to bid. So we corrected that, and Webber's won the bid. They bid very low and did a great service to us and installed the furniture, which is most that we have today (1980's)."

Actually, there was a bit more to the Webber's transaction. Simon called his old friend, **Julian Harris,** who was in charge of public affairs for UT, and told him, "I'm sore as a boil, and I'm going to cause trouble!" Julian asked him to be patient while he got in touch with UT's purchasing agent. Late in the week, the purchasing agent called Simon, informing him that purchasing officials had not been aware that Webber's could handle such an order. Bid invitations had called for more than thirty thousand dollars worth of furniture, he said, and bids were due on the following Monday. There would not be time to send specifications to Simon. Sorry.

Simon knew that the specs called for furniture equivalent to that supplied by the **Alma Desk Company** of North Carolina. It so happened that Webber's had been doing business with that firm. Simon called Alma and asked the sales manager for the specifications, instructing him to "Wire me the cost price of all the furniture that they are ordering. I'm going to bid, and I'm going to get it."

Webber's bid cost price on everything. **Manager Artie Golden** drove to Knoxville with the bids.

"They said UTSI wanted better furniture but couldn't afford it," Simon recalled years later. "So, I bid one price on the specifications and one price on the better furniture." Webber's price on the better furniture, for use in the top executive's offices, was below the lowest offer on the **cheaper** furniture by all Knoxville bidders.

"The Knoxville bidders had a big profit figured in," Simon said. And what about Webber's? "We got a two percent discount for paying within ten days," Simon said. "I went to the bank and borrowed money and paid the bill," and gave the profit to Golden, who had spent a lot of time on last-minute figuring.

Julian Harris called, saying: "You taught our purchasing department a good lesson."

"I was interested in UTSI having good furniture," Simon said.

Webber's had sold UTSI more than thirty thousand dollars worth of furniture, and nearly a quarter of a century later, Dr. Young said, "It has served us well."

Morris Simon, too, would serve the Space Institute well in various ways over the next twenty-five years.

✦

Chapter Nine

A NEW KIND OF SUPPORT

DESPITE inconveniences of continued construction, the Space Institute quickly moved forward in 1966 with several new faculty members, dedication of the first building complex, formation of the **UTSI Support Council,** and the start of a public lecture series. This also was the year that Dr. Goethert came home from Washington to become full-time director of the Space Institute. He brought with him specific ideas for research areas, including hypersonics, VSTOL flight, holography, MHD, scramjets and materials, and soon UTSI was involved in academic and research efforts in each of these areas.

In February, **Dr. Eugene Carl Huebschman,** who had been associated with Teledyne Inc. of Alexandria, Virginia, was appointed as professor of electrical engineering, and **Dr. Frederick M. Shofner,** a native of Bedford County, was named as an assistant professor of electrical engineering. Their appointments gave the Institute a total of fourteen full-time faculty members. In May, **Dr. Shen C. Lee,** a native of China, joined the faculty as assistant professor of aerospace engineering, and in July, **Dr. Firouz Shahrokhi,** a native of Iran who had earned his doctorate from the University of Oklahoma, assumed duties as assistant aerospace engineering professor. That same month, **Dr. K. C. Reddy,** a native of India who had been an instructor in aerospace engineering at the University of Maryland, came to the Space Institute as assistant professor of math. In December, **Dr. Sidney M. Harmon,** formerly an associate professor of math at Oklahoma State University, assumed that position with UTSI.

Three part-time professors—**E. K. Latvala** and **Dr. James W. L. Lewis,** both employees of ARO at AEDC, and **Dr. Demetrius Zonars,** an Air Force scientist at Wright-Patterson—also were appointed in December.

In June, an operating budget of five hundred and ninety-five thousand dollars was approved for the Institute, and enrollment for the summer classes totaled one hundred and seven, an increase of fifty-seven percent over the first summer session, with nearly half of those students enrolled full time. A major milestone, however, was the dedication ceremonies on **July 11, 1966,** as national leaders in education, government, the military and industry filed into the seven-hundred seat auditorium. This auditorium was part of the administration building that was being dedicated and named for **Frank G. Clement,** three-time Tennessee governor. Officially, it was this A Wing that was named for Clement—a plaque bearing his likeness adorns the wall near the first-floor entrance—but some newspapers reported that the entire Space Institute had been named for the famous orator.

Clement, the principal speaker at the dedication, said, "The UT Space Institute stands astride the path of the future, uniquely qualified in a very special way to be a vital contributor to man's search for new knowledge."

"We are pioneering in space today," he continued, "and this Institute, in partnership with the Arnold Engineering Development Center and the United States Air Force, is one of the volunteers leading the way. It is quite evident that nowhere else in the free world is there an installation that can match this one right here in Coffee and Franklin counties. There are others somewhat similar in talent in their own special ways, but none that has the particular capabilities of (Arnold Center)."

Clement, who was running for the U.S. Senate, left the ceremonies early to make a hand-shaking tour of Manchester and to meet with supporters at the law offices of **Garrett, Shields and Rogers.** He told the group that he had made a mistake in his unsuccessful Senate campaign against **Ross Bass** of Pulaski in 1964 by making too many speeches and shaking too few hands. Shaking hands, however, worked no magic for the man who had pushed to get the Institute established; he lost the election in November to **Howard Baker.** (Four area counties got new heads of government that year. At that time, they were called county judges, later county executives. New judges were **John W. Ray,** Coffee County, **Roy Crownover,** Franklin County, **Otis H. Templeton,** Moore County, and **Taylor Crawford,** Lincoln County.)

Dr. Frank P. Bowyer, chairman of the engineering committee of the UT board of trustees and the trustee who had been the Space Institute's chief advocate in the crucial days of getting the board's blessings for its establishment, was master of ceremonies for the dedication. **The Rev. W. C. Link,** pastor of Manchester's First Methodist Church, gave the invocation. In addition to Clement, other speakers included **Dr. Goethert, General Schriever, W. Harold Read,** UT vice president for finance, **Daniel J. Houghton,** president, Lockheed Aircraft Corp., and **John L. Sloop,** deputy administrator, advanced research and technology for NASA. Other dignitaries on stage included **R. M. (Bob) Williams,** managing director of ARO, Inc., **James L. Bomar Jr.,** former lieutenant governor, **Dr. Alvin Nielson,** dean of the UT College of Liberal Arts, **Dr. Hilton Smith,** dean of the UT Graduate School, **Brigadier General Lee V. Gossick,** AEDC commander, **Keith Hampton,** state personnel commissioner, **State Senator Ernest Crouch, John Neely,** UT director of physical plants, **Tullahoma Postmaster W. D. (Tut) Parham, James H. Alexander,** state treasurer, and **Ralph W. Emerson,** director of the state industrial development division.

Dr. Goethert seized the opportunity to issue a challenge for the future: "I assure you that the name of the University of Tennessee Space Institute is already well known in many foreign countries, and we can have students from abroad, including from Japan, more than our finances permit us to handle at this time. . . . It is within our reach to continue to build up a unique academic institute of excellence, known and respected throughout the entire world." He pledged to "work hard to carry Tennessee's flame of academic studies and engineering excellence beyond the boundaries of our state, with the aid of the Air Force, NASA, and industrial partnerships, to the next generation of young scientists and engineers who will be called upon to shape the world of tomorrow."

General Schriever called the Institute a "unique educational establishment, which offers distinct advantages to the U.S. Air Force, to the University of Tennessee, and to its students." He said life at Arnold Center "is clearly more attractive for scientists and engineers because they have a nearby graduate school where they can continue their studies. Likewise, the students at the Space Institute have access to many research and test facilities which are not found elsewhere." The Institute, he said, was a good example of what could be done when the Air Force and a civilian university "pool their resources to meet a common need," and he hoped it would serve as a pattern to be followed by other educational institutions. "If there is one thing that is clear today," Schriever said, "it is the overwhelming need for highly trained people. The rate of scientific and technical advancement means that the old four-year concept of higher education is no longer enough."

With General Gossick looking on, the governor presented Schriever with a certificate citing him for his "inspiration, leadership, cooperation and counsel" in working toward establishment of UTSI. Schriever would retire in a few weeks (August 31, 1966) after thirty years service in the Air Force, but he would continue to be actively interested in the Space Institute during its first quarter of a century.

UT Vice President Read noted that Arnold Center "has brought together one of the heaviest concentrations of aerospace brain power anywhere on the face of the globe" and with this concentration of brain power, its potentialities in the field of education were evident.

"Now we have a two-million dollar structure in which the Space Institute can begin a new era of service in the aerospace sciences," Read said. "But I want to remind you that this building is only a small part of our aerospace teaching facilities at the AEDC. Our students continue to have the advantages of the entire four hundred million dollar research and testing center for their studies, and this . . . is quite a laboratory!"

Robert Parson of Sewanee, who many years later would be named acting manager of the physical plant, began working as a custodian at the Institute on June 15, 1966; **Wilson Hill,** his brother-in-law who worked at UTSI, drove to Lynchburg where Robert was working on a project for Monteagle Silo Company and told him **Dewey Vincent** needed a custodian. Parson worked evening shift three months before going to the midnight till morning slot.

"It was lonesome out here," he said, "with nobody else around." **Normally** no one else was around, but Parson remembered almost running into **Dr. Lloyd Crawford,** professor of chemical engineering, in a restroom doorway about three o'clock one morning. Crawford was coming out as Parson started in, and "We almost scared each other to death," Parson said.

(In 1966, while in his third year of teaching at Tennessee Tech in Cookeville, Dr. Crawford spent some time at UTSI as a consultant working with Dr. Dicks. A native of Canada, Crawford joined the UTSI faculty in June, 1967, as an associate professor of chemical engineering. In addition to his teaching duties, he was to work with Dicks on problems concerned with high speed propulsion. Crawford received his undergraduate and master's degrees for the University of Saskatchewan in Canada and his doctorate from the University of Cincinnati. He would have a long career with UTSI and be heavily involved in MHD research later.)

In September, UTSI, expecting about sixty-five full-time students for the 1966 fall quarter, also was making long-range plans for a student dormitory and student center. **Harrie Richardson** said the UT administration was being asked to place those additional facilities in the University's capital outlay plan for which funds would be asked of the legislature. The proposed dormitory and student union (one building) would be located immediately behind the existing semi-circular Institute beside Woods Reservoir. Richardson estimated that the largest number of students the proposed facility ever would house would be about one hundred and fifty, saying that no more than two hundred and fifty full-time students was anticipated, but he noted that many of those would be married students with families and housing needs.

The UT trustees in June had approved a building and improvements program that allocated two hundred thousand dollars for the Institute over the next four years. These funds were expected to be used to complete second-floor portions of the Institute's library and laboratory wings, and for buying laboratory equipment. About ten thousand dollars was diverted for early construction of a coffee shop in the Institute's basement. Richardson said two rooms on the second floor had been left unfinished so that at least one could be developed later for use by an industrial or research firm that might have contracts with UTSI. Goethert had frequently cited the need for student dormitory facilities as well as faculty housing as part of long-range plans, and during the dedication ceremonies, he said that the Institute had more inquiries from qualified students than it could accommodate.

It was an ambitious proposal and one that would run into heavy opposition on the hill in Knoxville as well as from the Tennessee Higher Education Commission. It would be three years before ground-breaking ceremonies for the Industry-Student Center—including dormitories, dining facilities and conference rooms—would in fact occur. And three more years would pass before approval came for completing the third phase of the complex, the recreational wing. This might not have happened at all except for the persuasive power of **Morris L. Simon** who, by a fluke, got to speak on behalf of the proposal. But that, too, was yet to come.

✧ ✧ ✧

Joel W. Muehlhauser came home to Shelbyville that summer after four years in the Navy. He planned to go to graduate school, but where he did not know. He had received a bachelor's degree in Engineering Physics from Auburn University in 1962 before entering the Navy. One day in late summer, his mother suggested that he drive over to visit his high school classmate, **Dr. Frederick M. Shofner,** who had started teaching at UTSI in February. Joel took his mother's advice and as a result found himself enrolled at the Space Institute for the Fall Quarter, 1966. One day Shofner suggested that his friend check with **Dr. Shahrokhi** or **Dr. Dicks** about a possible Graduate Research Assistantship (GRA). Shahrokhi referred the young man to John Dicks, who offered him a GRA, and Muehlhauser was on his way to a Ph.D. in physics and a career at the Space Institute. It was a career that would put him in the thick of the MHD research that was to play such an important role at the Institute. Muehlhauser and Shofner had graduated from Shelbyville's high school in 1958. Two of their classmates, **John I. Shipp** and **J. D. Trolinger,** both Physics majors, were about to make history at UTSI.

In October, 1966, UT's vice president for finance, **W. Harold Read,** reported to the UT trustees that UTSI had led UT in research contract gains during the 1965–66 fiscal year. The Institute had research contracts valued at two hundred twenty-nine thousand, three hundred twenty-one dollars, Read said, which was a three hundred and sixty percent increase over its first year. Most of these contracts were for Air Force projects, but some were with the Chrysler Corporation's space division, and others were with Union Carbide Company. Colonel Richardson estimated that UTSI's contracts for research would exceed three hundred thousand dollars in the 1966–67 fiscal year. UT showed a twenty-one percent gain with contracts valued at more than twelve and a half million dollars. Second behind UTSI in percentage of gains was the UT Medical Research Center.

December of 1966 was a time of sadness for residents of Manchester and Coffee County with the death of two prominent citizens. **Coy St. John,** who had been Manchester's postmaster for twenty-two years, died on the first Monday of December. **Chancellor Robert L. Keele,** a former Manchester mayor, died four days later, on December 9. But Manchester and Tullahoma newspapers also reported good news in early December.

On Pearl Harbor Day, The Tullahoma News reported that a twenty-member Support Council with representatives of Bedford, Coffee, Franklin, Lincoln, Moore and Warren counties had been formed "to promote a close relationship between the Space Institute and leaders of lower Middle Tennessee, to keep these leaders and other citizens better informed about the institute's status and progress, and to help the institute to make this area an educational research and industrial center." The group's first meeting was scheduled for the following Tuesday (December 13) at the Officers Open Mess at AEDC. After the meeting, the council members would have dinner with **Scott Crossfield,** the first person to pilot the X-15 aircraft. Crossfield would lecture on December 13 in the Space Institute's auditorium, marking the beginning of UTSI's public lecture program.

Initial stories about the Support Council reported the following "temporary" officers: **State Senator Ernest Crouch,** chairman, **G. Nelson Forrester,** a Tullahoma attorney and civic leader, vice chairman, **Richard C. Robertson,** also an attorney from Tullahoma, secretary, and **Dr. Ewing J. Threet,** Manchester dentist and former state senator, treasurer. Six members were to be designated by residents of the following Chambers of Commerce: Manchester, Tullahoma, Franklin County, Fayetteville-Lincoln County, McMinnville-Warren County, and Shelbyville-Bedford County. Other members were listed by counties as follows:

COFFEE: Hatcher Meeks, president of MCA Sign Co. of Tennessee and president of the Manchester Chamber, and **Morris L. Simon,** newspaper publisher from Tullahoma; **BEDFORD: Franklin Yates,** publisher of the Times-Gazette, and Circuit **Judge William Russell; FRANKLIN: I. J. (Ike) Grizzell,** president of First National Bank of Franklin County, Winchester Mayor **Clinton Swafford,** and **Dr. Robert Lancaster,** professor of political science and dean of the University of the South, Sewanee; **LINCOLN: William R. Carter,** president of CFW Construction Company and president of the Tennessee Road Builders Association, and Fayetteville **Attorney Don Wyatt; MOORE: State Senator Reagor Motlow; WARREN: Henry Boyd,** McMinnville nurseryman.

On December 22, 1966, the Manchester Times reported that the same officers had been elected permanently. However, Dr. Goethert, writing about formation of the council, listed **Morris L. Simon** as the first vice chairman—a post that he held for many years. In fact, **Crouch, Simon** and **Threet** were elected as life-time officers years later (after the death of the first secretary, **Dick Robertson**). In 1989, **Nelson Forrester** confirmed that he had served as vice chairman only during the organizational period. Simon's memory was that Crouch accepted the office of chairman only on condition that Simon serve as vice chairman. Members representing the various chambers were **James H. McKenzie** of Tullahoma; **Rita Johnson** of Manchester; **Ben Anderton** of the Winchester-Franklin County chamber; **Floyd Nelson**, Fayetteville-Lincoln County chamber; **Bill Mittwede,** Shelbyville-Bedford Chamber, and **Bob Jones,** McMinnville-Warren County chamber. Other members named in the December 22 story were the same as previously reported except for two changes: **Hatcher Meeks** was no longer included as a Coffee County member, and **Cecil Shofner,** a Bedford County farmer, had replaced **Judge Russell** as Bedford's representative.

Remembering the formation of the Support Council, **Ernest Crouch** in 1989 said it came about because UTSI continued having problems with parental Knoxville. "UT just couldn't see the light," said Crouch, so original supporters of the Space Institute formed the Support Council. UTSI had no alumni, and wouldn't have for a few years, Crouch said. "The idea was to establish a support group to serve as an alumni association to support UTSI's growth and development. It just grew and grew and developed into an influential group. I believe it was **Morris Simon's** original idea to form the Support Council. It really paid off for the Space Institute." Crouch served as chairman of the Council for seventeen years.

Two days before Christmas, 1966, the Manchester Times reported that **Bob Williams,** president and managing director of ARO, Inc., would become a partner in the parent firm of Sverdrup & Parcel on January 1. **Edward M. Doughtery,** son of **Dr. Nathan W. Doughtery,** retired dean of UT's College of Engineering, had been elected vice president and assistant treasurer of ARO. He had been deputy to Williams since 1963 and a member of the board for two years.

The public lecture program was a particular source of pride to Dr. Goethert. When it began with a lecture by **Scott Crossfield** on December 13, 1966, Goethert said, "This is to show the people of Middle Tennessee our sincerity in the desire to become an active member in the surrounding communities while at the same time sharing with them and the Institute's faculty, students and staff and with our friends at (AEDC) the opportunity to hear some of the most expert people in their chosen fields on a wide variety of subjects." Later, Goethert said he had been "particularly delighted to see so many young people from the surrounding area, from the high schools, private schools, the military schools, the colleges and universities attend the lectures. It is evidence that the minds and imaginations of our young people are strongly attracted by the challenges of the unknown, by their devotion not to small everyday programs, but to the achievements of 'giant steps of mankind' which are calling this generation at a historical period of time."

Crossfield, who had flown the rocket-powered X-15 at more than twice the speed of sound, chose as his topic for the first lecture: "The Educational Social and Industrial Relationships." **Lieutenant General John W. Carpenter III,** commander of the Air University, Maxwell Air Force Base, Montgomery, Alabama, spoke on January 30, 1967, and the last speaker in this first series was **Marine Corps Major Gerald Paul Carr** of Denver, an astronaut in training at Houston for Project Apollo, who spoke February 28, 1967, on "The Manned Space Program." **Carr was the first astronaut to visit either the Space Institute or Arnold Center.** Among those who heard his lecture was **Dr. Frank P. Bowyer,** UT trustee.

John Stack, eminent aerodynamicist airplane designer of international fame, lectured on April 17, 1967, and afterwards presented the Institute with a sizable check for assistance in building the UTSI library. A few hours earlier, Phillippe Poisson-Quinton, head of the Applied Aerodynamics Division of ONERA of Paris, France, spoke on "From Wind Tunnel to Flight, the Role of the Laboratory in Perfecting Airplanes and Rockets." He had given this lecture earlier in New York as the 1967 Wright Brothers Lecture. Other early lecturers included **Alexander H. Flax,** assistant secretary of the U.S. Air Force for research and development, December 4, 1967, **General Schriever,** who gave a review of the first decade of the military and civilian space programs on February 23, 1968, and **James Webb,** chief administrator for NASA, who on June 13, 1968, spoke on the civilian space program.

Relations with the people in Knoxville continued to deteriorate after the Institute opened, according to Dr. Young, particularly at the departmental and college levels. He said a "key point of controversy between us and Knoxville was the residency requirement in Knoxville for our Ph.D. students. Most of the department chairmen we were involved with felt that with only about five faculty, we had little business awarding Ph.D. degrees in aeronautical engineering, mechanical engineering, electrical engineering and physics, but Goethert had sold us as a Ph.D.-granting institute."

Young said that Goethert, through Schriever, wrote harsh letters to **Andy Holt,** pointing out their understanding that UTSI would grant Ph.D.'s and that the added time for residency was not suitable for Air Force Institute of Technology (**AFIT**) students.

"So Goethert, whose office (for two years) was next door to General Schriever, would write General Schriever a letter pointing out the tremendous advantages of Ph.D. studies at UTSI with particular emphasis on the AEDC research equipment available but noting the controversy over our right to grant Ph.D. degrees. **Schriever** would write to **Andy Holt,** copy to **Hilton Smith,** expressing his amazement and disappointment that the Ph.D. problem existed, and Holt would talk to Smith in preparation of a diplomatic response."

Young said that Holt indeed was shrewd in his replies, and the upshot was that the possibility of waivers was arranged by **Hilton Smith.** Young was convinced that most of the Goethert-Schriever-Holt correspondence was written by one man—**Jack Shea.** Thus, he thought, Goethert was "writing to himself. But it accomplished the goal and soon we could get waivers for individual cases in aeronautical engineering, mechanical engineering, physics and electrical engineering, **Hilton Smith** being the judge. As expected, some of the department heads at UTK were appalled by this arrangement, and I got a fairly hard time."

None of the first AFIT students was working toward a doctorate, but soon afterwards a few were. Prior to the arrangement that was finally worked out, Young often found himself in an awkward position.

"No AFIT students went to Knoxville, and few others did," he said, "but it was awkward for us because prior to their coming (to UTSI), we could not guarantee a waiver for a given student."

In March, 1967, UTSI awarded its first doctoral degrees to three physics students: **J. I. Shipp, J. D. Trolinger,** and **Walter L. Powers.** Trolinger's mentor was **Dr. Arthur Mason;** Shipp and Powers both completed their work under **Dr. John Dicks'** supervision. It was a major victory for Dr. Goethert and a milestone for the Space Institute. In December, 1967, UTSI's first engineering doctorate—a Ph.D. in mechanical engineering—was earned by **Gerald S. Dzakowic** with **Dr. Walter Frost** as professor. By the close of UTSI's twenty-fifth year, one hundred and forty-eight Ph.D. degrees had been earned there beside Woods Reservoir.

Also early in 1967, UTSI received two new research grants from the Air Force at AEDC. Twenty thousand dollars was for renewing a contract for research directed by **Dr. Snyder.** The other grant, for $98,870, was for research by **Dr. Dicks** on the effects of waste contamination on supersonic combustion systems.

The fifth speaker in the public lectures series was Air Force Brigadier **General R. C. Richardson III,** who spoke May 9 on the role of technology and atomics in national security. He was deputy commander of Headquarters Field Command for the Defense Atomic Support Agency at Sandia Base, New Mexico. In October, **Dr. Edward C. Welsh,** executive secretary of the National Aeronautics and Space

Council, lectured on "Change Through Space Technology."

Dr. Gerhard Braun remembered the summer of 1967 as a time when Goethert was too busy to visit with an old friend from Germany. Goethert had invited **Professor Dr. Heinrich Hertel** to spend his sabbatical leave at the Space Institute. They had worked closely together during World War II as Hertel designed a new fighter airplane, which was planned to replace the Messerschmitt fighter. At that time, Goethert was chief of the only transonic wind tunnel in operation in Germany, and Hertel was the chief of design at the Heinkel Aircraft Company at Warnemunde, Germany.

"When Hertel arrived at the Space Institute," Braun recalled in 1988, "he expected several days of discussions about present scientific problems and old memories. But Dr. Goethert had no time for anything else but a proposal he had to present in a few days in Washington. Then he left for the presentation without finding time for his friend. Unfortunately, Dr. Goethert had right away to go on another business trip. Dr. Hertel had fallen in love with the Tennessee countryside. He took a UTSI car and rode along the 450 miles of the Natchez Trace Parkway, singing and enjoying the beautiful weather."

One evening in June, a convocation honored eleven students who had completed work for master's degrees: **Captain Kenneth L. Anderson, Robert A. Cassanova, Captain Charles J. Chase, Ned E. Clapp Jr., Frank L. Crosswy, James R. Lancaster, Captain Roy C. Lecroy, Richard K. Matthews, Joseph C. Powell, Burley H. Shirley and Joseph A. Sprouse.** Also honored were eight of UTSI's part-time faculty: **Dr. William A. Dunnill, John Stanley Ingham, Dr. Richard Kroeger, Eino Latvala, Dr. Clark Lewis, Thaddeus Lockhard, James Wallace and Dr. Wendell Norman.** Preceding the convocation was a lecture by Air Force Colonel **Raymond S. Sleeper** on "The Soviet Technological Threat." A 1940 graduate of the U.S. Military Academy at West Point, Sleeper would come to UTSI as a professor of cybernetics on April 1, 1970, and during his eight-year tenure he also would serve as UTSI's aviation education resource director. Sleeper was a native of Laconia, New Hampshire, with both a master's degree and doctorate from Harvard University.

Dr. W. T. Becker, assistant professor of metallurgical engineering at UTSI, was awarded a small contract in July by the Air Force Materials Laboratory to study development of a new material for aircraft "skin." Becker held three degrees from the University of Illinois. **Drs. Susan Wu, James Wu and Arthur A. Mason** were promoted from assistant to associate professors.

A new firm, **J. B. Dicks & Associates,** was established that summer by Dr. John B. Dicks. Dr. Goethert, noting that the firm grew out of research performed at UTSI, suggested that its establishment might lead to more significant industrial development in Middle Tennessee. The new company was working on a government research project by a $500,000 subcontract from Chrysler Corporation. The new firm was to design a generator to produce electricity from heat. **Dicks** was president, **Uwe Zitzow** was vice president, **Mrs. Betty Jones,** secretary-treasurer, and **Richard V. Shanklin III,** chief engineer. Zitzow was an Air Force officer at AEDC for three years before becoming a research assistant at UTSI in 1965. Mrs. Jones also was an employee of the Institute, and Shanklin was one of the first full-time students at the Institute, having received his master's there in 1965. Other employees included **Steve Lanier** and **Jack Garner,** engineers, **Charles Newman** and **Jimmy Bowling,** technicians, **Richard Chadwick,** draftsman, and **Mrs. Florida Smith,** secretary.

James A. Brantner, who had retired in April as an Air Force colonel after twenty-five years in the military, was named to the new position of assistant for management and security at UTSI in the summer of 1967. A native of Wisconsin, Brantner held a bachelor's degree from the University of Alabama and had done additional study at George Washington University and at Ohio State University. Shortly afterwards, in December, **Paul West** of Normandy was employed as Brantner's business assistant. West had graduated in August from Middle Tennessee State University with a degree in business management. He would be a mainstay in managing the Institute's business through the years, being named budget officer in 1987. He also was in charge of UTSI's book store. Through the changing seasons, West shared his home-grown flowers with the Institute. He and his mother, **Mrs. Mable West,** also were well known for their handmade quilts.

Two new professors were added to the Institute's faculty. **Dr. Heinz Dieter Gruschka** had come to UTSI for a year as a visiting professor from the German-France Institute in St. Louis, France. He had taken his doctorate in physics two years before from the Institute for Applied Physics in Germany, having received his master's from the University of Gottingen. **Dr. James R. Maus, Jr.,** who had been a research assistant at UTSI for more than a year, was appointed as assistant professor of aerospace engineering. Maus had completed undergraduate and graduate work at North Carolina State University, where he also had taught. He had worked as a mechanical engineer for the Chemstrand Corporation.

In November, **Dr. Edward Boling,** UT's development vice president, told the Support Council that the Nashville architectural firm of Waterfield & Associates had been selected to design a $350,000 Industry—Student Center.

The **Binkleys, Ruth and Freeman,** left the Space Institute on Nov. 30, 1967, as Freeman accepted a full-time job with the state of Tennessee in Nashville. However, they both returned to UTSI on June 1, 1969.

In the fall of 1967, a young engineering graduate from the University of Cincinnati enrolled as a student at the Space Institute and thus began an association that would one day include his service as executive assistant to the UTSI dean. **Gary David Smith,** son of Mr. and Mrs. **Norman L. Smith** of Madison, Indiana, would receive his master's degree in mechanical engineering on March 17, 1971, with **Dr. Robert L. Young** as his thesis adviser. Smith's thesis was on "Development of a High Temperature Probing System in an H2-O2-N2 Environment." Smith and **Roy Schulz**—as "co-op" students at Cincinnati—had driven together to their jobs at Arnold Center. Both continued "co-op-ing" as UTSI students, where Schulz earned both master's and doctoral degrees. Smith also did post-graduate work in fluid mechanics, combustion, propulsion and energy conversion. He, Schulz, and **Dr. Thomas V. Giel Jr.** all worked in the Engine Test Facility at Arnold Center under the supervision of **Dr. Carroll Peters,** who also had been Gary's work supervisor while he was a co-op student. **Carl Catalano** later joined this group, and like the others, he would eventually be lured by **Dr. Susan Wu** into becoming a member of the Space Institute family. Smith joined the UTSI staff on June 3, 1980, and he would become ECP's lab manager at the Coal Fired Flow Facility until going to Washington, D.C., for a year in 1985 as an ASME Congressional Fellow. During his stay at UTSI, Smith would become increasingly active with the American Society of Mechanical Engineers, and he would author more than twenty-five technical publications. Schulz, Giel and Peters all would become part of the Institute's growing faculty in the early 1980's.

Perhaps no member of the faculty generated more excitement, though, than the "owl man," who came in the Institute's fourth year.

✦

Chapter Ten

WHISPER OF A WING IN FLIGHT

ON JANUARY 1, 1968, **Dr. Tibor Charles Helvey,** a native of Budapest, Hungary, joined the UTSI faculty as professor of cybernetics and worked hard to make cybernetics a household word in the area. He frequently lined up guest lecturers, and he also liked to tell the cybernetics story in other places when he got the opportunity. For instance, a year later, he would speak at the 135th annual meeting of the American Association for the Advancement of Science in Dallas, where he praised Goethert's "forward looking" attitude, which he said was responsible for cybernetics being taught at the Space Institute. He was optimistic that the first graduate program in cybernetics might be started at the Institute. Helvey is best remembered, however, as the "**owl man.**" During his five and half years at the Space Institute, he aroused the most interest with his study of the silent flight of owls.

He had two huge owls shipped in from Florida and housed them in a large classroom in E Wing. **Dewey Vincent** remembered: "We had to go change locks on the doors and put covers over the windows because he claimed that people looking in the windows were getting his owls excited. He said that nobody should feed those owls but him. He went in there one day—he was feeding them chicken livers—he was going to hand-feed one of them, and it liked to have eaten his hand off."

Helvey was training the owls to sit on a perch high in one corner of the room so that they had to fly over a string stretched across the room and glide to a landing in the opposite corner. He was particularly interested in the glide. While the owls were not unusually fast, they were able to capture their prey because of their quiet glide. **Jim Goodman,** who was with the acoustical engineering group, worked with Helvey on the project, recording the noise, or lack of it, of the owl flights. One night they worked well past midnight, and **Dewey Vincent** was called back to the Institute.

"I had to shut down the motors, everything on this whole campus, to kill any noise, so they could record," Vincent said. "What they would do is take scissors and change the configurations of the wings to see what effect this would have on the sound that was generated by their flight." While Dewey found the experiments interesting, he recalled that the "room was a mess when we finally got it back."

Some faculty members said that Helvey once put a smaller owl in the room, and it mysteriously disappeared in about fifteen minutes—apparently being devoured by the bigger birds.

Charlie Burton remembered: "We had to clean that mess up! He had one side of the wing down there closed off. We'd have to cut off the power when he ran his owl tests."

Dr. Richard Kroeger, associate professor of aerospace engineering and a specialist in aerodynamics, and **Dr. Heinz Gruschka,** assistant professor of physics with a specialty in acoustics, were involved with Helvey in some of the research, part of which was financed by the Air Force Flight Dynamics Laboratory. The objective was to develop ways to eliminate noise from the passage of an airplane though the air, part of a plan to develop helicopters and short-range planes that could be used at small airports.

Helvey had caused a stir ten years earlier while he was at Cape Kennedy helping to design a "moon station" by suggesting that a woman scientist should be included in a three-person crew for a moon landing. A woman's presence might offset conflicts that he thought would arise from three men being confined in small quarters. Helvey also had designed an inflatable coffin for disposing of astronauts who might die in space. **The Rev. Jim Rhudy,** an Episcopal minister and an engineer at AEDC, took a leave from those positions in 1968 to work with Helvey.

Rhudy was fascinated by Helvey's desire to use cybernetics to help solve common problems such as polluted streams, housing and transportation and in general to improve living conditions. Both men were the subject of a feature story written by **Louise Davis** in the Nashville Tennessean Magazine December 29, 1968. Rhudy would later spend several years on the staff of the Space Institute.

Burton also remembered his first meeting with **Bob Kamm** in 1968. Charlie was mail clerk and "when I got to the post office, my truck was not big enough for the sacks of mail from California. I asked Dewey what to do, and he said, 'Store it until I get there.' We had half a mail room full of sacks from California. Dewey came, and I started carrying sacks of books and papers; it took awhile to get through. That's how I got to know Bob Kamm. He took over as Dewey's boss. Kamm is a nice man." Kamm had been in charge of NASA's western offices at Santa Monica, California, for nine years. One day he called Goethert and told him, "Doc, they're closing me down." He had some government reports that he thought would be good for the Space Institute. Goethert called in **Art Mason,** instructed him to go to California, meet with Kamm, and see if the reports would be useful. **Harold Ellison** retired in 1968, and on March 4, 1968, Kamm joined the Space Institute's staff as assistant to the director, his old friend **Dr. Goethert.**

(While at Santa Monica in January, 1960, Kamm had come near dying. His wife **Mary Ann** and their three daughters were driving west at the time, on their way to join Bob. Tests had not confirmed physicians' suspicions as to the cause of Kamm's illness, and he was in critical condition. NASA officials contacted the head of the Tropical Diseases Department at the University of California at Los Angeles who diagnosed the illness—regardless of what tests had or had not confirmed—as **amoebic hepatitis,** and effective treatment began. During the crisis, NASA put out a tracer, and Texas state police apprehended Mrs. Kamm and the girls at about two o'clock one morning as they were traveling across the Lone Star state. NASA sent a pilot from its Flight Research Center to fly Kamm's family members directly to California. The pilot? **Neil Armstrong**—whose fame as the first American to reach the moon was yet to come. So were the classes that he would take at the Space Institute. **Jim Caldwell,** a former security guard at AEDC whom Kamm had coaxed out of retirement to join NASA's security force at Santa Monica, drove Mrs. Kamm's car from Texas on to California.)

Dr. Young recalled that Kamm was a big help in getting students from various industries to enroll at UTSI. Two of these later became astronauts. As chairman of the Los Angeles Federal Executive Board, Kamm had gotten acquainted with **Major General Joseph Bleymaier,** deputy director of the Department of Defense's Manned Orbiting Lab (MOL) program. When Kamm got to UTSI, he learned of Goethert's public lecture series and arranged for General Bleymaier to speak at UTSI on June 5, 1969. Because of this, the Space Institute wound up with two astronauts as students—**Henry (Hank) Hartsfield** and **Don Peterson,** both of whom later flew manned missions, and both of whom later returned to the Institute as guest speakers. Hartsfield earned his master's degree in engineering science in March, 1971. His thesis was: "An Investigation of Spontaneous Emission Noise in Quantum Amplifier." Peterson did not complete his work for a degree at UTSI.

Kamm explained why the astronauts came. On Monday morning after Bleymaier had spoken at UTSI, he was driving to work and heard on the radio that the MOL program had been cancelled. This raised a big question as to what would be done with those astronaut trainees in the MOL program. NASA agreed to take some of them but could not take all of the trainees immediately. "Because he had just been here," Kamm said, "UTSI was fresh in his (Bleymaier's) mind. He decided that some should be sent to UTSI for graduate training while awaiting assignment to NASA. That's how Henry Hartsfield happened to come to the Space Institute." (Two other astronauts with the MOL program—**Major Robert Overmeyer** and **Lieutenant Colonel Al Crews**—were with General Bleymaier when he lectured at UTSI in June, 1969.) Kamm later was named assistant dean and remained active in Institute affairs even after retiring in 1988.

Dr. Thomas Gee assumed duties in January, 1968, as assistant professor of electrical engineering. He had worked for the U.S. Naval Research Laboratory in Washington and taught at Virginia Polytechnic Institute while completing requirements at VPI for his doctorate. He would be promoted to associate professor in January, 1971. UTSI registered a record 250 students (about 80 of these were full time) at the start of this winter quarter, and **Miss Virginia Richardson** expected more to sign up late.

Dr. Martin Summerfield, one of the nation's leading authorities in the field of jet propulsion, spoke at a banquet marking the close of a two-week short course on that subject at UTSI. Summerfield discussed Dr. Theodore von Karman's work in aerodynamics. Summerfield, professor of jet propulsion engineering at Princeton University since 1951, had known von Karman during Summerfield's early scientific career.

Orrin J. Greenwood, who had retired after seventeen years at AEDC, was named a part-time assistant to Dr. Goethert in the spring of 1968. His duties were to include helping with the organization and arrangements for seminars and symposiums. UTSI hosted the Ninth Symposium on Engineering Aspects of Magnetohydrodynamics with **Dr. John Dicks** and **Dr. Leon Ring** serving as arrangements chairman and program chairman respectively. More than 200 attended. In May, **Dr. Eliot Morris,** a U.S. Geological Survey scientist connected with the "Surveyor" television moon probe, spoke at UTSI.

The Space Institute was assured of a six hundred thousand dollar grant from the Department of Defense for basic research on magnetohydrodynamics—a three-year project directed by **Dr. John B. Dicks.** He said previous MHD contracts at UTSI had totaled about $330,000. MHD research would, in fact, eventually provide a major source of revenue—and notoriety—for Dicks and UTSI.

The UT Trustees approved a sixty-million-dollar four-year construction program that included $250,000 for UTSI. UT officials said they would go ahead and build initial portions of an industry-student center, using about $350,000 available. **Bob Kamm** thought that bids might be taken in the fall with work starting early in 1969, but progress was considerably slower.

Dr. Arthur Mason, associate professor of physics, was granted tenure in June, 1968, and four of his fellow faculty members were promoted from assistant to associate professor. These were: **Dr. Sidney M. Harmon,** mathematics, **Dr. Firouz Shahrokhi,** aerospace engineering, **Dr. Fred M. Shofner,** electrical engineering, and **Dr. Walter Frost,** mechanical engineering.

The Institute got its third annual five thousand-dollar grant from the Department of Health, Education and Welfare to buy library books.

Effective July 29, **Rita B. Johnson** was named to the Institute's administrative staff, where she was to work with **Bob Kamm** in establishing "closer liaison with news media and other organizations." She had been secretary of the Manchester Chamber of Commerce for several years and was a member of the UTSI Support Council. This was, in essence, the start of the Space Institute's organized public relations efforts.

Kenneth K. Roberge, a candidate for a master's degree in mechanical engineering, left in late July for Technische Hochschuile in Aachen, Germany, to study for his doctorate. He was the first UTSI student to study in Aachen under a student and faculty exchange agreement signed by both schools in September, 1965. Aachen had sent eight students and two faculty members to UTSI during this time. Roberge was a native of Connecticut and had lived for years in Nashville.

Dr. Eugene Huebschman, president and treasurer of Oceanography and Pollution Research Corporation, said the new research and development firm would soon open an office in Manchester. The company had sold an initial offering of three hundred thousand shares of stock. Other officers, all UTSI faculty members, were **Dr. Gerhard Braun, Dr. Frederick M. Shofner,** and **Dr. W. T. Snyder.**

Len Dolan, who had served as the first manager of the Institute's short course program, resigned, and **Miles Maynard,** retired AEDC engineer, succeeded him in August, 1968. **Dolan,** a former city manager of Johnson City, Tennessee, worked for ARO, Inc. at Arnold Center for fourteen years. In September, 1967, Dolan retired as chief of the General Services Division, and a few weeks later assumed his duties with UTSI. **Maynard,** a native of Holyoke, Massachusetts, had first come to Tennessee to work in 1942 on the atomic bomb project at Oak Ridge. Ten years later, he had joined the staff of ARO, and in 1966 he had retired as a staff engineer in the engineering support facility.

Dr. Marcel K. Newman, coordinator of research and professor of mechanical engineering, directed a course in modern materials technology. A banquet at the conclusion of the course featured **Dr. Robb M. Thomson,** director of materials science for the federal government, as speaker. He cited the need for more sophisticated materials in building and medicine.

Colonel Jean A. Jack, former deputy chief of staff for tests at AEDC, wound up thirty years' military service with his retirement that summer. Jack, who had played a role in getting UTSI established, was a veteran of sixty-seven combat missions in the Southwest Pacific during World War II. He would become a student at the Institute and later teach at Middle Tennessee State University in Murfreesboro.

William F. Kimsey, a student associate at UTSI, went to Toulouse, France, in September to present a paper at a meeting of propulsion specialists from thirteen NATO nations. Later, as **Doctor** Kimsey, with two degrees from UTSI, he would hold a top position with one of the AEDC contractors (Sverdrup Technology). **Robert Jones** of McMinnville was elected secretary of the Support Council to serve with **Chairman Ernest Crouch, Vice Chairman Morris L. Simon, and Treasurer Ewing J. Threet. Rita B. Johnson** was named alternate secretary. New members on the council were **Bill Crabtree** and **Thornton Taylor,** both of Fayetteville, **Hubert Nicholson** of Decherd, **James Bomar Jr.** and **William D. Copeland,** both of Shelbyville, **Joe Powell Shelton**

of McMinnville, **Joe Majors** of Tullahoma, and **Billy Mason** of Franklin County.

In October, 1968, work was started on the first separate laboratory building. This lab, built beyond a wooded area in front of the main building, served two purposes. Half of it was devoted to MHD research, and the other half housed Goethert's wind tunnels.

Professor Lenoid Ivanovich Sedov, a top Russian space scientist ("Mister Sputnik") lectured in October on "Science and Society." He was accompanied by another scientist, **Dr. G. G. Chernyi.** Sedov said Soviet space flights were a rehearsal for space probes to Mars and other planets.

Dr. Maurice A. Wright, a native of England, joined the UTSI faculty late in 1968 as associate professor of metallurgy. He came from Norton Research Corporation, Cambridge, Massachusetts, where he was senior metallurgist, involved in research concerned with the mechanical properties of materials. He had taken his Ph.D. from the University of Wales in 1962.

On January 16, 1969, **Dr. George E. Mueller,** NASA associate administrator for manned space flight, packed the auditorium at the Space Institute with his lecture on "The Promises of Space." **Dewey Vincent** said this was the only time he remembered the auditorium being filled. Mueller showed a film, "**The Eagle Has Landed.**" This was the first public showing of the moon landing, and as Vincent remembered, it was unplanned.

"Somehow the news media got hold of it and broadcast it, and we had people come in here from as far away as Cookeville. We didn't have a cafeteria, and the group had all gone to the Officers Club for dinner. When they came back, you could hardly get down to the door. They were parked all the way to the main access road. The lobby was full of people, the halls were full of people, the auditorium was full. They were sitting on the stage. Dr. Goethert was always striving to have a full house, but I thought he was going to faint!" Another major lecture that year featured **Dr. Lewis Larmore,** vice president and director of Douglas Advance Laboratories.

Things were buzzing in early 1969. **Bob Kamm,** in January, was approved as consultant to **Harold B. Finger,** NASA associate administrator for organization and research. **Brigadier General Gustav E. Lundquist** announced plans to retire later that year as AEDC commander. In July, it was announced that **Brigadier General Jessup D. Lowe** would succeed Lundquist. **Dr. Sam Ingram** had been chosen as president of Motlow State Community College, which would open in the fall. On February 4, Ingram set up temporary administration offices in the education building of First Baptist Church, Tullahoma, and began building his staff. **Kenneth B. Slifer** would become the first dean of the institution and **Dr. Don C. England** the college's first dean of students. **Richard Morris** had been named new administrator at Harton Hospital in Tullahoma. Also in February, **Dr. Heinz Gruschka,** assistant physics professor at UTSI, was heading up a study of ways to reduce jet noise under a NASA grant. Goethert was named as a consultant to FAA for its new noise abatement program.

On March 1, **Dr. Fletcher Donaldson,** a pioneer in the computing field, was named professor of computer science—the first for the Institute. He came to UTSI from Lockheed in Sunnyvale, California, where he had served as a senior consulting scientist in the Computer Department. Donaldson was a man with a mission, and eventually he would call on one of the most powerful men in the U.S. Senate to help him accomplish it.

After completing his Ph.D. work at the University of Texas, Donaldson joined the University of California at Los Alamos in October, 1951. At Los Alamos, he computed the yield of atomic bombs before they were tested in the Nevada desert, using one-dimensional hydrodynamics (spherical symmetry) on IBM punched card equipment. He spent a year on a ten-man programming team writing library programs for converting his hydrodynamics problem to the IBM-701, IBM's number one assembly line-produced, stored program computer.

In accepting Dr. Goethert's offer to join UTSI, Donaldson secured a promise that he could pursue his primary interest—the use of computers in **medical care**—along with his teaching. This interest, or obsession, persisted, and in the years ahead, Donaldson would realize his dream to see the medical community—doctor's offices, medical laboratories, pharmacies—served by a comprehensive computerized system. While on the UTSI staff, Donaldson would lead a long but successful fight to force MUMPS (Massachusetts General Hospital Utility Medical Programming System)—a computer language and operating system developed originally for medical applications—into the public domain.

Thirty-three graduate students, all from AEDC, reported for Donaldson's first class at UTSI, but it was quickly divided into junior and senior groups. Donaldson informed the senior students that they were going to design a Regional Medical Information System, and, according to Donaldson, "thus began three quarters of study of the best medical system efforts in the country, the best minicomputers available for such a system, the best programming language, (and) the medical care system in the five-county, five-city area surrounding Tullahoma."

✧ ✧ ✧

On March 26, 1969, a man who had pushed hard to get the proposed von Karman institute established, **Major General Samuel Russ Harris,** died in Tucson, Arizona, at age 66. Two days later, **Dwight D. Eisenhower,** the general with the boyish grin who served from 1953 to 1961 as president of the United States, died in Washington, D.C.

Charles Wilson Bray in March completed work for his Ph.D.—the first UTSI student to get a doctorate in electrical engineering. **Dr. F. M. Shofner** was his professor.

AEDC was planning an eight-million-dollar expansion of a test facility with the addition of an Airbreathing Propulsion Test Unit (APTU).

Four speeches on the impact of the Space Institute were scheduled in the area by **Dr. Mason, Dr. Young** and **Dr. Goethert** as part of a drive to raise from ten to fifteen thousand dollars to develop a recreational area and picnic grounds at the Institute. **Rita B. Johnson** of Manchester, liaison between the community and the Institute, was a member of the committee that included **Dr. Threet** of Manchester, **William Crabtree** of Fayetteville, **Joe Powell Shelton** of McMinnville, Dr. Goethert and **Colonel Harrie Richardson. Jain-Ming Wu** and **Ying-Chu Lin Wu** (Jimmy and Susan to their Institute friends) became U.S. citizens in a naturalization ceremony in April. On April 4, the Manchester Times reported that Goethert had said UTSI would ask bids soon on a new industry-student center for which Goethert and others had been fighting for years.

"Goethert probably started pushing for the student center early in 1966," Dr. Young said. "Folks at the University were appalled, saying that no student center had ever been built from state funds, and that unless we could get donations or gifts, we couldn't have it. They said we didn't need it with the Officers Open Mess close by. Luckily, we found that the student center and even the gym at Motlow College was to be built with state funds, so (Ed) Boling's objections were silenced when we brought this to his attention, and our Support Council helped get a sufficient appropriation for the center."

The Times story said the unit would cover more than thirty thousand square feet, and that it had been in the planning stage for about two years. To be built were a dormitory wing with twenty double occupancy efficiency type living units and a student center and a dining hall to seat four hundred. It had been designed in three sections, and the third section—a recreation building—would be built later when money was available, according to the story. (The third wing would encounter more opposition from some quarters, including the Tennessee Higher Education Commission, than the first units did.) Estimated cost of the center was six hundred and fifty thousand dollars, and UT reportedly had about three hundred and seventy-six thousand dollars in appropriations and borrowing authority to help finance the structure and equipment. It was estimated that it might take a year to fourteen months to complete the center, which had been designed by Cooper and Waterfield, Nashville architects, under the direction of **Malcolm Rice,** UT architect. UTSI officials wanted Harwood as architect, Young said, but Boling was unhappy about "one flaw" in construction of the auditorium: After completion, it was discovered that the first few rows of the balcony did not provide a view of the stage. "We felt that this error could be overlooked," Young said, "because we knew that Harwood has spent some of his own money to assure that the auditorium acoustics were very good. He had hired as consultants the famous MIT firm of Bolt-Berenack at his expense to advise on the acoustics. But Boling was adamant, and Harwood did not get the . . . center."

A critical stage in the planning of the center involved Goethert's desire for a four-hundred seat cafeteria. How to justify a cafeteria of that size in the woods, and could it at least break even financially were two critical questions.

"I knew, based on AEDC experience, that cafeterias have to be subsidized when only one meal is served," Young said. "There were also questions about the ability to keep the dorms full in order to support themselves." The manager of UTK cafeterias and dorms was consulted and extensive surveying was done. To Young's astonishment, the manager said the dorms could be self-supporting and that the cafeteria would break even. This evaluation overcame Boling's objections.

It was to be called an Industry-Student Center because Goethert had the idea that the offices overlooking the lake might be rented for the use of aerospace company personnel involved in testing at AEDC. A good idea, Young thought, but it never worked because AEDC provided these offices at the testing site. Neither, Young said, did the cafeteria ever break even although for awhile Goethert insisted on serving evening and weekend meals.

"This proved to be a financial disaster," Young said. "Goethert had many ideas, all basically good, but some not feasible, and he pursued them with glee and gusto in such a dogged manner that most were attempted."

In May, it was reported that UTSI had accounted for the largest share of space research done by UT in the past four to six years. Space Institute officials were saying they needed more labs. AEDC opened its new airfield. Haircuts were going to two dollars a head in Manchester. **Andy Holt** announced that he would retire in 1970 as president of the University of Tennessee. **Olney H. Anderson,** former Tullahoma mayor, died May 16 at seventy-one, and two days later, **Prentice Cooper,** three-term governor of Tennessee, died at seventy-three. (His son, **Jim,** would one day become a U.S. congressman and prove to be an outspoken supporter of the Space Institute.) Then on May 15, 1969, bids for the Industry-Student Center were opened, and **Hardaway Construction Company** of Nashville had submitted the apparent low bid of seven hundred seventeen thousand, five hundred dollars. Action was delayed since this was more than the six hundred and twenty-five thousand dollars available. A few weeks later, Dr. Holt said that Hardaway would be awarded a six hundred seventy-three thousand dollar contract to construct the center. The building and executive committees of the UT Trustees had approved a financing plan that involved deleting some partitions and built-in equipment.

Dr. Franz Fett, a faculty member at the Technical University of Aachen University and an expert on jet engines, was appointed as a visiting assistant professor of mechanical engineering at UTSI.

In June, UTSI's library got a five-thousand-dollar grant from the U.S. Office of Education, and the student branch of the American Institute of Aeronautics and Astronautics at the Space Institute had won first prize in AIAA's Bendix Award Contest for its project to develop a visual aids center at UTSI.

Dr. S. N. Chaudhuri joined the Institute as associate professor of aerospace engineering in July. **Aldrin** and **Armstrong** were on the moon (July 20, "Apollo Eleven"), and Manchester officials moved into a new City Hall for which **Mayor C. V. Myers** had battled. Goethert was lecturing at Aachen.

At 1 p.m. on July 25, 1969, ground was broken for the Industry-Student Center. Among those attending from Knoxville were **Dr. Holt, Chancellor Weaver, Dr. Boling,** and **Dr. Hilton A. Smith,** vice chairman for graduate students and research. And they would not have to eliminate partitions or built-in equipment after all. The State Building Commission had approved eight hundred twenty-seven thousand dollars to complete the whole project—**except** for the third phase, the recreational building.

UTSI added a faculty member that fall—**Dr. Yaun-Siang Pan,** a native of Taiwan and specialist on sonic boom propagation, ionospheric plasma flow and supersonic and hypersonic flow. He had taken his Ph.D. in engineering mechanics from Brown University in Providence, R.I.

In August, The Tullahoma News editorialized that the new student-industry center should be named for **General Lief J. Sverdrup.** ARO, Inc., founded by Sverdrup, had contributed "more than $1.2 million in operating funds" to UTSI and to the UT graduate program that preceded it, according to the editorial. And, The News insisted, "Gen. Sverdrup originated the idea for a technical institute here, and in so doing laid the foundation for what is now UTSI." Morris Simon would always remember Sverdrup's early efforts to secure private funding to establish a von Karman space institute. However, the industry-student center would bear the name of UT's president, **Andy Holt,** as recommended by faculty, staff and students. (The Support Council recommended that the center be named for Holt and that **next** Institute building be named for Sverdrup.)

Undergraduate evening classes began in September with **Miles Maynard** as coordinator. In late December, **Dr. Edward J. Boling,** vice president for development and administration at UT, was named to succeed **Andy Holt,** who would retire as UT president in September, 1970. Goethert said Boling was a UTSI committee's first choice for the position.

On Sept. 1, **Dr. Kenneth Kimble** assumed duties as assistant math professor at UTSI. He had just received his Ph.D. from Ohio State University where he also had earned his undergraduate and master's degrees and had held several instructorships. He had hesitated about coming to UTSI because of its weak computer program. In the years ahead, Kimble would play a big part in getting the Vax system installed at UTSI and would serve as manager of the Institute's Computer Center.

Things were going well for the Space Institute. The fall of 1969 provided an opportunity for a joint celebration by the University of Tennessee and the Space Institute as UT celebrated its one hundred seventy-fifth year, and UTSI marked its fifth birthday. But sadness would soon follow as the state mourned the death of a man who had shared Goethert's dream for a space institute and convinced the legislature to put up the money to pay for it.

The Space Institute held its "Fifth Anniversary and Founder's Day Celebration" on October 5, 1969. A brochure, recapping the celebration, was published later, and it featured a picture of **Frank G. Clement** in memorium to the former governor who died in a two-car accident in Nashville on November 4 of that year—seventeen years to the day since he had first been elected governor of Tennessee. The tribute acknowledged Clement's role in getting the Institute established and stated that the "Frank G. Clement Administration Building of the Institute, dedicated in July, 1966, would be an ever-living tribute to his contribution to education in the state of Tennessee."

A host of speakers participated in the ceremony on October 5 with **Dr. Charles H. Weaver,** chancellor of UT, presiding, and an invocation by the Reverend **Delwyn Fryer,** pastor of First United Methodist Church, Manchester. **Dr. Andy Holt** was the key speaker. Others included **Dr. Goethert, General Schriever, Joseph Gavin Jr.,** vice president, Grumman Aircraft Engineering Corporation, **John Stack,** vice president, Fairchild Hiller Corporation, **E. M. James Jr.,** director of University Grants and Contracts, NASA headquarters, Washington, D.C., and Brigadier General **Jessup D. Lowe,** Arnold Center commander.

In addition to the ceremony, a tea was held in the UTSI lobby, tours were conducted, and a Founder's Day banquet was held at the Officers Open Mess at AEDC with Dr. Weaver as key speaker. Dr. Goethert welcomed guests to the banquet, and remarks were made by **Colonel Harry L. Maynard,** director of plans and technology at AEDC, **E. M. Dougherty,** vice president of ARO, Inc., **Senator Ernest Crouch,** Support Council chairman, **Dr. John B. Dicks Jr.,** UTSI physics professor, **Dr. Robert L. Young,** deputy director for educational programs at UTSI, and Air Force **Captain Roger Crawford,** president of the Student Government Association at UTSI. Goethert read a telegram from **Governor Buford Ellington,** expressing regrets that he could not attend and congratulating "all who make this observance possible."

During the anniversary ceremony, Goethert noted that at the first convocation five years earlier, "the land of the present campus was a wooded wilderness, with barely a wagon trail leading into it. Now five years later, there are buildings of a most graceful architecture. There is a separate laboratory area, which opened its first research building recently. There is a Student-Industry Center under construction, with student and guest housing, a dining hall and conference rooms for business and social functions and provisions for recreational activities. **And above all, there is still the unspoiled beauty of the campus' natural landscape, of the lake, of the mountains in the background, and the wildlife around us.** "

In those five years, Goethert said, the faculty had grown from three to thirty full-time professors, all with doctoral degrees, and the student body had grown beyond the "half-mark goal" of five hundred students, all on the master's or doctoral study level. (Actually, the Institute had opened with five full-time faculty members, but Drs. Young and Newman also had duties outside the classroom.)

"Our research activities, an indispensable and most important part of a graduate school program, has grown beyond expectations," Goethert said, "spanning a broad spectrum of aerodynamics, turbojet-engine dynamics, ramjet propulsion, advanced materials, MHD electric power generation, heat flux, radiation spectroscopy, laser applications, aircraft noise, sonic boom . . ." He acknowledged that there had been serious problems, including unpredictable cuts and "shifts of support," but he expressed pride that "our built-in flexibility of operation, and the broad basis of our support sources enabled us to maintain a healthy growth of our faculty, students and research." Progress in the end, was made by people and not by money or outside support, he reminded his listeners. Goethert also paid tribute to the late **General William Rogers,** whose widow attended the ceremony. As commander of AEDC, Rogers had been an avid supporter.

General Lowe also praised the "foresight" that had been shown by General Rogers and General Schriever, and he said without question "the Space Institute has more than justified all of the arguments that were put forward for its formation and support by the Air Force . . . when we were selling this program."

Dr. Holt, after quickly reviewing four ingredients for a "truly great university"—significant curriculum, competent curriculum, eager students and adequate physical facilities—emphasized that the fifth ingredient was "dynamic leadership." This really involved a good bit of arm-twisting, Holt said, adding that "Dr. Goethert is the best arm-twister I have ever known." As "main victims" of Goethert's arm-twisting, he cited General Schriever, the Air Force, NASA, **Joe Gavin** and his industries, and **Ed Boling, Herman Spivey, Ram Raulston, Hilton Smith** and **Charlie Weaver.** "Sometimes we thought this arm-twisting was a little bit painful," Holt conceded. "Sometimes I'd just as soon not had it, but after it is all over, I'm awfully glad he did it because otherwise we wouldn't have this tremendous institute we have. Really and truly, this great institution . . . is an eloquent tribute to the vision, to the dedication, and to the perservance of Dr. Goethert."

Another special part of the celebration was the raising of a new flag of the University, designed to pay tribute to UT's one hundred seventy-five years and, in Goethert's words, "to mark our devotion to carry on towards a dynamic future." He asked Dr. Boling to entrust the flag to **Captain Crawford,** SISGA president. Accepting the flag, Crawford noted that one hundred seventy-five years before "this great nation was not even well-explored, so we had the frontier of the West to look forward to. The next one hundred seventy-five years, I think I can safely say, our frontier will be space. And so UTSI looks forward one hundred seventy-five years in pride, and back one hundred seventy-five years with pride."

During the banquet, **Ed Dougherty** cited UTSI's contributions to ARO, noting that ARO employees had taken thirty-two hundred student courses at UTSI, and that during the first five years, the Institute had awarded one hundred degrees to ARO employees, including five Ph.D.'s. Captain Crawford, speaking for students during the banquet, said, "We have a very fine mixture of civilian, military and foreign students . . . UTSI has just the right mixture of the kind of people who come from various academic backgrounds that supplement the educational program." He cited the "great opportunity" for UTSI students to participate in the growth of a school.

"The faculty and Dr. Goethert and Mr. Kamm and Dr. Young have been very receptive to almost all the suggestions that the students have made," Crawford said, "and we try to make those suggestions that are appropriate to the growth and success of the University. They have a very progressive academic program, they have a very large research program, which has been emphasized many times, and we can't underscore the very close relationship here with Arnold Center and ARO, Inc."

Crawford, who was the second SISGA president (**Hrishikesh Saha,** later a teacher at A&M University, Huntsville, was the first), was working on his doctorate and years later would return to UTSI. He had earned his master's degree in the early 1960's from the Air Force Institute of Technology, Wright Field. Before enrolling at UTSI, he was assigned to the Air Force Space Systems Division in Los Angeles, where he worked on the Titan III space booster. After earning his Ph.D. at UTSI in 1971, Crawford remained with the Air Force until retiring as a lieutenant colonel in 1982. In March of that year, he began working with Dr. Goethert as an assistant, and in the fall of 1982, he joined the UTSI faculty as associate professor of aerospace and mechanical engineering.

Chancellor Weaver, in his banquet address, said UT was outstripping all other organizations in two places in the state. One was at Oak Ridge "where they are making and will continue to make, tremendous progress in working with the National Laboratories, and the other is here, where we continue to make tremendous progress in working with the AEDC." Weaver hoped to see the day when UTSI would have a faculty of one hundred and a student body of a thousand.

(In six years, Weaver would play a more active role in the life of the Space Institute when he would be named dean upon Dr. Goethert's "retirement." However, at the fifth anniversary celebration, Goethert was serving as director. Dr. Weaver would appoint him as dean in July, 1971.)

Dr. Young gave a quick review of progress, noting that graduate course offerings had increased from eleven to more than thirty courses per quarter and that the full-time student body had grown to about ninety, and that one hundred sixty-eight students had earned degrees from UT through the Space Institute. Fourteen of those degrees were Ph.D.'s while the others were master of science degrees in engineering, physics or mathematics. Young said that ninety-four of the graduates were sponsored by either ARO, Inc. or the Air Force at Arnold Center. Young also emphasized that UTSI's academic programs had been strengthened and broadened to include "all regimes of flight from subsonic to hypersonic, propulsion systems from jet engines to rockets, computer science and technology, physics, cybernetics, and so forth." During the past year, he said, UTSI had housed and assisted in UT's evening extension activities in the area. About three hundred and fifty persons from surrounding communities had enrolled in a wide range of credit and non-credit courses at the undergraduate level in those evening classes. (**Bob Clement,** son of the former governor and himself a future congressman, had pushed for the evening classes while coordinator of field services at the UT Nashville Center.)

"Through the extension program," Young said, "it is possible for certain selected students to obtain junior and senior year **undergraduate work.**" With the new Motlow College in its second month, Young expressed an eagerness to cooperate with Motlow on the undergraduate work. Since 1964, he said, the Space Institute had offered one or two-week intensive, highly specialized, short courses for the benefit of industrial and governmental organizations.

"From a beginning of two short courses in Supersonic Ramjet Technology in the 1964—65 period," he said, "this short course program of continuing education has grown to some twenty short courses offered last year."

During its five-year history, UTSI had offered thirty-eight short courses with a total of nine hundred and forty persons attending.

It was a fitting birthday party for the Space Institute, and officials had done a commendable job of sharing its accomplishments with the public, but the celebration was not over. At the urging of the Support Council, six different Chambers of Commerce sponsored "UTSI Day" on June 14, 1970. **William A. Anders,** Apollo 8 astronaut, was the featured speaker for this occasion.

Astronauts **Donald H. Peterson** and **Henry W. Hartsfield Jr.** began classes at UTSI in January of 1970. On February 5, an "agreement of cooperation" had been signed with Von Karman Institute for Fluid Dynamics, Brussels, Belgium. **Robert O. Dietz** of Manchester was on leave from his Air Force civilian post at AEDC and serving as director of the von Karman Institute, which was operated by an aerospace scientific group of the North Atlantic Treaty Organization. Under this agreement, several UTSI professors, including **Drs. Jimmy and Susan Wu,** would spend about one academic quarter each at Brussels as guest lecturers and researchers. Also, several UTSI short courses, primarily in the field of space booster technology, would be presented at the institute in Belgium with **Drs. Goethert** and **Shahrokhi** as co-directors. UTSI also had entered into an agreement with NASA's Marshall Space Flight Center on March 12, calling for reciprocal use of certain equipment, library and other facilities. Goethert later recalled that this agreement, administered by **Marion Kent** of NASA, was especially useful for UTSI by allowing access to costly instrumentation and computers at the Marshall Center, which were superior to those available at the Space Institute.

(Later that year, Mr. Dietz received the Air Force's highest civilian decoration for public service, the Exceptional Civilian Award, for his "exceptional service" while serving as director of the von Karman Institute. **Dr. John L. McLucas,** undersecretary of the Air Force, made the presentation in September in ceremonies at AEDC. Dietz was an ex-officio member of the NATO advisory group for aeronautics research and development and chaired a special working group devoted to wind tunnel testing.)

Dr. William H. Pickering, director of the Jet Propulsion Laboratory in California, in January lectured at the Institute on "The Grand Tour of the Solar System." **Dr. T. Charles Helvey** was named chairman of a new thirteen-county South Central Comprehensive Health Planning District. **Dr. and Mrs. B. H. Goethert,** accompanied by **Hrishikesh Saha,** a doctoral candidate at UTSI, were on a month-long tour of six Indian universities and research centers. **Governor Buford Ellington** signed into law a bill creating a seat on the UT board of trustees for the Space Institute area. **Don Shadow,** a 29-year-old Winchester nurseryman, was appointed to the seat, the youngest person to serve as a trustee. He had graduated from UT in 1963. **Thornton Taylor** of Fayetteville, a member of UTSI's Support Council and a state representative for ten years, died of lung cancer March 5, the day after his 59th birthday.

Bell Aerosystems of Buffalo, N.Y., gave UTSI an air cushion vehicle, and Institute engineers began working to modify the underside of the vehicle in an effort to improve its efficiency. **Dr. Richard Kroeger,** professor of aerospace engineering, said the Space Institute was the only American university involved in air cushion vehicle research. Especially interested was **Dr. John Koester,** a new faculty member from California Institute of Technology, where he had received both master's and Ph.D. degrees. Koester had built a similar vehicle while an undergraduate at Notre Dame University.

Dr. Frank Collins was one of four aerospace engineering professors from the University of Texas visiting the Institute in March; he later would join UTSI's faculty. **Luther Wilhelm,** Student Government Association president, and **Dr. Frederick M. Shofner** were included on a statewide advisory group named by Dr. Boling. A chapter of Sigma Pi Sigma, physics honor society, was established at UTSI in early May, 1970, with **Dr. Arthur Mason** as faculty advisor. Retired **Colonel Raymond S. Sleeper** was appointed as cybernetics professor at the Institute, and the new Industry-Student Center opened. A former Air Force officer student at UTSI, **Major Richard L. Menninger,** died at Hollomon Air Force Base of injuries suffered in a light military plane crash. **Governor Ellington** spoke at a dinner at UTSI in June climaxing a national aerospace education leadership symposium at Middle Tennessee State University.

Dr. William T. Snyder, one of the five full-time professors when UTSI started classes, resigned in the summer of 1970 to become professor and head of the Engineering Mechanics Department at UT in Knoxville and to serve as chairman of the graduate program in engineering science within the College of Engineering. A fellow professor from those first days, **Dr. Marcel K. Newman,** was conducting a short course on the use of nuclear power in transportation systems of the future. And another of their peers, **Dr. John B. Dicks,** was pushing MHD as "the only answer" to increased power demands. He was doing magnetohydrodynamic research at this time under a five-year, million-dollar federal contract. **Dr. Jimmy Wu,** who had come to UTSI in its first year, received the General H. H. Arnold Award, top honor given by the Tennessee Section of the American Institute of Aeronautics and Astronautics.

The six Chambers that sponsored "UTSI Day" were the same ones with representatives on the Support Council: Manchester, Tullahoma, Franklin County, McMinnville-Warren County, Shelbyville-Bedford County, and Fayetteville-Lincoln County. **Robert Marbury,** who was president of the Tullahoma chamber when the idea was conceived, was chairman for the day and coordinated the planning activities. A resolution, dated June 14, pledging support and appreciation for the Space Institute, was signed by the following Chamber presidents: **J. C. Eoff Jr.,** Tullahoma, **R. Richard Baldwin,** Manchester, **James F. Cunningham Jr.,** Franklin County, **Charles L. Smith,** McMinnville, **William R. Carter,** Fayetteville, **J. W. Bowman,** Shelbyville. This resolution expressed appreciation to Dr. Holt and to **Dr. Edward J. Boling,** president-elect of UT, and to Dr. Goethert, "without whose personal dedication and great drive the Space Institute would not, in just five short years, be recognized nationally and internationally as a 'Center of Excellence' in Aeronautical, Aerospace and related academic and research programs."

Promising continued active support, the resolution concluded that "in particular, we will endeavor to ensure that the status, accomplishments, long-range goals and contributions to our communities of the Space Institute located in our midst, are known and appreciated by all citizens in our respective communities."

More than a thousand of those citizens got a first-hand look at the Institute when they turned out for the special day. A respectable crowd filled the stage of the auditorium. Among the speakers, in addition to the former astronaut, were **Dr. Hilton A. Smith,** vice chancellor for graduate studies and research, UT, **Dr. Charles H. Weaver,** UT chancellor, **William R. Carter** of the Support Council and president of the Fayetteville-Lincoln County Chamber, and **Dr. Goethert.** Marbury introduced the speakers and special guests and read the resolution adopted by the various chambers.

Others seated on stage were **Charles Wilson,** minister of Main Street Church of Christ, Manchester; **Marion Kent,** director, University Affairs, NASA; **Dr. Young; Don Shadow** of Winchester, a UT trustee; **Dr William Britt,** UT assistant vice president for development and administration; **Robert M. Williams,** managing director of ARO, Inc; **Colonel Ed Feicht,** vice commander, AEDC; **Jim Cunningham,** Franklin County Chamber of Commerce; **J. Stanley Rogers** of the Manchester Chamber; **Charles Smith,** president of the McMinnville-Warren County Chamber; **Dr. Sam Ingram,** president of the newly formed Motlow College; **Dr. Edward McCrady,** vice chancellor, University of the South, Sewanee; **Coffee County Judge John W. Ray,** Manchester; McMinnville **Mayor J. W. Anderson;** Shelbyville **Mayor H. V. Griffin;** Decherd **Mayor J. M. Shelley;** Tullahoma **Mayor Jack T. Farrar; Frank Pearson Sr.,** mayor of Cowan; Lincoln County **Judge Taylor Crawford,** Fayetteville; Winchester **Mayor Herman Hinshaw,** and **Vice Mayor E. W. Jernigan** of Manchester.

Visitors saw a flight demonstration program that included a short takeoff and landing of the STOL aircraft. Use of a helicopter as an air-ambulance also was demonstrated. **Professor Koester** demonstrated an air-cushion vehicle coming off Woods Reservoir onto land. Dr. Goethert thanked **John Stack,** "the famous aeronautical pioneer who sent us for this day from New York two advanced aircraft for demonstration flights." Goethert also told the visitors that "We are greatly honored by the presence of Mr. Anders, currently the executive secretary of the U.S. Space Council, and one of the famous Apollo 8 astronauts, who as one of the first men in history circled around the moon and saw with human eyes the craters, the boulders, and the vast extent of the lunar landscape."

Expressing pride in the Institute's progress in its first five years, Goethert said, "We are proud of the confidence shown in us by numerous federal and private groups who have given us the financial support so necessary to pursue our research ideas. We will need your help also in the future. Any idea or program can survive in the long run only if sufficient young people such as our students of today—the engineers and managers of tomorrow—are properly selected, supported, and educated."

Goethert said the "young Institute is undertaking to kindle the flame of enthusiasm among faculty and students, not for taking the easy way of 'just living' or 'hunting for money,' but for devoting their professional lives to exploring the unknown and to the creation of new generations of aircraft and space vehicles.

Mr. Carter, representing the Support Council and its chairman, Senator Crouch, who could not attend, cited the impact the Institute had had on the economy in the area.

"Of a budget of one million, two hundred thousand dollars," he said, "more than eight hundred thousand dollars is for salaries and stipends and most of it is spent in the local area. During the last year, more than four hundred people have been here for short courses and were housed in this area during their stay. The Institute has more than one hundred employees, and this is the equivalent of an industry itself."

While emphasizing that the Institute was "most important in and of itself," Carter said he believed that the research-type industry that would be attracted by the presence of UTSI "is of even greater importance to us." He said three research firms had been established in the area since 1964—two by personnel of the Institute and one because of the research environment of the Space Institute.

Dr. Weaver, commending the "fine relationship" between UTSI and the surrounding area, said the Institute "now truly has an interface with the entire free world . . . I look on the interface right here as the anchor point . . . for one end of an operation that expands all over the world."

Mr. Anders said he could think of no more valuable contribution to the nation "than providing talented graduates in the aerospace field," adding, "A strong graduate education such as that being offered (at UTSI) is invaluable for the young men entering the future aerospace world." He referred to contributions to aerospace programs, saying that a "major source of new technology, new ideas, and new processes have been institutes, graduate schools, and the facilities associated with them, such as you have here.'' He then declared ''the most valuable role of the graduate institution in the United States'' to be the education, stimulation and development of talented and creative engineers, scientists and managers for the nation's aeronautical and space needs.

''Apollo was wholly dependent on a team of government, industry, and university people,'' Anders said. ''More importantly, universities and institutes were the source of nearly all the technical and management people in Apollo, as they were in Gemini and Mercury before them.''

As an example, Anders said astronauts averaged more than one graduate degree per man, all acquired at institutes such as UTSI.

The special day was a boost to UTSI personnel because it demonstrated that the unity and support of communities throughout the area, which had played such an important role in getting the Institute established, remained strong after five years. Another boost came just before Christmas that year (1970) when UTSI was accredited by Southern Association of Colleges and Schools. This came on December 23, about a month after the Space Institute's first international short course at Brussels in conjunction with the Von Karman Institute, on ''Space Shuttle Vehicles,'' conducted by **Dr. Goethert** and **Dr. Firouz Shahrokhi.** UTSI was talking with Indian officials about working out a similar exchange agreement with India.

Two new professors—**Ronald H. Kohl** and **Dr. Harry L. Deffebach**—had joined the Institute's staff for the fall, 1970, quarter. Kohl would teach physics courses in electromagnetic theory and Deffelbach would teach electrical engineering courses in communication systems. Kohl, a native of Jackson, Mississippi, was a graduate of Cornell University and was finishing his dissertation for his Ph.D. from Ohio State University. A native of Birmingham, Deffebach held three degrees in electrical engineering from Auburn University.

Dr. Lloyd Crawford was heading up a nine-month investigation of combustion processes in a turbojet engine afterburn system under a $44,000 contract with the Air Force Propulsion Laboratory. **Dr. Chung-Ming Wong,** director of the U.S. Department of the Interior, speaking at the conclusion of a short course on engineering aspects of environmental control, said half the people in the United States were drinking unsafe water. **Dr. Robert A. Frosch,** assistant secretary of the Navy for Research and Development lectured on ''The Sin of Simulation,'' focusing on neglect of the art of engineering in systems engineering. Dr. Goethert took pride in noting that UTSI was almost ''paying its own way'' with non-state funds totaling $1,030,000 for the year compared to operating costs of $1,232,000

In March of the new year, the Space Institute, with the Technical University of Aachen, would conduct the first international short course at Aachen, Germany, in ''Aircraft Noise.'' And in July, 1971, Dr. Goethert would become **dean,** rather than director, of UTSI.

Goethert was getting close to sixty-five years of age, but he had no intention of quitting. UT, as it turned out, saw it differently. And **Morris L. Simon** would get inspired again.

✦

Chapter Eleven

TUG OF WAR

GOVERNOR Winfield Dunn was on hand at UTSI in the summer of 1971 for the start of a month-long workshop on air transportation, which was combined with a four-day conference on aviation in the U.S., held at Middle Tennessee State University in Murfreesboro. **Professor Raymond Sleeper** directed the workshop, which examined a wide range of technical, social, economic and geographic factors affecting air transportation. **John L. Baker,** assistant administrator of the Federal Aviation Administration, spoke at the dinner launching the events. Close to one hundred persons—many of them professors—from throughout the nation attended the workshop, which featured a host of speakers, including **James M. Beggs,** under secretary of transportation.

UTSI was being ranked above most other universities in the Southeast in the number of graduate degrees conferred in aerospace and mechanical engineering for the past two years. Dr. Goethert said of the seventy-two advanced degrees earned by Space Institute students over two academic years, nine were in aerospace and mechanical engineering. UTSI and the University of Aachen combined efforts in the spring of 1971 to present an international short course on aircraft noise, held at both institutions. **Grant L. Hansen,** assistant Air Force secretary for research and development, lectured at UTSI, keynoting the sixth national thermophysics conference hosted by AIAA.

On June 17, 1971, **Dr. Charles Weaver,** UT chancellor, announced during a meeting of the UT trustees in Knoxville the promotion of Dr. Goethert from director to dean effective July 1. **John Paul Grantham** succeeded **Charles Nitter** in June as manager of the industry-student center. Six months later, **A.J. Zazzi** replaced Grantham in the position. The new UTSI Automobile Club held its first rally in July with **Trevor Moulden**—a graduate student and future UTSI professor—as course marshal. **Jules Bernard,** a journalism graduate of Louisiana State University, became the Institute's short course director and information officer in the summer of 1971. He had worked in public relations with the Boeing Company in Huntsville and had formerly operated his own photography and public relations business in Louisiana. **Lee B. James,** former manager of the Saturn V launch vehicle that had sent Apollo missions on their way to the moon, was appointed associate professor of cybernetics. A retired colonel, James had been among the first Army officers assigned to the field of guided missiles and space vehicles in 1947. Completing the recreational wing—third and final phase of the Industry-Student Center—turned into a tough tug-of-war between legislators, UT and Space Institute officials, and the Tennessee Higher Education Commission (THEC). In 1971, area legislators got the General Assembly to approve completion of the additional wing, using some unobligated bond money, but due to a miscalculation, there was no money for it. Furthermore, members of the commission were cool toward the whole idea. **Dr. John Folger,** executive director, took the position that it was inconsistent to recommend building the unit with general obligation funds, and the commission had assigned the project a low priority, which meant that it probably could not be built before 1975, if that soon.

On the last Friday in September, 1971, THEC members voted 6 to 2 to **defer** action on a proposal to establish a nursing program at Motlow College and directed their staff to make recommendations in November as to the feasibility of the proposal. A large delegation, including **G. Nelson Forrester, Richard Morris, State Senator Reagor Motlow, Dr. Sam Ingram,** and **Gloria Calhoun** of Manchester, who was expected to head up the program, had urged approval. Folger was against it, saying that within a very short time, there would be too many nurses in the state.

The commission met on November 29 at an amphitheater in Chattanooga. Among those attending this meeting to speak for the **nursing** program was **Morris L. Simon.** As it turned out, however, Simon's best speech was on behalf of finishing the student center.

"I had worked with Dr. Goethert, trying to get funds for the center," Simon recalled, "but I was there that day to speak on behalf of the nursing program."

Senators Crouch and Motlow and several other legislators and Support Council members also attended the meeting and were seated at various places surrounding the commissioners. Attorney Forrester, who had lobbied the commissioners on behalf of the nursing program, Simon, and **Dick Morris,** Harton Hospital administrator, were seated just behind **Dr. Edward Boling,** who was then the UT president. Simon was chatting with **Bo Roberts,** a former newspaperman and UT vice president for development. Boling made a pitch for the student center project, but THEC members told him they had decided to postpone actions on the center. Boling took his seat.

Apparently in jest, since he knew Simon was there to speak for the **nursing** program, the chairman—**John Long** of Springfield—looked at Simon and asked: "Do you have anything to say about the student center, Mr. Simon?"

Simon rose like a trout to the bait.

"Yes indeed I do, Mister Chairman!" he said. "I certainly do. Dr. Boling hasn't told the whole story about our needs for this student center. It just happens that I am vice chairman of the Support Council. . . . I know you have decided to hold the center in abeyance, but I think you are making a very serious error in doing this. I think we need it now. Why? Because our students are fifteen miles from town. They have no place to go between classes, no place to go for recreation. . . . You will never have an institute that will be meaningful unless you give them the facilities to support it, and this is one very desperately needed facility."

Simon then eased into the second phase of his argument, calling attention to the fact that "a number of very prominent legislators" and other citizens were members of the Council.

"It is my understanding that you get your support from the state legislature," he said. "I would like to call your attention to some people here who are very important to this organization, to UT, and to the state." Simon then looked at the elevated section and proceeded to point out various legislators who were seated there, including **Senator Ernest Crouch,** chairman of the Support Council.

"These eminent citizens are interested in what you do and don't do," Simon said. Before singling out **Senator Reagor Motlow,** whose family owned Jack Daniels Distillery in Lynchburg, Simon slipped in a little humor, saying, "I suggest you give heed to one man who is present today because he can cut you off from all your whisky. . . ."

As the laughter died down, Simon continued: "**Senator Reagor Motlow** is on the Support Council. You all are aware of the part he has played in education during his years in the Senate. All you have to do is to look up at him and ask him if he doesn't think this is a very definite need of the Institute."

One member of the commission quickly said, "I think we ought to reconsider our decision." Another chimed in: "I think you are right."

As Simon sat down, amidst applause, **Bo Roberts** whispered: "They will never believe that you and Boling didn't plan all this!"

The commission agreed to reconsider its action. Later that day, the commissioners also unanimously approved plans for a nursing program at Motlow (which by then was fully accredited by the Southern Association of Colleges and Schools.) The battle wasn't over, though. State **Representative J. Stanley Rogers** and **Senator Ernest Crouch** had a lot of work to do.

Money to complete the center was still an issue when the legislature convened in January, 1972, and on January 7, THEC members refused to recommend spending two hundred thousand dollars asked by UT for building the recreational wing to complete the center as designed. **Dr. Archie Dykes,** UT chancellor, had urged them to reconsider. Chairman Long commented that there had been a "full presentation" on the subject made in November. Dykes disagreed with Dr. Folger, who had repeated his stand against using general obligation money for the building. Dykes said the wing was important to the educational program. Plans were for the wing to have facilities for squash, handball, billiards and table tennis.

THEC took no action on the project. Its position was that all recreational facilities were financed by users—that revenue from gate receipts and student fees were pledged to pay off bonds. UT's position was that setting up student fees at UTSI would be prohibitive because of the small enrollment.

After the meeting, Senator Crouch said, "What they say isn't the final say by any means. There will be a lot of changes made in their recommendation." Crouch was a member of the state fiscal review committee, which would soon meet.

In August of 1971, the U.S. Department of the Interior and the Tennessee Valley Authority gave UTSI a $299,000 research grant for MHD research. Dr. Dicks headed up the research team. **Luther R. Wilhelm,** former president of the student government association, joined UT's faculty that fall as an assistant professor of agricultural engineering. He had earned his master's degree in mechanical engineering at the Space Institute and was completing work at UTSI on his Ph.D. **Dr. Fletcher W. Donaldson** was addressing the Congress of County Medical Societies in Anaheim, California. At this meeting, he was elected a fellow of the Society for Advanced Medical Systems. Donaldson also was appointed to a special committee on problem-directed medical systems of the society.

Dr. Jack T. Farrar, who as Tullahoma mayor had presided during the pressurized days of change brought about by the decision to locate AEDC at the Camp Forrest site, died on the night of November 8 at the age of sixty-five.

Meanwhile, UTSI acquired a twin-engine Cessna 310 (''Flying Laboratory'') Aircraft—the first of two such planes coming to the Institute—and an experimental flight mechanics program was initiated.

On January 6, 1972, Dr. Dykes, during a dinner meeting of the Support Council, said the Space Institute held the key to much of the future of the whole UT system ''because of its dedication to graduate study and research.''

''The Space Institute is possibly the most eloquent manifestation of graduate training that we have because it is **the only institution that I know of in the state completely graduate and research-oriented,''** Dykes said. ''We would be hard pressed to find anywhere in the country scholars superior to those found right here.'' He was accompanied by **Hilton Smith.** Among legislators present were **Senators Crouch** and **Motlow,** and Representatives **Rogers** of Manchester, **Ed Murray** of Winchester, and **W.R. (Spot) Lowe** of Lewisburg.

Attorney Clinton Swafford of Winchester, a member of the Support Council, described the Institute's new program of undergraduate fellowships for a select group of students who had completed their freshmen and sophomore years and who were believed suitable to complete undergraduate work and go on to seek graduate degrees in aerospace science at UTSI. (This was an experiment by Goethert, who saw it as a means of ''seeding'' the graduate program.)

Simon reviewed the history of how the Support Council was formed in 1966 and told of the Institute's work, including cybernetics and MHD, as it applied to many vocations.

''One of the nice things about the Institutes,'' he said, ''is that they not only ask for the support of the citizens, but then they ask the communities 'How may we help you?' '' As an example, Simon said UTSI had offered the expertise of its faculty and students in helping develop Northern Field into a municipal airport and industrial park.

Also in January, 1972, UT awarded a $19,876 contract for construction of a library wing at UTSI. Amacher Construction Company of Tullahoma was chosen for the project, which would add 3,840 square feet to the library. This was the second of three major construction projects eyed for the Institute. (UT had asked THEC for $530,000 in building expansions for the next two years). Construction was already under way to double the MHD lab. The third one, of course, was the recreational wing of the Industry-Student Center, and a major push to resolve that problem was shaping up in the legislature.

The Institute got a $25,087 earth survey grant from NASA. Dr. **Shahrokhi,** who in a couple of months would direct a short course on remote sensing of earth's resources, was in charge of the program.

On January 27, 1972, **Mrs. Jessie Ruth Jackson Myers,** a Manchester artist and wife of the perennial mayor, died, and three days later, **Ridley Hickerson,** a former Manchester mayor and banker, died at eighty-nine.

Early in February, **J. Stanley Rogers** said that the Space Institute would get back some of the money authorized by the legislature the previous year for completing the recreational building. Rogers said if necessary he would introduce an amendment to **Governor Winfield Dunn's** budget to put in the full two hundred thousand dollars. He said State Finance Commissioner **Russell Hippe** was working to include about eighty-eight thousand dollars that was unobligated out of the funds the legislature had approved in 1971. When UT appropriated the money, Rogers said, UTSI wound up short because of a calculating error.

On February 25, the Manchester Times reported that Governor Dunn was including two hundred thousand dollars in his twenty-six-million-dollar building budget for state colleges and universities to be allocated for the student center wing. This pretty well cinched it.

Joe Carr, Secretary of State, about fifty legislators, and other state officials toured AEDC in March and attended a dinner and reception at the Space Institute. On April 14, the General Assembly passed an omnibus appropriations bill containing two hundred thousand dollars for the recreational building. UTSI officials said they would seek bids ''this year.'' In June, the higher education commission, meeting in Nashville, gave final approval for Motlow and Middle Tennessee State University, Murfreesboro, to launch a two-year program to train registered nurses, starting in the fall. **Miss Barbara Ann Rowden** of Nashville was named as director of the nursing program since **Gloria Calhoun** had resigned.

Retired Colonel Edgar J. Masters joined the UTSI staff that spring (1972) as assistant to Dr. Goethert in charge of student personnel services.

In May, **Dr. William D. Jackson** of Lexington, Massachusetts, was appointed as a professor of electrical engineering at UTSI. Born in Edinburgh, Scotland, Jackson had taken his Ph.D. from the University of Glasgow in 1960. His major responsibilities at UTSI would involve helping **John Dicks** with MHD research. A former lecturer at MIT, Jackson left UTSI at the end of August, 1973, to join the senior staff of a newly created Electric Power Research Institute with responsibility for energy conversion research. A year later, he joined the Office of Coal Research in Washington, D.C., and from that office a few years later he would negotiate a contract to locate the Department of Energy's Coal Fired Flow Facility at UTSI.

In June, 1972, UTSI received a three hundred and forty-five thousand dollar research grant from the Federal Aviation Agency to study jet noise. **Dr. Heinz D. Gruschka,** associate professor of physics, headed up this project.

The recreational wing of the Industry-Student Center would open in April of the following year, and on October 6, 1974, it was dedicated as "The Andrew Holt Industry-Student Center." The UT board of trustees named it in honor of **Dr. Andrew David Holt,** the "distinguished and beloved 16th president of UT who guided the institution from 1959 to 1970." A plaque erected on the front wall of the building noted that creation of UTSI was "one of the outstanding developments in the Institution's phenomenal growth under his inspiring leadership."

Dr. Roger Crawford, SISGA president when the first work began on the Industry-Student Center and later a professor at UTSI, said Dr. Goethert was "very student-oriented. He was better to the students than he was to the faculty."

In February, 1972, UTSI received a $19,760 grant for a management study to be conducted by **Lee James,** associate professor of cybernetics at the Institute. On March 30, **Burt C. Kendrick**, a man with a voice like a bass fiddle and one of the most colorful police chiefs in Manchester's history, died. And on the following Monday, April 3, **Buford Ellington,** two-time Tennessee governor, suffered a fatal heart attack while playing golf in Boca Raton, Florida. **President Richard Nixon,** riding the crest of his popularity before the pitfall that lay ahead, made his historic visit to China that spring. **Dr. Wernher von Braun,** deputy administrator of NASA's space program, spoke at UTSI on April 28, highlighting a national congress on aerospace education involving more than three hundred junior and high school teachers from around the nation. He predicted that a baby would be born on the moon by the year 2000. **Mrs. Mary Anderson,** former head of the state aeronautics commission, acting on behalf of **Governor Winfield Dunn,** made von Braun an honorary citizen of Tennessee. (Mrs. Anderson's cousin, **Madge Gibson,** later would serve as secretary for Dr. Joel Muehlhauser and others at UTSI.) In a couple of months, von Braun would retire and join Fairchild Industries.

George Wallace handily won another term as governor of Alabama in early May, and two new tennis courts at the Space Institute were "bombed" during dedication ceremonies. Construction of the two "Laykold" champion tennis courts had started in the fall of 1971. Paverite Inc. of Knoxville built the courts adjacent to the student center. They were financed by UTSI's share of student activities fees collected by UT, and by special donations. Goethert was an avid tennis player, and when the courts were officially opened in May, the match of the day featured the UTSI dean challenging **Dr. Walter Herndon,** associate chancellor for academic affairs at UTK. Herndon showed up with a racket "bent like a scoop," according to **Bob Kamm.** "We had line judges," he said, "and everything that Herndon hit was 'in' and everything that Goethert hit was 'out.' Herndon was the guest!"

There was another guest for the dedication of the courts, but he was not invited. He came roaring out of the sky, zoomed in a few feet above the tennis courts, and released his cargo: rolls of toilet paper! Tissue streamers freely decorated the court and interrupted the Goethert-Herndon match. With the crowd cheering, Goethert managed to smile. Old-timers said his diplomatic display of appreciation of the humorous situation almost certainly was not conveyed to the bombardier, one **Sean Roberts,** the first pilot for the Flight Mechanics Program, which had been initiated in January after the Space Institute bought a surplus Cessna airplane from the Department of Defense for five hundred dollars.

Dr. John Dicks was named Alumni Distinguished Professor at the UT Chancellor's honors banquet. Three other UTSI professors—**Drs. Marcel K. Newman, Firouz Shahrokhi, and Jimmy Wu**—were chosen for the 1972 edition of "Outstanding Educators of America." Wu was promoted to full professor in July. A year later his wife, **Dr. Susan Wu,** was promoted to professor of aerospace engineering—the first and only woman engineering professor at UTSI. She also was a member of J.B. Dick's MHD research team.

One morning in the summer of 1972, **Jules Bernard** walked into the registrar's office and told **Miss Virginia Richardson,** "I would like for you to meet **Neil Armstrong.** He wants to register for credit in a short course."

"It was really a treat!" Miss Richardson recalled years later. She had known that **Neil Alden Armstrong,** the famous astronaut who had been the first man to the moon, was attending a short course on campus, but "I thought I would never get to meet him." Armstrong was residing in Lebanon, Ohio, and teaching at the University of Cincinnati. **Sean Roberts** led two short courses that summer: Special topics in aerospace and mechanical engineering. Armstrong audited one and took the other for credit (he mad an A). He had his master's from Southern California, and Purdue University, where he had completed his undergraduate studies, had awarded him an honorary doctorate.

"The students all took his picture," Miss Richardson said, "and I've kicked myself ever since for not having my camera."

Miss Virginia Richardson

Gabriel D. Roy, who was an assistant professor at Kerala University in India before studying at UTSI, became president of a new Arts Club. **Dieter Kowak** of Wurselen, Germany, was secretary-treasurer. Faculty advisor was **Dr. Jimmy Wu,** and **Miss Richardson** and **Mrs. Mary Lo** were executive committee members. **Senator Howard Baker** spoke at a conference on aviation. **Dr. Richard V. Shanklin III,** one of the first students at the Institute, in July, 1972, joined the faculty as an assistant professor of mechanical and aerospace engineering. **Dr. James Reginald Maus Jr.,** who had been appointed to UTSI in 1967 as an assistant professor, was promoted to associate professor of aerospace engineering, Maus, a research assistant at UTSI in 1966, got his Ph.D. from North Carolina State University the next year. Fifteen students were studying in a new program leading to a master's degree in engineering administration. More than eight hundred people turned out at UTSI in September for a "This Is Your Life" tribute to **Congressman Joe L. Evins.**

Jerry Coble, a young engineer at AEDC and a research assistant to **Professor Raymond S. Sleeper,** died of "natural causes" in November. He obtained a master's degree from UTSI in aerospace engineering and was working towards a doctorate in engineering science. Coble was a member of Tullahoma Airport Commission.

Just before Christmas, 1972, the U.S. Department of the Interior granted the Space Institute a $1,180,000 extension of its contract for MHD research. This extension would take the work through June, 1974.

A major milestone was reached in January, 1973, with the establishment of the Energy Conversion Research Division (ECD) at UTSI with **Dr. John B. Dicks Jr.** as director. Dicks and UTSI had been involved in MHD research from the start with the first MHD contract signed in 1964 with the Air Force Systems Command. Various other faculty members and graduate research assistants (**Joel Muehlhauser** for one) had participated in the research, but it was not a full-time commitment until 1973. Nevertheless, early efforts had been fruitful. While testing an MHD generator with a solid rocket propellant combustor under the first contract, UTSI scientists found that alumninum oxide deposits on the electrodes of the generator not only did not detract from the power generation but actually protected the electrodes from erosion. This discovery, coupled with the projected energy supply position for the U.S., led to the conclusion that **coal-fired MHD was a potentially attractive option for electrical power generation.** This led, in 1971, to development and testing of coal-fired combustors and generators at UTSI—the first time anyone in the world had done this. MHD would provide the backbone of research at the Space Institute over the years and account for more than seventy million dollars in research contracts by 1989. (ECD later became known as the Energy Conversion Research and Development Programs (ECP). The first MHD lab building was referred to as the Energy Conversion Facility (ECF).

Dr. Goethert returned from a month-long lecture series in Germany where he had been presented the Aachen University's highest award bestowed on a faculty member: the "Plaque of Honor."

In March, 1973, the Space Institute was given authority to establish multidisciplinary programs. Also in March, the "UTSI Flyers" were established, and in May, a Club of the Society of Sigma Xi was chartered at the Institute.

A new master's of science program in Aviation Systems was approved in July, 1973, and the next April a second airplane—a big, single-engine DeHavilland "Otter," capable of carrying eight to ten passengers—was obtained from the U.S. Department of Health, Education and Welfare (HEW) for use in UTSI's research programs. Construction of an Aeroacoustic Laboratory Building was completed at UTSI in September, 1973. The second joint short course with the Technical University of Aachen was held in April, 1974, in Aachen and in Tullahoma.

Dr. T. Charles Helvey, planning to retire, organized a private educational Institute for Information and Control Systems, also known as the Kappa Institute, with offices in Tullahoma. **Dr. Gerhard H.R. Reisig,** a Huntsville space scientist and cybernetics expert, was associated with Helvey in the venture. Helvey had graduated from the Humanistic College in Budapest, Hungary, in 1920. Later he earned his master's degree at the Braunschweig Technical University in Germany and his doctorate at the Kaiser Wilhelm Institute in Berlin.

UTSI's faculty got a major boost in September, 1973, with nine new full-time professors and four new names added to the part-time staff. Added full-time were **Dr. Kenneth E. Tempelmeyer,** professor of engineering science; **Dr. Terry Feagin,** assistant professor of aerospace engineering and computer science; **Dr. Roy D. Joseph,** assistant professor of electrical engineering; **Dr. Tien Chen Lieu,** research associate and professor in mechanical engineering; **Dr. Neil L. Loeffler,** assistant professor of electrical engineering with a specialty in ceramics engineering; **Dr. Alan Milner,** associate professor of metallurgical engineering; **Dr. Thomas Conley Powell,** assistant professor of mechanical engineering; **Dr. Grant Patterson,** assistant professor of aerospace engineering, and **Dr. Manfred Whittman,** visiting assistant professor of aerospace engineering from the Technical University of Aachen. These additions gave the Institute a total of thirty-three full-time faculty members.

Dr. Trevor Moulden, post-doctoral fellow, became a part-time assistant professor of aerospace engineering. Moulden had just returned from the tenth International Symposium on Space Science and Technology in Japan, where he had presented papers by **Dr. Jimmy Wu, Dr. Donald Spring,** and himself. Other new part-time professors were **Dr. Kennedy F. Rubert,** professor of aerospace and mechanical engineering, **Dr. John C. Vaughn III,** assistant professor of mechanical engineering, and **Dr. Carmelo L. Battaglia,** assistant professor of sociology and psychology. (**Rubert,** an early Scramjet authority, had been with NASA at Langley Field.) UTSI also promoted **Dr. Maurice A. Wright** to full professor of metalurgical engineering.

On November 19, another faculty member was added. **Dr. Willi F. Jacobs** of Sandy Springs, Georgia, was employed as a professor of aviation systems. Jacobs taught aircraft performance as well as aerospace engineering courses during the next ten years, commuting from his home near Atlanta. He retired on August 31, 1983. Jacobs had been a student of **Ludwig Prandtl** in Germany.

Dr. John Dicks was in Russia that fall, working out details of an exchange program with staff scientists at the Soviet High Temperature Institute in Moscow. Five Soviet scientists visited the Space Institute in December. **Terry Wenzlick** was the latest manager of the Industry-Student Center.

As the curtain was closing on 1973, both the House and Senate approved a supplemental appropriations bill with two million dollars in extra energy research funds for UTSI.

The Space Institute had survived for almost a decade. It was time to celebrate.

On Sunday, June 2, 1974, the combined chambers of commerce from Manchester, Tullahoma, Bedford, Franklin, Lincoln and Warren countries and the UT alumni chapters from Bedford, Coffee, Franklin, Lincoln and Warren counties sponsored "UTSI DAY" in recognition of the Space Institute's tenth anniversary.

It was a good day for a party. After two days of rain, the sun came back, and hundreds of visitors turned out. The tennis courts got plenty of action that day, but the main event was a challenge match between Dean Goethert and **Dr. Maurice Wright,** and UT's new **Chancellor Jack Reese** and **Fred Peebles,** dean of the college of engineering at UTK. Reese and Peebles were the victors.

A major attraction—despite necessary "surgery"—was the manned hot air balloon flight by **Eric Soesbe** of Tullahoma. It was scheduled for late in the afternoon, and weary spectators gravitated to the site of the launch. Soesbe, in blue jumpsuit and cap and sporting an old-fashioned moustache, spread the limp fabric of the balloon over about a fifty-foot area. Then he discovered a five-foot rip in the canopy, and the cry went out: "Is there a doctor in the. . . .?" There was. **Dr.**

Charles H. Webb of Tullahoma, with eager assistance from **Larry** and **Deidra Horn** of Tullahoma and Dr. **Goethert,** applied sutures, and the balloon eventually soared. (Spectators who had pressed close to watch could hear Dr. Webb compare the "messy" procedure to a hysterectomy.)

Numerous displays and posters called attention to activities of various student organizations, including the boat club, auto club, the AIAA, the Institute flyers, tennis club, and the arts club, which sponsored a "Fancy Dress Competition."

Ray Ricco of Tullahoma, president of SISGA, addressed the visitors on behalf of the students and expressed appreciation of the students to "Dr. Goethert for his hard work and untiring efforts in making the Institute the success that it is today."

UTSI Flyers and the Tullahoma Bunch presented a "flyby" with **Jules Bernard,** short course director and public affairs officer for UTSI, **Stephen Landers, Bill Landers,** and **Bill Baker** of the Tullahoma Bunch serving as ground crew and announcers. NASA's Marshall Space Flight Center provided several exhibits, including space shuttle, solar energy and Skylab displays.

There were many speeches, with the **Reverend Garnett R. Smith,** rector of St. Barnabas Episcopal Church, Tullahoma, praying ". . . that knowledge may be increased and good learning flourish and abound to the benefit of man."

Other dignitaries seated on the stage included **Dean Goethert, Dr. Robert L. Young,** associate dean, UTSI, **Dr. Andy Holt,** UT president, **Dean Peebles, Chancellor Reese, Dr. Hilton Smith,** vice chancellor for graduate studies and research at UTK, **Senator Ernest Crouch,** chairman of the Support Council, who presided over the program in the auditorium, **Morris L. Simon,** vice chairman of the Support Council, **Don Shadow,** a UT trustee, **Ike Grizzell,** and **William R. Carter,** members of the Support Council, and **Ray Ricco.**

"Ten years have gone by since we first assembled . . . as a small group of five professors and a handful of full-time graduate students," Dr. Goethert said. "At that time, the Institute was merely a plan, a dream, and we are fortunate that so many people had faith in us and gave us their unwavering support. And most importantly, they let us go to work, let us operate on our own, and experiment with new ideas of graduate education, research and community services. At the end of the first decade, we of the Institute believe we lived up to your confidence!"

Goethert contrasted conditions in 1964 with those ten years later:

"When ten years ago there was a wild area, barely accessible by foot or car, there is now a thriving campus. . . .

"When ten years ago we had merely a handful of five professors, there are now more than 30 resident professors and more than 20 adjutant professors from research laboratories, government and industry around the country.

"When ten years ago the academic program . . . was in its infancy, there is now a broad spectrum of academic opportunities with an ever-growing number of graduate students. . . .

"When ten years ago only very few sponsors ventured some small sums of research money for the young unknown team at UTSI, there is now a broad basis of research sponsors, with a total commitment of $1.2 million in this year. . . .

"The steady growth . . . is well reflected in the funds for operation which grew from $180,000 in 1964 to $2.2 million in 1974. . . ."

Goethert expressed pride that "with the exception of a fourth to a third fraction, the funds of the Institute operation were earned by our people from outside sources, mostly from out of state, that without a drain on the resources of the University."

He said the struggle for growth "in the face of tremendous nationwide competition was not easy, and a few of our group could not stand the heat, but those who stayed became stronger and harder and thus better prepared for the struggle in real life, outside the sheltered confines of the University campus."

Looking ahead, Goethert envisioned a doubling of the faculty within the next decade, further strengthening of research programs, and further development of the campus. Prophetically, he visualized establishment of a UTSI research park, and a "high-technology vocational training center jointly with the surrounding communities, in which high-technology industrial groups will join hands with the researchers of the Space Institute and thus establish in mid-Tennessee an academic-industrial technological base, which will stimulate high-quality life far beyond the bounds of this area."

In "due time," Goethert thought the University would give the Institute authority "for establishing a small **undergraduate** program on the junior and senior level, for further utilization of the Institute's resources and for providing unique opportunities for gifted students to grow up in the advanced research and applied engineering atmosphere of the campus." (This was Goethert's idea for providing "seed" students for UTSI. "Above all," Goethert said, "we foresee that the University planners will recognize that the young growing Space Institute has come a long way from childhood to adolescence and should be entrusted with adequate authority and local responsibility for uninhibited development and utilization of its resources."

With top UT officials present, along with many community leaders, alumni and Support Council members, Goethert was not one to let pass an opportunity to plug for more independence. It would come slowly and grudgingly, but it would come.

Goethert concluded with a pledge to make the name of the Institute "known not only in Tennessee and the U.S.A., but make it continue to ring all over the world!"

Dr. Young called attention to "three hundred and five people, some tall, some short, some thin, some fat, different races, including men and women, a few nearly normal, but all—I emphasize **all**—great." He referred to the number of "engineers and scientists who have earned graduate degrees from the University of Tennessee through our academic research programs in our ten years of existence." Saying that none took advantage of all the opportunities for self-development at the Institute, Young added, "These three hundred and five grasped enough of what is here to earn **two hundred and forty-five Master of Science degrees** and **fifty Doctor of Philosophy degrees.**"

Dr. Holt said the first ingredient of a truly great institution of higher learning is "to have stem-winders at the top," explaining that a stem-winder is one who gets things done.

"Well, we've been blessed at this institution by having more stem-winders at the top than any place I know," the retired UT president continued. "I guess it has to start with the Air Force. That'd be **General Arnold** and **General Schriever** and **General Rogers** and all the rest of them. The next bunch of stem-winders I'd suggest would be the ARO big-wigs. That'd be **Bob Williams** and **Ed Dougherty** and all the rest of them who said: 'We can help you do it. We can provide the facilities, and we'll provide the management.' The next bunch I'd suggest would be the University of Tennessee big-wigs, **Dr. Brehm, Hilton Smith, Herman Spivey,** UTSI's own **Bob Young,** your board of trustees, and all the rest of them who said: 'We'll provide the training; we'll take it over.' And the other big-wigs, of course—**Ed Boling** is a big-wig; he's carrying it on right now. Then there's **Jack Reese,** who is making a great contribution to this institution. I'd have to add . . . the legislators and the congressmen who provided the money to do the job."

Holt saved the biggest "big-wig"—**Dr. Goethert**—for last, saying: "He is undoubtedly the stubbornest big-wig I have ever seen. He knows absolutely nothing about the meaning of no, and if he did, this institution wouldn't be here right now. He gets an idea, and he tells you what it is. He listens, but he doesn't understand what you said. He never takes no for an answer. He's got more determination, more vision, and more dedication than any educator I've ever seen."

Chancellor **Reese** said the Institute's future was "extraordinarily bright" for several reasons, including its capacity to respond to societal needs.

Space Institute officials also seized the birthday party as an opportunity to express their gratitude to two members of the Support Council who had provided two new recreational areas at UTSI by naming the facilities for them. "Carter Park" was named for **William R. Carter,** president of CFW Construction Company of Fayetteville, who had arranged for his firm to clear land near the Industry-Student Center on which the tennis courts were located. Picnic tables were installed later. Carter, a UT alumnus and former member of UT's engineering faculty, was a charter member of the UTSI Support Council, and in 1982 he would become its second chairman. He was presented with a photograph of the park. "Mullican Beach" was named in honor of **J.D. Mullican,** president of J.D. Mullican Construction Company, McMinnville, which had constructed the beach. Mullican was ill and unable to attend the party, but **Richard Smith,** a UTSI student, accepted a photograph of the beach on Mullican's behalf.

It was a full day, and as the sun went down on Woods Reservoir, everything seemed peaceful with little trace of the wrangling that had occurred a few months earlier after UT insisted that Dr. Goethert retire because of his age. Those waters had been temporarily calmed, but rough sailing lay ahead and a period of transition that would bring some major changes to the Space Institute.

✦

Chapter Twelve

IN SEARCH OF A DEAN

EARLY in January, 1974, the Space Institute registered one hundred and seventy-seven students for the winter quarter—an increase of twenty-two. UTSI had offered a site for a three-county vocational school on the Institute's campus; Tullahoma's Chamber of Commerce passed a resolution urging that the school be built. It would have served Coffee, Grundy and Franklin counties, but it was not to be.

Dr. Harvey M. Cook Jr., who had been Dr. Goethert's assistant at ARO, resigned at the end of 1973 as chief of the technical staff at ARO, a position he had held since 1966. Only forty-eight years old, Cook quipped that his body couldn't keep up with his mind. It was not a joke. Cook's back was broken in 1971 when he was the victim of a hit and run accident in Boon, West Germany, and he was still paying the price. Dr. Cook had played a very active role in the long drama to establish the Space Institute. He also had been very active in professional societies, and he was the first non-German to be appointed as a scientific member of the West German Aeronautics and Astronautics Organization; he had been technical advisor to that organization for eleven years. Cook made his first appearance at AEDC in 1951, when the center was under construction, as a project engineer for one of the contractors. He started working for ARO in May, 1953, and three years later he became manager of ETF at AEDC. Dr. Cook had served as president of the Tennessee section of the American Institute of Aeronautics and Astronautics (AIAA) and also had headed up the Tennessee section of the American Rocket Society. (The American Rocket Society and the Institute of Aeronautical Science merged into AIAA in 1963.)

Late in January, 1974, the news that would shock UTSI supporters broke in area newspapers: **"UT RETIRING DEAN OF SPACE INSTITUTE."** According to news reports, **Dr. Hilton Smith,** vice chancellor for graduate studies at UTK, dropped the bombshell on Tuesday, January 22, during a first visit to the Institute by the new UT chancellor, **Dr. Jack E. Reese.** Among UT officials present was **Mrs. Margaret Perry,** dean of graduate studies for UT.

Manchester and Tullahoma newspapers reported that "Top officials of the University of Tennessee have begun a move to bring about the retirement of **Dr. B.H. Goethert. . . .**" The "surprise announcement" by Smith came during a briefing arranged for Reese by the UTSI faculty and administrative staff. Smith said a University committee was being formed to seek a successor to Goethert, and a decision might come in the fall or later. The newspapers reported that Goethert declined comment "but it was learned that he was told the day before," and that Smith acknowledged the retirement was involuntary on Goethert's part.

While Goethert may have "seen it coming," the announcement caught the faculty by surprise, and some of them reacted angrily. **Senator Ernest Crouch,** Support Council chairman, said he was "surprised and concerned" and pledged to urge that Dr. Boling, Reese and Smith reconsider.

Goethert, who was born October 20, 1907, in Hanover, Germany, was sixty-six—a year beyond the normal retirement age required by UT. He had been granted a one-year extension and had requested another, which had been turned down. UT's policy permitted a faculty member to continue teaching on a year-to-year basis until the age of seventy. Requests for extensions normally went from faculty member to immediate supervisor, who in Goethert's case would have been Smith. The Manchester Times and Tullahoma News reported that Smith, in response to a question, said that Goethert had sought another extension "but didn't get the request in on time," and that the decision was being made in January to allow plenty of time to locate a successor later in the calender year. (After the initial stories were published, Smith quickly denied saying that Goethert did not get his request in on time.)

The first news stories noted that "Despite differences with the (UT) administration," Goethert had been praised on public occasions by Dr. Holt, Dr. Boling, and **Dr. Archie Dykes,** who had left Knoxville as UT chancellor the previous September to become chancellor at the University of Kansas.

Crouch contended that Goethert's expertise would be needed the next three years because Congress had just voted two million dollars to step up research at UTSI. He was, Crouch said, "surprised and concerned that the administration of Knoxville would serve such short notice on Goethert." Goethert's leaving would mean a hardship on the programs and further developments of the Space Institute, the senator said, and Goethert "should be given the same consideration given other deans over the years . . . and be allowed to serve until he was seventy." Crouch continued that "Due to his (Goethert's devotion to duty, and due to the fact that he provided the leadership and the concept to create the Space Institute to begin with, and due to his broad background, it will be most difficult to go out and find a successor to him. . . ." Crouch said he thought it was the feeling of all Support Council members that the UTK administration "may have taken a step which will be injurious to UTSI, and therefore the entire area and state, in its decision not to allow Dr. Goethert to continue in the role in which he has brought prestige to the Institute and to UT. . . ."

Within a week, students and faculty at the Space Institute formally requested that Goethert be allowed to postpone his retirement. **Walter Harper,** Student Government Association president, said all three SGA officers and the other members unanimously supported the action. **Dr. Arthur Mason,** as faculty secretary, said a letter was being sent to the administration on Goethert's behalf. Meanwhile, UT officials were countering the criticism by saying the decision on Goethert had been made considerably earlier. Chancellor Reese said the decision came about as a result of an "agreement" with Goethert in October, 1972. Dr. Smith said the decision to deny Goethert's request for a second extension was made in October, 1973, by Knoxville authorities because it was thought that "the search should go forward" as it would be lengthy since it would be a major task to find a capable successor to Goethert. Smith said he had asked Goethert if he wanted to make the announcement, and Goethert declined and asked him to make it. However, Smith added, Goethert had been aware of the decision since 1973 and could have announced it at any time.

Reese cited a letter that he said Goethert wrote to **Hilton Smith** in October, 1972, expressing a desire to serve through August, 1974. He quoted Goethert as writing in that letter that, "as discussed before," he wished to retire "not at the end of the current year, but after one more year of service, on 31 August, 1974." "The University was pleased to grant a one-year extension to allow him to round off a decade of service to the institute which he helped establish and which he has served so well," the chancellor said.

Continuing in this conciliatory vein, Reese recognized Goethert's "loyal and dedicated service to the University and his tremendous achievement in building UTSI from a dream into one of the nation's primary aeronautical research centers."

Dr. Young remembered that Chancellor Reese had become perturbed at barbs from Young and Goethert just before the matter of Goethert's retirement came up on that January 22. The state university system had gone on "formula-funding" wherein the primary factor was the number of student credit hours produced at various levels. UTSI officials had been told that UTK was supporting UTSI at the same rate for full-time equivalency graduate students as it cost UTK, but what that figure actually was formed the subject of endless arguments. The formula gave some definite figures, with specific credit hour rates for master's degree students, and much higher for doctoral candidates. Goethert and Young got a copy of the formula and according to their calculations, their appropriations from UTK through the years had been much too low.

"At budget time," Young recalled, "the people at UTK were not impressed with our figures and claimed we had made several mistakes and were much too high in our estimates. They said we would get X number of dollars regardless of what the formula showed, which was the way it had always worked anyway."

Then the state, on the philosophy that all funds coming to a university were state funds, decreed that overhead income would be subtracted from formula-based state appropriations. This did not make a lot of difference to UTSI, but it hurt UTK because the Space Institute's overhead income even at that time was significant, and with the possibility of major funding in areas such as MHD, the overhead subtraction clearly posed a threat. Goethert could hardly believe the ruling about subtracting the overhead income because he felt that it destroyed initiative for both an institution and its faculty to obtain sponsored research. Young shared this opinion, and undoubtedly many in Knoxville did, too, but there was little they could do about it.

The chancellor's meeting with the faculty got under way with UTSI reporting on research projects, academic programs, and hopes for the future, but shortly afterwards, Young recalled, the tenor changed. Young and Goethert made strong statements about the subtraction of overhead.

"Reese had heard it before, and he became perturbed," Young said. "It was not completely fair because he couldn't do anything about it himself. He became tired of hearing it, and after about an hour, Reese turned to Hilton Smith and said, 'Do you wish to tell Dr. Goethert, or shall I tell him?"

According to Young, Smith answered: "You tell him," and Reese announced that Goethert would be asked very shortly to give up his position as dean. The chancellor said he had asked Smith to make arrangements for a search committee.

While some of the faculty, who had been "ridden pretty hard," may have welcomed the announcement, Young was speechless and registered his protest later. Returning from lunch with Young, Dr. Smith asked him if he would be chairman of the search committee.

"Hell, no!" snapped Young. "I'm a candidate for the position."

Later, Young concluded that Smith's asking him to chair the search committee was a good indication that he had no real prospects of getting the job, but he eventually did apply for it, as did **John Dicks.**

Goethert showed no interest in assisting in the search; he especially wanted to be dean at the time of the tenth anniversary, and as it turned out, he would be.

At the tenth-year convocation in January, Goethert said enrollment had reached one hundred and eighty-eight students, including sixty-eight full time, and there were forty-four faculty members, twenty-nine of them full time. **Dr. Maurice Wright** had spent the previous quarter as a visiting professor at the University of Aachen. **Dr. Susan Wu** was given a faculty adviser's award, which her husband accepted for her since she was attending an MHD meeting in Washington. **Dr. Terry Feagin,** assistant professor of aerospace engineering and computer science, received a twenty-four-thousand-dollar NASA grant for a study pertaining to the earth's orbit. The Support Council adopted the following four goals for the year: To obtain more adequate state funding for operation and capital outlay, to upgrade the administrative status of the Institute in relation to UTK, to establish a research and development park, and to establish a high-tech training center in the research park. And a young (thirty-three) **Lamar Alexander** jumped into the race for governor of Tennessee.

Melvin Edward Partin, a machinist technician at UTSI's MHD lab, was killed in an accident in Hillsboro in January. While he was working under a car, a chain hoist broke, and the vehicle fell on the 29-year-old Decherd man.

By mid-February, 1974, some members of the Support Council had met in Nashville with **Dr. Edward J. Boling,** UT president, and several other top UT officials to protest Goethert's retirement. Boling told them that there would be a "careful search" for Goethert's successor and that Goethert might remain as dean for a long time, depending on when a successor was found. The council members unsuccessfully tried to get the president to specify a definite time that Goethert would serve.

Morris L. Simon, vice chairman of the Support Council, made the main presentation at the meeting with Boling, questioning the "abrupt" way the matter had been handled and urging him to name a successor to work with Goethert. Senator Crouch and **William R. Carter,** also spoke. Other council members present included State **Senator Douglas Henry** of Nashville, and State Representatives **J. Stanley Rogers** of Manchester and **Ed Murray** of Winchester. **Ray Ricco** also attended. **Charlie Alexander** was present, having driven to Nashville with Simon. Accompanying Boling from UTK were **Charles Smith,** administrative assistant, **Dr. Joe Johnson,** executive vice president, **Chancellor Reese** and **Dr. Smith.** This meeting—held in a motel on the north side of Nashville near Interstate 24—bore more fruit than was evident in the stories published about it because within two weeks, **Dr. Charles Weaver,** vice president for continuing education at UT, was being considered as a successor to Goethert—and reaction was usually favorable, especially from members of the Support Council.

(As an engineering "co-op" student at UT in 1940, **Charlie Weaver** was told: "There are going to be some soldiers in the woods at a place called Tullahoma, and they need a telephone." The "place" was Camp Forrest, and Weaver was sent down to install a ten-pair terminal for the telephone company. In subsequent quarters, he worked throughout the Camp Forrest area as an installer-repairman.)

Actually, Weaver's entry into the picture at UTSI seems to have resulted not from the Nashville meeting, but from a conversation **after** the meeting. Dr. Boling, lingering after the session ended, walked over to where Simon and Alexander were sitting and joined them.

"Do you know someone who would make us a good man?" he asked Simon.

"Absolutely," Simon said.

"Who?"

"Charlie Weaver."

"I'd take him in a minute," Boling responded, "if he would have it. I know he wouldn't."

"Do you mind if I talk with him?" asked Simon.

"No."

"Will you tell him that I would like to talk with him and what I want, so that he can come down?"

"I certainly will," Boling said.

A few nights later, Dr. Weaver and Simon had dinner at a restaurant in Manchester. It quickly became evident that the two held different ideas about the amount of autonomy the Space Institute should have, and this drastically changed the course of the conversation. Weaver felt that the UT engineering department should have more, not less, control over the Institute.

"If that is your position," Simon said, "then you are the wrong man for the job. I don't want you here."

Caught off guard, Weaver smiled and said, "You really mean that, don't you?"

"Absolutely," Simon answered. While he had very much favored Weaver for the dean's post, he said, he would be totally opposed to having Weaver come, given his stand against autonomy for UTSI.

"Well," Weaver said, "maybe I can change that position. It's something to talk about."

Indeed it was.

Faculty and students agreed to invite Weaver down to talk; **Art Mason** said a date would be set soon. Mason headed a six-member search committee that was to meet soon with Dr. Smith. Other members were **Dr. Susan Wu,** professor of aeronautical engineering, **Dr. William Bugg,** head of the Physics Department, **Dr. William T. Snyder,** head of the engineering science and mechanics department at UTK, and **Raymond J. Ricco Jr.,** new SGA president.

The search committee learned at its first meeting in February that Weaver had given permission for his name to be proposed as a successor to Goethert. Boling and Reese, recommending Weaver, told the committee that a big advantage in getting Weaver was that he was a professional engineering educator with administrative experience, and he was already in the UT system. Weaver had told top officials that if he were named dean of the Institute, he would wish to continue as vice president of continuing education at UT, and that he would not wish to assume duties until shortly before the 1975–76 academic year. This, of course, would mean that Goethert could remain until that time, which was just what the Support Council wanted. Immediately, council members lauded Weaver as a successor to Goethert. Crouch thought Weaver would be "great," and Goethert could stay and work with him. **Billy Bob Carter** said he had known Weaver for thirty years, and he would be "an excellent man" for the job. **Dr. Ewing J. Threet** said he thought it was always good to grow within the system. **G. Nelson Forrester** was well pleased. **Colonel Charles B. Alexander,** by then retired from the Air Force and serving as executive director of the Tullahoma Chamber of Commerce and Industrial Board, said Weaver had a "top notch reputation."

It seemed that **Dr. Charles Weaver,** who had been born in Murfreesboro fifty-four years earlier and had grown up in Nashville, was just what the doctor ordered. But on March 4, Weaver told the search committee that he was withdrawing as a prospect for the UTSI dean's position. **Art Mason** and **Walter Frost** had driven to Knoxville to discuss the Institute's future with Weaver. Instead, he told them that he was withdrawing. His chief reason, Mason said, was concern that he might not be able to give proper attention to both jobs. "Now we will begin a full search," Mason announced.

While members of the committee may not have been fully aware of it, the idea of the UTSI faculty and students questioning him about his "position" on the Space Institute had not set well with Weaver. He felt this was out of line since he had been closely associated with the Institute, especially while he was engineering dean and chancellor at UT. This was a factor in his withdrawal, and some say it was the "chief reason." Clearly, Weaver did not intend to go through the routine procedure, nor did the administration in Knoxville intend to subject one of its vice presidents to the scrutiny of the search committee. Another factor in Weaver's decision to withdraw was that he got wind that some of the faculty did not want him. Their opposition, however, was largely not to him personally—he was known as a good administrator although he was criticized for his militant handling of student demonstrations—but it was a reaction to what many perceived as intentions of the administration to dissolve the search committee and to appoint Weaver. A strong feeling prevailed that if Weaver wanted to be a candidate—and perhaps the favored one—he should participate through the normal search committee route. There was concern, too, about having a dean who would spend part of his time being vice president in Knoxville.

Toward the end of March, **Bob Kamm** said the Space Institute expected to get a million dollars in federal grants and contracts by the end of the fiscal year, compared to $913,000 the previous year.

On April 4, **Dr. Henry Tompkins Kirby-Smith,** retired Sewanee surgeon and chief of staff at Emerald-Hodgson Hospital, died after a long illness. He was sixty-six.

More than two hundred engineers and scientists attended a symposium on MHD in early April. **Susan Wu** headed the committee for the event. A reporter at the meeting wrote that a consensus expressed at the symposium was that development of a commercially feasible MHD generator was at least ten years away. Meanwhile, Dr. Dicks was doing one of the things he did so well—testifying before a Senate subcommittee in Washington on April 9, asking for an additional three million dollars in MHD research funds for the Institute. Congress had voted two million in December to speed up MHD research, but UTSI had not received a penny of it.

General Lowe announced that he was retiring as AEDC commander. **Leith Potter,** deputy director of the von Karman facility at AEDC and one of the part-time instructors when UTSI opened, received his Ph.D. in mechanical engineering at Vanderbilt.

In May, the Space Institute had its own banquet, celebrating its tenth year, with **Jess Safley** of Nashville as speaker and **Morris L. Simon** as master of ceremonies. The big event, though, was the party thrown by the area chambers on June 2. And in October, the Space Institute would host a "Ball" with flaming tapers and a dedication ceremony for the Industry-Student Center.

In mid-June, UTSI students garnered a bit of publicity by using the Institute's twin-engine plane as a "flying-lab" to gather data for the Air Force. **Dr. Ronald Kohl,** assistant physics professor at UTSI, directed the project, along with **Dr. William Dunnill,** associate physics professor. Two physics students involved were **Robert Partin** of Estill Springs and **Michael Flaherty** of Bell Buckle.

A lot of attention also was focused on MHD in June. Dr. Dicks, director of the Energy Conversion Research Division was the main speaker on June 21 at an energy rally in Springfield. **Dr. Kenneth Tempelmeyer,** professor of mechanical engineering and a member of Dicks' staff, and other UTSI faculty members also attended.

A House-Senate conference committee gave its blessings to a bill containing $12.5 million for beefing up energy research. The House earlier had passed a bill containing $7.5 million for MHD research. Then the Senate added five million and accepted the committee's report that MHD work at UTSI should be accelerated. **Congressman Joe L. Evins** supported the additional five million dollars and said the committee was disturbed at delays in the MHD program.

Newspapers reported that if Congress approved the additional five million, this would make $12.5 million available for MHD research at UTSI. But **Mike Mansfield** later would cast a shadow on the appropriation for UTSI by insisting that the committee meant for the money to be spent at the Montana College of Mineral Science and Technology.

On June 27, the Manchester Times reported that the House had concurred with adding five million, and the Senate was expected to take final action to add the money that day. **John Dicks** was pleased, but he also was disappointed that three million dollars had not been earmarked specifically for UTSI as he had requested. Congressman Evins said none of the five million would be spent in Montana although the Montana institution was mentioned specifically in the conference committee's report. Dicks was disappointed that the Senate's language—specifying UTSI—was not kept in the report. The Space Institute still had not received the two million dollars approved by Congress in December, and the U.S. Office of Coal Research was saying it could not release the money until a new contract was negotiated. Dr. Dicks said those negotiations were going "fairly well."

Ross Bass of Pulaski, who once had defeated **Frank Clement** in a race for the U.S. Senate, was running for governor.

It was going to be a long hot summer with Dicks battling Mansfield's ploy to steal UTSI's MHD thunder, UT turning up its nose at the search committee's recommendations for a dean, and the Support Council fighting mad over persistent reports that UT was maneuvering to put the Space Institute under the control of UT's College of Engineering.

Tennessee Senators **Howard Baker** and **Bill Brock** apparently had other things on their minds when the Senate Appropriations Committee met early in July, 1974, because neither mentioned UTSI while **Mike Mansfield** bulldozed through his recommendation that up to five million dollars be specified for work at Montana's College of Mineral Science and Technology. The Senate majority leader read only the portion of the Conference Committee's recommendation that pertained to Montana, ignoring the committee's plea that MHD research be accelerated at UTSI. He also got **John McClellan** of Arkansas, chairman of the appropriations committee, and **Senator Alan Bible** of Nevada, head of the interior subcommittee, to agree with his contention that the original recommendation had been for the additional five million to go to Montana.

In voting for the 1974–75 energy bill, the House had followed the recommendation of the U.S.

Office of Coal Research and allotted $7.5 million for MHD research, the same as for the previous year.

Baker, who spoke for more atomic energy funding, and Brock, who spoke on behalf of added research funds at UTSI, did not mention that UTSI had requested a special appropriation of three million dollars (for research in MHD) and also had hopes of sharing the extra five million.

Dr. Dicks said Montana had done some research in the chemical processing of coal and in coal liquification processing. It seemed that **Mike Mansfield** and his crowd were making some progress in Montana's bid to compete for the pilot MHD plant, but the Space Institute had a reputation in MHD research, and **John Dicks** was not about to let anyone forget it.

Thomas A. Wiseman Jr., formerly of Tullahoma and two-time state treasurer, jumped into the race for governor.

UTSI promoted four faculty members to full professor. They were **Dr. Arthur Mason** of Manchester, and **Dr. Walter Frost, Dr. Firouz Shahrokhi,** and **Dr. Lloyd Crawford,** all of Tullahoma. Shahrokhi was holding UTSI's first classes in remote sensing under a $210,000 grant from the Department of Health Education and Welfare. Thirteen students from eight different colleges were attending.

Late in July, the Supreme Court ordered **President Nixon** to release some of the tapes from his office, and there was growing talk in Washington of impeaching him.

ARO announced the promotion of **Dr. Jack Whitfield,** director of the von Karman Gas Dynamics Facility, **Michael Pindzola,** head of the Propulsion Wind Tunnel, and **Joel Ferrell,** head of the Engine Test Facility, to vice presidents.

Colonel James A. Brantner resigned on July 31, 1974, as assistant for management at UTSI. The 58-year-old Wisconsin native and University of Alabama (1941) graduate had first assumed the position at UTSI on June 1, 1967. Because of illness in his family, he had taken a temporary leave from September 10, 1971 until March, 1972. During this time, **Colonel Otis Brooks Thornton** had held the job. Thornton left in March, 1972, to take a job at Motlow College. After Brantner resigned, another retired Air Force colonel, **John F. Stubbs,** succeeded him, effective September 1, 1974. He left in October, 1976, to take a management position with the Clinch River Breeder Reactor Project. Effective October 22, **Bob Kamm** assumed duties of the office on a "temporary basis." Thus the title of "assistant for management" faded away.

In July, **James N. Chapman,** a retired Army colonel residing in Manchester, had entered UTSI as a graduate research assistant. He not only would complete work for his Ph.D. and become a research associate professor but also would move into responsible positions of management within the Energy Conversion Programs. Chapman, whose U.S. Army career stretched from 1951 to 1974, had earned his bachelor's degree in electrical engineering from the University of Illinois in 1950 and his master's in the same field from the University of Michigan in 1959. He also had earned the equivalent of an M.B.A. degree with advanced management study at the Industrial College of the Armed Forces in 1972 and had completed all work toward a doctorate at the University of Michigan except for the writing of a dissertation. He completed this work toward a Ph.D. in engineering science at UTSI in 1977 with **Dr. Lloyd Crawford** as his major professor.

In 1977, Chapman was named a temporary administrative assistant with ECP and in April of the next year the position was made full-time until July 24, when he became a section manager. He subsequently became program manager, Energy Conversion Division, deputy manager, ECP Research and Development Laboratory, and finally, manager, Advanced Energy Conversion Department. In 1979, he was named a research assistant professor of electrical engineering, and then years later he became research **associate** professor.

Chapman's Army career had included serving as operations research analyst in the office of the Secretary of Defense. He also had commanded an eleven-hundred-man battalion responsible for providing communications service to two provinces in the Republic of Vietnam.

By August, 1974, the Search Committee had seven candidates—including **Dr. Robert L. Young and Dr. John B. Dicks**—to succeed Dr. Goethert as dean, and students and faculty were scheduling dates to interview them in the next couple of weeks, after which the candidates would go to Knoxville for interviews. Other candidates were **Dr. Victor W. Bolie,** chairman of the electrical engineering and computer science department at the University of New Mexico; **Dr. Salamon Eskinazi,** professor and former head of the mechanical and aerospace engineering department at Syracuse University, New York; **Dr. George Leitmann,** vice chairman for graduate study and research, mechanical engineering department, University of California, Berkeley; **Dr. Gene Marner,** vice president for engineering, McDonnell/Douglas Electronics Company, Chesterfield, Missouri, and **John D. Nicolaides,** professor and former head of the aerospace and mechanical engineering department, Notre Dame University, South Bend, Indiana.

Young, who thought that he and Dicks stood head and shoulders above the other candidates, said that Goethert asked **Dr. Barney Marschner** to apply for the job, and that Marschner said he did apply. Young said search committee members said they did not consider Marschner as a candidate because they never received all the information on him.

"This is a great shame and a great error on somebody's part," Young said. "He would have been a fine replacement."

In early September, the committee would recommend **Young, Dicks, Nicolaides** and **Eskinazi.**

Democrat **Ray Blanton** and Republican **Lamar Alexander** emerged from the primaries in August as contenders for the governor's post in the General Election in November. Blanton would win, but Alexander would take his revenge the next time around. **J. Stanley Rogers** of Manchester was nominated for his second term as state representative, and in the fall he would be re-elected as House majority leader. Rogers, always a friend of the Institute, would be of special help within the next two years in getting UTSI on a line-item budget.

Late in July, the FAA added $73,000 to UTSI's contract to study jet noise. This made a total of $498,000 in FAA research funds allocated to UTSI since June of 1972.

A few days later, **John Dicks** said UTSI and the U.S. Office of Coal Research had reached an "agreement in principal" on a seven-million-dollar, three-year contract for energy research. Details were being ironed out in Washington talks. Final approval of the deal would mean that the Space Institute within the coming year would be able to start the design of a special MHD energy test plant in UTSI's lab section. Dicks said additional staff members were being added to the energy research staff in preparation for the expansion. Dicks was planning to have two million dollars available for the first year and to be able to operate into 1980 at least. He foresaw the MHD plant being available for special research tests by industrial firms and other institutions in the country. Dicks was especially enthused that the Office of Coal Research would furnish a super-conducting magnet that would go with the lab equipment, rather than allocating money for its purchase. This would make the whole deal worth about eight million dollars, Dicks said. The Space Institute's MHD unit was the only one in the U.S. that was fed by coal, Dicks reminded. And he said the contract was recognition that "we are the only people who have any experience in direct coal firing." He **expected** to have the contract signed by September 1; it would take a bit longer than that.

UTSI had had a $550,000 contract with the Office of Coal Research and TVA for MHD researching during the past three years, carrying on research with a small generator located in the main academic building at UTSI. With the seven-million-dollar contract in the works, Dicks could see great things coming for the Institute. He noted that the funding for this contract was coming from a "different package" than the Montana appropriation. Montana was still in the ball game, however. Congress a second time specified that the five million be spent at the Montana institution, using basic research and key design developed at UTSI. **Joe L. Evins** was confident that not all of the appropriated money would go to Montana, however.

With the MHD contract in sight (actually it would total $8.1 million), **Dr. Goethert** called his friend, **Henry H. Sherborne,** a retired colonel, and discussed the possibility of Sherborne setting up a purchasing department for ECD. After about three calls, Sherborne agreed. But until the contract was signed, there was no money for the position. Soon afterward, however, Goethert asked Sherborne to help set up a flight operations office.

Sherborne's thirty-two-year military career began as a buck private in the U.S. Army. After finishing cadet training in 1935, he was commissioned as a second lieutenant. He got his bachelor's degree in Mechanical Engineering from Illinois Institute of Technology at Chicago, and in 1940 transferred to the Army Air Corps (later Air Force) just in time for World War II. Born in Chicago and reared in California, Sherborne had attended junior college in California and later would get a master's degree in Aeronautical Engineering from the Air Force Institute of Technology.

In 1955, he drew his first assignment to Arnold Center where he became friends with **Goethert** and with **Bob Kamm.** At AEDC, he headed up a design engineering group. "Only ETF was finished," he recalled years later. "Everything else was under construction." He then was sent to California to build a missile site in the Sacramento Valley, but two years later he returned to AEDC, where he retired in 1966. Now, in the fall of 1974, Goethert had persuaded him to go back to work. As Sherborne recalled, **Sean Roberts,** UTSI's first pilot, had left just before the colonel joined the staff, and UTSI was getting another "Otter" for its skimpy fleet. Early in 1975, with the MHD contract signed, Goethert called him to a meeting also attended by **John B. Dicks, Hilton Smith and Colonel John Stubbs,** and discussed the purchasing job. It was a heated session with Dicks expressing the view that he should have something to say about who was hired. (Dicks was anxious to get something worked out so he would not be dependent upon Knoxville's purchasing department headed by **William O'Toole.**) Sherborne got the job, but part of his time was still to be devoted to flight operations.

About this time, **Wilma Kane** was put in charge of UTSI's first purchasing office. About five years later, both purchasing offices were combined with Sherborne as manager and Kane as agent.

Bob Williams, who had first joined ARO in 1954, and who had lent strong support through all the dark days of getting the graduate program and then the Space Institute established, was retiring on September 15 as chairman of the board of ARO. **General Leif J. Sverdrup,** who had founded the company and been ARO's first president, would succeed Williams, who would remain on the company's board. Williams had been named chairman of the board on July 1 at which time **Ed Dougherty** succeeded him as president of ARO, Inc.

In the fall of 1974, **Dr. Virgil K. Smith, III,** a native of Hattiesburg, Mississippi, joined UTSI as an associate professor of aeronautical engineering and mechanical engineering. He would stay about five years before going to work at AEDC. **Dr. Frank G. Collins,** a native of Illinois, began his long career with the Institute that fall as associate professor of aerospace engineering. Collins got his undergraduate degree in civil engineering from **Bob Young's** alma mater, Northwestern University, and he took his Ph.D. in mechanical engineering from the University of California at Berkeley.

The Support Council got riled in September, and its top officials headed to Knoxville on the 27th to protest rumored plans to downgrade the Space Institute. **William R. Carter** of Fayetteville, who headed up the special delegation, was accompanied by Support Council Chairman **Crouch,** Vice Chairman **Simon, Clinton Swafford,** legal counsel, **G. Nelson Forrester, Ed Murray, J. Stanley Rogers,** and **Senator Reagor Motlow.** They met with **Drs. Ed Boling, Jack Reese** and **Hilton Smith.**

Before the trip, Carter said the council was "deeply disturbed by the inadequate financial support of the Institute this last year and most recently by the persistent reports that with the upcoming change in the Institute's deanship, this transition period might be exploited to administratively downgrade the Institute and thus hamper its future growth." Carter, who had once taught engineering at UT, referred specifically to reports that top UT officials wanted to put the Institute under the control of UT's College of Engineering. To do this, Carter said, would mean "losing sight of the fact that the Institute's academic programs are of an interdisciplinary nature and include important elements of more than one University of Tennessee Knoxville college."

Such control also would violate recommendations made two years earlier by a committee of the Southern Association of Colleges and Schools, Carter noted. That committee, he said, had recommended more administrative independence, more state funds and a broadening of the curriculum at UTSI. He said the committee also called for having the Institute report directly to the Knoxville chancellor rather than to the vice chancellor, "and preferably to have it have a vice chancellor of its own."

Carter had put his finger on a very sensitive issue and one that would simmer for fourteen years.

President Boling assured the group that he would not downgrade the Institute. He was, he emphasized, desirous of seeing the Institute get bigger and better.

On September 8, 1974, **President Gerald Ford** pardoned **Richard Milhous Nixon** of all wrongs associated with the Watergate scandal. **John Dicks** represented Tennessee at a convention of the Federal Energy Administration in Atlanta in late September. The search committee sent its recommendations for a successor to Goethert on to Knoxville, and folks at UTSI briefly put their problems aside and donned their fancy clothes for an anniversary "Ball" on October 5. **Johnny Gidcomb's** orchestra from Columbia provided music for dancing that Saturday night. The next day, Dr. Boling came down to dedicate the Andrew David Holt Industry-Student Center. The UT Singers also participated in the week-end celebration. A proclamation from **Governor Winfield Dunn,** citing the Space Institute's contributions to "the betterment of mankind" due to the "wise leadership" of Dr. Goethert, was presented to Goethert on Saturday night at halftime of the UT-Tulsa ballgame by **Gil Thornton,** state commissioner of agriculture. The governor proclaimed that October 5 as "UTSI Day."

Dr. Boling and the new AEDC commander, **Colonel Webster English,** signed an agreement the next afternoon, allowing each to use the other's "specialized equipment and facilities" for research. This was the first such agreement. Boling also signed an "Agreement of Cooperation" between UTSI and Alabama Agricultural and Mechanical University in Huntsville.

Perhaps the most historically significant event of that weekend was the signing of an agreement with the Technical University of Aachen, Germany, that established an "honorary lecture series" in honor of **Dr. B.H. Goethert** and **Dr. A.W. Quick** of the University of Aachen. Dr. Boling and **Dr. Burkhart**

Mueller, chancellor of the Technical University of Aachen, signed the agreement. Thus the **"Quick-Goethert"** lecture series, which would become an annual tradition, was established.

Mr. and **Mrs. Robert Choate** of Tullahoma won the fourth annual auto rally sponsored by the Institute's Auto Club.

Two groups offered support in October of the Institute's MHD program by calling upon the next governor (**Blanton**) and legislature to make five million dollars available for UTSI's energy research. The first was the Tennessee Electrical Cooperative Association, during a meeting in Nashville, after **J.C. Hundley,** executive director, urged passage of a resolution to this effect. Shortly thereafter, the Elk River Development Association followed suit.

Young and Dicks both were interviewed in Knoxville regarding the dean's job. In a meeting with Smith, Boling and Reese, Young was offered the job of acting dean. Reese was anxious to get the matter of Goethert's retirement settled before **Hilton Smith** retired since Smith was perhaps the single UT official most knowledgeable about the Space Institute. Young saw the offer as simply a means to get the search under control for awhile. The offer offended him. He felt that he had been "on trial" for nearly seventeen years. He thought that the officials doubted whether he could handle the job of dean.

"I couldn't have," he said many years later, "**if** they expected me to completely reverse all that Goethert had done and become completely subservient to our leaders in Knoxville." He told them he wanted to think it over. As acting dean, he thought, he would be pretty much "under their thumb," and privately he felt that it was an insult. He later wrote back that he would be pleased to accept the Dean position, but not the Acting Dean. One thing that influenced his decision was a remark that **Hilton Smith** privately made to him after the meeting—that he might **not want** to accept the position. Young wasn't sure what he meant by this remark, but it set him to thinking.

Hilton Smith notified the Search Committee during the last week of October that officials were unable to make a decision on the successor to Dr. Goethert and asked for more nominations even though he said that the persons previously recommended were still under consideration. **Dr. Mason** acknowledged that this took the committee by surprise.

In Manchester, **Robert Wilce Casey Sr.,** a former mayor and for twenty-eight years clerk and master of Chancery Court, died on Halloween, seven days before his 95th birthday.

Ray Blanton from Adamsville, Tennessee, defeated **Lamar Alexander** by a margin of 125,000 votes to succeed the Memphis dentist, **Winfield Dunn,** as governor. State Representative **J. Stanley Rogers** and **Congressman Joe L. Evins**—nominated by the Democrats in August—were returned to office in the General Election.

That fall, **Dr. Jack H. Hansen,** a specialist in remote sensing, joined the Space Institute faculty. He had taught at California State University at Humboldt and did research in ecology at Berkeley and forest management with the Bureau of Land Management. He had received bachelor's and master's degrees in forestry at Berkeley, and he took his doctorate in remote sensing at the University of Missouri.

On November, 7, 1974, Arnold Engineering Development Center celebrated its twenty-fifth anniversary.

As UTSI continued negotiations for the seven-million-dollar MHD contract, Arnold Center's commander, **Colonel W.C. English Jr.,** wrote a letter to Dr. Boling, declaring AEDC's opposition to plans for a fifty-acre research park at UTSI. English, who reportedly also noted that **Ed Dougherty,** ARO president, concurred, said the park would violate terms of the transfer of the 365-acre site of the Institute to UT in the early sixties. He also expressed concern that it might have a negative effect on the environment. The commander suggested that the research park be located on 791 acres of AEDC property that had been declared surplus and was at that time in the hands of the General Service Administration. This land, adjacent to Interstate 24, had been declared surplus by the Air Force so that it could be used by Coffee County Industrial Board for developing an industrial park. The Space Institute fired off a point-by-point rebuttal.

A new governmental scientific study hailed the Space Institute's work, finding that UTSI was setting the pace throughout the world for the largest energy research. The study summarized American and Soviet research in MHD, concluding that the Space Institute facilities have the "longest record of continuous operation and have yielded the most detailed study of MHD generators' performance anywhere in the world." The report also noted that "UTSI's generator has been operated for more than an hour on coal and is the only installation thus far that operates using coal as a fuel.'' The 104-page report was based on a comprehensive study made for the Defense Advanced Research Projects Agency by the Rand Corporation of Santa Monica, California.

The U.S. Department of Interior announced intentions to lease about thirteen acres from UTSI as a site for a new facility to conduct government-sponsored energy research. On December 7, the UT trustees approved the leasing plans. Two days before Christmas, the U.S. Department of Interior in Washington signed a $8.1 million contract that had been approved over the weekend by UT and the Department of Health, Education and Welfare. This was the contract that **John Dicks** had been working for, only instead of seven million dollars, it had been increased to $8.1 million because of work that had been added into the three-year pact. The amount of the contract was eight times UTSI's annual budget, and it was believed to be the greatest single research contract UT had ever received. And with the $1.3 million magnet that was supposed to come with the deal, the overall value of the contract would have been $9.4 million. As it turned out, however, it was just as well that the value of the magnet was not figured in.

It was a nice Christmas present.

Governor-elect Blanton and Congressman Evins visited the Space Institute for a briefing. The year was winding down, but there was one other present under the tree.

Hilton Smith said the search committee had submitted another four or five candidates for the deanship. However, this didn't seem too important after newspapers reported just before Christmas that **Charlie Weaver** was again considering accepting the offer. A story said that members of the UTSI faculty planned to meet with him in hopes that he would indeed reconsider. On December 26, the Manchester Times reported that Weaver would visit the Space Institute on January 6 and 7 to talk about the position of dean, which **Dr. B.H. Goethert** still held as 1974 ended.

✦

Chapter Thirteen

A Time of Change

DOMESTIC workers got a dime-an-hour raise as a new minimum wage took effect on January 1, 1975; farm workers were still getting $1.60 minimum. The payroll at Arnold Center had set a new record in 1974, cresting at $50,922,000—an increase of $3,027,000 over the previous year.

John Dicks said he had been assured that UTSI might get thirty-five million more in federal funds for stepped-up energy research if only the Tennessee Legislature would commit five million dollars to the cause. **Lieutenant Governor John Wilder** was among the first to say he favored appropriating the five million, which was to be provided over a period of years and matched on an eight-to-one basis with federal funds.

Builders in Coffee and Franklin counties had met at the Boar's Pen restaurant on the last Monday in 1974 and organized Highland Rim Builders. **Bob Whaley** of Tullahoma was first vice president, **Dennis Maccagnone,** Manchester, second vice president, **Joe Hagan,** Tullahoma, secretary, and **Tom Ballou,** Tullahoma, treasurer. Selection of a president was delayed and shortly thereafter, Hagan became president, and **Giles Bennett** of Franklin County replaced him as secretary.

The Space Institute was offering five evening classes—a non-degree program offered for the convenience of undergraduates. AEDC awarded a $61,000 grant to UTSI for a special study of aircraft models used in wind tunnels with **Dr. Trevor Moulden,** assistant professor of aerospace engineering, in charge. **Dr. Terry Feagin** was conducting an $18,000 project for NASA to help spacecrafts plot their own orbits with greater accuracy.

Then began the "Weaver Era."

Manchester Times, on January 23, 1975, reported that the 54-year-old Weaver would become "dean-elect" of UTSI on February 1 and on June 1 would assume the full duties as successor to Dr. Goethert.

Chancellor Reese said Weaver would maintain "fully operational offices" on both campuses; he would continue serving as UT's vice president of continuing education. Reese said that Goethert, in a letter to Hilton Smith, had reported that both faculty and staff at UTSI had unanimously endorsed Weaver. The chancellor acknowledged that the dual role was unusual, but he said he and Dr. Smith believed the appointment would be successful "because of Dr. Weaver's technical competence and deep commitment to the various units of the entire University." Reese expressed gratitude to Dr. Goethert and said he looked forward to a "continuing association with him." In fact, Goethert would continue at UTSI as professor of mechanical and aerospace engineering.

Dr. Boling, UT president, said he expected the appointment of Weaver to help with UTSI's growth and expansion. He said in all matters pertaining to the Space Institute, Weaver would report to Dr. Smith, vice president for graduate studies and research; in his other role as vice president at UT, he would report to Boling. Reese noted that Weaver had had administrative authority over UTSI when he served as the first UT chancellor for three years. He had experience, too, as an electrical engineer and an energy researcher at Oak Ridge, Reese said, and as a "top University administrator."

A young **Charlie Weaver** had entered Vanderbilt College, but in 1940 he transferred to UT to take advantage of cooperative training, and as a "co-op" student working for Southern Bell Telephone Company, he had gotten acquainted with Camp Forrest. He received his bachelor's degree in electrical engineering from UT in 1943 and worked as an engineer and supervisor in atomic research at Oak Ridge. After the war, he re-entered UT, earning his master's degree in 1948, and began teaching at UT, becoming assistant and associate professor and then full professor, dean of engineering, chancellor and ultimately vice president for continuing education—a position he had held for four years at the time he was chosen as dean of UTSI. (He received his Ph.D. in 1956 from the University of Wisconsin.) From 1959 to 1956, Weaver was Westinghouse professor of electrical engineering and chairman of the department of Auburn University. He returned to UT in 1965. Weaver was among those considered for the presidency of Georgia Tech shortly before the chancellor announced his appointment to UTSI.

Weaver said he looked forward to assuming his duties at the Space Institute, especially since it meant a "return to engineering." He said he wanted to "develop to the fullest" the Institute's ties with AEDC.

Before June 1, folks on "the hill" in Knoxville would get nervous about John Dicks' success in shaking the money tree for MHD research, and some UT eyebrows would be raised as plans for a UTSI research park materialized. Dr. Weaver would have his hands full, juggling not only two jobs, but also two campuses and two philosophies.

Late in January, 1975, the UTSI Research Park Inc., was formed and chartered as a not-for-profit organization to support the development of a fifty-acre park on the UTSI campus. **I. J. (Ike) Grizzell,** president of First National Bank of Franklin County (and a long-time member of the Support Council) was temporary chairman. **Walter Wood,** head of the South Central Development District with offices at Motlow College, was temporary secretary-treasurer. **Charles B. Alexander,** executive director of the Tullahoma Industrial Board, was temporary executive director. It was agreed that the initial board of directors would include representatives from Bedford, Coffee, Franklin, Lincoln, Moore and Warren counties as well as a representative from UTSI.

Charlie Alexander, Morris L. Simon and **Dr. Ewing J. Threet** represented Coffee County; **Grizzell, Karl Mawhorter,** and **Clinton Swafford** represented Franklin County; **William R. Carter** of Fayetteville, represented Lincoln County; **Franklin Yates** of Shelbyville was Bedford's representative; **J.D. Mullican** of McMinnville represented Warren County, and **Truman Ashby** was Moore's member.

This organizational meeting had been called by representatives of the Tullahoma Industrial Board and Greater Franklin County Inc. to name temporary officers, decide on selecting the board of directors, and to talk about raising money. It was urgent, they agreed, that they raise sixty thousand dollars by March 1 to match a hundred-thousand-dollar grant from the Appalachian Regional Commission. (The State Department of Economic and Community Development had helped get the ARC grant offer.) An executive (fund-raising) group was formed, consisting of **Grizzell, Wood, Alexander,** and **Mawhorter,** who was executive director of Greater Franklin County Inc.

Goethert said UTSI already had one company leasing space in its laboratory area that had expressed interest in the proposed park. The idea was to locate special research offices apart from a production facility to work on prototypes and ideas for new product development, using the Space Institute campus and giving students practical experience.

On August 28, 1975, at the fifth meeting of the Research Park Board, **Dr. Goethert, Dr. Robert L. Young, and Robert Kamm** were appointed as UT directors. However, all three later resigned on advice of a UT attorney to avoid a possible conflict of interest.

In February, **Dr. Harry Wagner,** a vice president at Middle Tennessee State University, was named president of Motlow College, succeeding **Dr. Sam Ingram,** who was the new Tennessee Commissioner of Education.

Bob Kamm—an associate fellow of AIAA—received a certificate of appreciation from that group for his "sustained contribution to and continuous membership in the Institute and its predecessor organization (American Institute of Aeronautics) since 1939."

The Space Institute was training sixteen students to be electric technicians under a federal employment program administered by Motlow. **Dr. E.C. Huebschman,** UTSI professor, was in charge of the thirty-nine-week course.

In mid-February, clear signs of a developing schism between Dicks and Knoxville (and ultimately with Weaver) became evident. The UT administration at Knoxville was opposed to the formation of a separate organization for the Space Institute to carry on MHD research, which they said should be administered through Knoxville. Weaver made this position clear to visiting legislators during a briefing at UTSI on February 18. (Tennessee Electric Cooperatives Association sponsored the tour to acquaint about twenty key legislators with the energy research under way at UTSI.)

Dicks said UT's position could jeopardize the whole MHD program. The federal government, he argued, would insist on dealing with a relatively independent organization. He contended that an entity known as the Tennessee Energy Institute should be set up to receive money from the state, or that the MHD program should be operated by a separate division with more autonomy. If the state would raise the extra funds, Dicks said, the $8.1 million contract could be enlarged to a total $40 million project.

"At this time," **Weaver** maintained in a memo read to the legislators, "there is no need to set up any other organization. . . . In particular, the University does **not** endorse an energy research institute since such an organizational unit would encompass all energy research being conducted throughout the entire University." Weaver said UT welcomed additional funds for MHD research and "is pleased with the work of Dr. Dicks and his group in obtaining the $8.1 million contract."

Dicks' campaign for state funds was obviously a source of worry in Knoxville. Weaver said if state funds were to be allocated for MHD work, the University "must consider the effect such an allocation would have on all other budgets of the state, both educational and non-educational." The University, he said, "must preserve its fiscal integrity with regard to both its own budget and that of other state agencies. . . ."

"What that means," Dicks told a reporter, "is that they don't want us going directly to the legislature for money. I am afraid it's going to kill this program if they insist on running it through Knoxville. There is no way of selling this large a program without an organization here. I don't think we can get federal money without an organization."

An even larger result of "this kind of control," Dicks warned, was that the present energy research group—about fifty and growing to two hundred—"may have to leave UTSI and move to industry or possibly to another state." Afterall, he said, most of their salaries were paid through the federal government and not by the University.

"The University, by its opposition, can very well kill this operation in Tennessee," Dicks declared. "It's a very distinct possibility."

It seemed to Dicks at the time that most of the lawmakers were in sympathy with him, but he said they urged that he and the administration in Knoxville try to work out a compromise. **Ayers Merrill,** a Harvard graduate, friend of Ulysses Grant, and ambassador to Belgium, **may** have been strong on compromise, but that was not necessarily the dominate trait in his great-great-grandson, **John B. Dicks Jr.**

Dicks called attention to a memorandum from the U.S. Office of Coal Research written the previous year, questioning whether UTSI's administration was "adequate" for an $8.1 million contract.

"We finally won out over them on that," he said, "but if they are doubtful about running it through the University on an $8.1 million contract, you can imagine what their attitude is going to be on a forty million one. They are going to insist on a separate organization."

State Representative **Ed Murray** of Winchester said he believed the legislators were in full support of the MHD program. He emphasized that the five million dollars would be spread over the next several years and that each dollar the state put up would be matched by eight from the federal government.

Dicks said about six hundred thousand dollars would be needed from the state that year, which he said would generate almost five million dollars on the eight-to-one ratio.

(The Office of Naval Research awarded a third $39,800 contract to UTSI to do research on materials as light as aluminum and as strong as steel; **Dr. Maurice Wright,** professor of metalurgical engineering, was to direct it. Manchester Mayor **Clyde V. Myers,** at the last minute, entered the race for another term, facing **Buster Bush,** Manchester insurance man, as challenger.)

Late in February, Dicks was in Washington, checking out a new Mike Mansfield bill that would provide several million more dollars for MHD research. He said he feared that if Tennessee legislators did not come through with the five million dollars, all of the "Mansfield" money would go to Montana.

On February 20, the UT trustees unanimously gave approval for UT to award a contract for the design and construction of a facility to carry out the $8.1 million contract for the federal government. UT was negotiating with four companies: Bechtel Power Corp., San Francisco, Burns and Roe Inc., Oradell, New Jersey, Stone and Webster Engineering Corp., Boston, and United Engineers and Contractors, Philadelphia.

Governor Ray Blanton, chairman of the trustees, questioned whether UTSI and AEDC were doing duplicate work on MHD. He proposed that an attempt be made to coordinate efforts at the two places. The governor suggested that the trustees might take into consideration investigations by the Congress of duplicating projects. Chancellor Reese and Dr. Weaver assured the governor that there was no duplication of efforts, but Weaver agreed to arrange a meeting with AEDC officials regarding cooperation.

Arnold Center's MHD work used hydrocarbon fuel, and this research was focusing on short bursts of power. UTSI, burning coal, was focusing on long sustained runs. Both were financed by the U.S. Office of Coal Research, but each had different aims.

Robert P. Rhodes III, a graduate of Tullahoma High School with a bachelor's degree in business administration from the University of Denver, became the new manager of the Industry-Student Center, replacing **Walter Eads,** who had left several months before. **Ramon Cintron,** a native of Puerto Rico, joined the staff as cook in 1975, and the next year became manager of the cafeteria. Cintron had been **General Jessup Lowe's** personal aide when Lowe was AEDC commander, and in 1971 he had transferred with the general to California. After retiring from the Air Force with more than twenty-two years service, Ray returned to Tullahoma. He continued as manager of the cafeteria in UTSI's twenty-fifth year. **Mrs. Odie Mann** of Estill Springs joined the staff as chief cook in 1979.

Dr. Kenneth Kimble, assistant professor of math at UTSI, was elected president of the Manchester Rotary Club. In later years, he would direct UTSI's computer center. And he would continue as an active Rotarian.

Ernest Crouch and **Ed Murray** introduced companion bills in the legislature asking for five million dollars to support MHD work at UTSI. Crouch said six hundred thousand dollars would be available in the next fiscal year from the sale of bonds. **U.S. Rep. Joe L. Evins** pledged to help get a share of the "Mansfield" money for UTSI. The next week, **Charlie Weaver** said that a major part of new federal revenue for MHD research should be spent for work in Tennessee "in the vicinity of UTSI." He said UT believed state funds would add to the strength of UT's MHD position. UT was, Weaver said, "totally committed" to support the Space Institute's work in MHD" and also recognized the "equally importance" of research that was getting under way at AEDC. Apparently, some feelings had been rubbed raw. And the last thing needed at this point was a squabble between UTSI and AEDC.

Bids would be sought on an estimated two-million-dollar building for carrying on work under the $8.1 contract, Weaver said. Meeting in Nashville on March 14, the State Building Commission authorized UTSI to spend up to $1,071,000 in first-phase work on a new energy research facility. About $771,000 of this was for equipment.

On Monday afternoon, March 17, **Andre R. Ricou,** a Tullahoma retired Air Force major and a UTSI pilot, was injured when the Space Institute's Canadian-made DeHavilland Otter crashed shortly after takeoff from the Winchester airport.

Dr. Firouz Shahrokhi was preparing to host an international conference in remote sensing for about two hundred visitors.

In the spring, **Thomas D. Benson,** state commissioner of economic and community development, supported the research park, but the Air Force at AEDC was still **against** locating it at UTSI; however, this obstacle would soon be overcome.

Buster Bush swept to victory over **Clyde V. Myers** as Manchester mayor. ARO Inc. celebrated its 25th birthday with an open house on April 20. It had been in June, 1952, that **Dr. John H. Wild,** director of engineering for ARO, wrote to the AEDC commander, proposing a graduate program for technical personnel at AEDC. ARO, then, was "on record" as having supported that program, and on its 25th anniversary, ARO confirmed its support for the research park at UTSI.

The Space Institute got $62,410 from Marshall Space Flight Center to continue wind research. **Dr. Walter Frost,** professor of mechanical engineering, who was in charge of the project, said the four-year total for the continuing contract was $241,000.

On April 19, 1975, **Governor Ray Blanton** spoke at the dedication of Old Stone Fort State Park in Manchester. (**Frank Clement** had turned the first spade of dirt when the "fort" was first designated as a state park under his administration.)

The state Senate's finance, ways and means committee was recommending approval of the five million dollars for UTSI's MHD program.

On April 24, **Dr. Richard V. Shanklin III,** one of the first full-time students at UTSI, said in Tullahoma that commercial use of MHD would not be feasible before the late 1980's. Dr. Young introduced Shanklin, who spoke to members of the Tullahoma chapter of Tennessee Society of Professional Engineers. Shanklin, acting director of the MHD program office of the Federal Energy Research and Development Administration, said he believed that placing an energy test facility in Montana would be a major step toward meeting commercial power demands.

At 4:30 p.m. on Monday, May 5, 1975, **Colonel Harrie Richardson**—a man who had shared the dream of having a space institute and who had served as administrative assistant to the first director—collapsed at this desk in his home, 1507 Sycamore Circle, Manchester. His death was attributed to a heart attack. He was seventy-six. Sadly, his passing occurred even as plans were under way to honor his old friend, Dr. Goethert, with a dinner on May 28.

Early in May, Dr. Dicks testified before a Senate subcommittee in Washington, requesting that Congress appropriate $10 million for MHD research. A total of $3.5 million was sought for design and site preparation on the pilot MHD plant at UTSI, and the rest was to continue MHD work at UTSI and AEDC. (Arnold Center had the largest MHD generator in the world.) A few days later, **Senators Howard Baker**

and **Bill Brock** jointly sponsored an amendment to the National Energy Research Bill to provide for this money to be spent during the next year for MHD research at UTSI and AEDC. Of the $10 million, $6.5 million would be to continue existing contracts at the two facilities. Meanwhile, a state House committee was considering the bill to provide five million in state funds, which the state Senate had already overwhelmingly approved. If it cleared the House, it was expected that $40 million in state and federal funds would be available, and at least a fourth of this was expected to be spent in Middle Tennessee.

The U.S. Farmers Home Administration approved a $65,000 loan to help develop the research park. Pending was another $95,000 grant from the Appalachian Regional Commission for site preparation, and providing utilities and roads to the park.

Dr. James R. Connell, associate professor of physics, was put in charge of a project for research on mountain airflow under a $23,700 contract with the National Science Foundation. Legislative leaders from fifteen southern states were planning to tour UTSI's MHD facilities in the fall. **Dr. Firouz Shahrokhi** was named special advisor to Governor Blanton for the governor's program to develop Tennessee trade contracts in the Middle East. And UTSI threw a party on May 28 to honor Dr. Goethert before he and his wife headed to Germany for a month's vacation.

More than two hundred people attended the dinner. Among these were top officials from UT, including **Dr. Boling, Dr. Joe Johnson,** executive vice president, and **Dr. Weaver;** legislative leaders, including Senators **Crouch** and **Motlow** and Rep. **Rogers;** community leaders; Support Council members; AEDC Commander **Webster England,** and faculty and staff of the Space Institute. They all praised Goethert for his leadership at UTSI. It was not an actual farewell because while Goethert was stepping down as dean, he would continue as professor, concentrating especially on research in aeroacoustics for the FAA.

The state House approved the five million for MHD research in late May and sent it to Blanton for his signature.

In a speech to the Tullahoma Rotary Club, Dr. Weaver outlined his goals as he prepared to assume duties of dean. He had outlined these goals—calling for new growth in UTSI's academic program and physical facilities and stronger ties with AEDC—in a memorandum to UTSI faculty and staff members. Specifically, he wanted UTSI to develop new independent academic programs, provide training for technicians and to move ahead with development of a research park. He said he would arrange for personnel at Arnold Center to take Space Institute classes either on the campus **or** at the center.

"If we are to continue to respond to the academic and academically related needs of AEDC and ARO," Weaver said, "we must perforce grow in a direction not directly related to AEDC and ARO."

He cited, as an example of an independent course, the aviation systems program at UTSI. As an example of establishing closer ties with AEDC, Weaver suggested setting up a shuttle bus service. Weaver also said he believed UTSI needed an on-campus student apartment complex. Later, UT architects drew up plans for locating the apartments across the inlet from the campus with an attractive bridge providing access.

The Appalachian Regional Commission came through with a $94,200 grant for the research park, which with the FHA loan, made $159,200 available for preparing the site. (Both the grant and loan were issued to Greater Franklin Inc.)

By mid-June, 1975, Governor Blanton had signed the bill providing five million for MHD research at UTSI, and UT revealed that it was studying a new location for the research park that would be agreeable with the Air Force at AEDC. Instead of locating the park on fifty acres in the northeast corner of the campus, adjacent to AEDC property, attention was switched to one hundred and fifty acres on the west side of the campus, across an inlet of Woods Reservoir, bordering on private property in Franklin County. This turned out to be a good move as the UT trustees on June 19 acted to clear the way for locating the park across the Rollins Creek inlet. The new location connected with the Space Institute's main academic area by a causeway crossed by a gravel road. Weaver said he was pleased with the new location, which he understood was agreeable to AEDC **and to the Girl Scouts.** (There had been concern that the other site would have encroached on Camp Tannassie.)

Thus, Dr. Weaver's tenure began on a harmonious note with one source of contention between UTSI and AEDC resolved. (At Arnold Center, **Dr. William Heiser,** former chief scientist of the Aero Propulsion Lab at Wright-Patterson, was succeeding **Donald R. Eastman** as chief scientist at AEDC.)

Reluctantly, Dr. Goethert had "turned loose of his baby." (Some of the sting had been taken out of his forced "retirement" by a special provision that allowed him to continue to draw his full salary.) But for the rest of his life, "Doc" would play an active role at the Space Institute, and in 1982, he would get to be "dean" again.

✦

Virginia Richardson, left, Bob Young, third from right, and Col. Masters, right, advise students registering for the 1974 term.

Hertha Goethert snips the ribbon dedicating the tennis courts while Hrishikesh Saha steadies it. Beside Goethert are Dr. Walter Herndon, vice chancellor for academic affairs, UT, and Don Shadow of Winchester, UT Trustee. That's Pastor A. Richard Smith on the left of Bob Young. Bob Kamm, Trevor Moulden, Frank Ianuzzi, Maurice Wright, Steve Littig, and Eugene Ball also are participating.

Mrs. Goethert and Dr. Marcel Newman visit with Senator Mary Anderson in 1968.

Visiting with Dr. Wernher von Braun are, from left Morris L. Simon, Don Shadow, Ernest Crouch, Dr. Goethert and Henry Boyd.

UTSI officials visit with state officials on Sept. 1, 1971. From left are Dr. John Folger, executive director, Tennessee Higher Education Commission, Dr. Goethert, Dr. Jerry Boone, assistant director, THEC, Bob Kamm, Ralph Disser, State Health Department, Bob Young, John Dicks, and Marcel Newman.

Visiting with Gov. Winfield Dunn during a trustees visit in 1972 are from left Bob Young, Barbara Perry, Dunn, Dr. Goethert, and Bob Kamm.

Dr. Goethert is surrounded by faculty members on the staff at the start of the 1970's. From left, front, are Bob Young, Lloyd Crawford, Firouz Shahrokhi, Susan Wu, Tom Gee, Goethert, Arthur Mason, Kenneth Kimble, Jimmy Wu and Lee James; second row, Franz-Rudolf Wagner, Kent Koester, and Fletcher Donaldson; third row, Virginia Richardson, Bill Dunnill, and Gerhard Braun; back row, John B. Dicks, Jim Maus, Heinz Gruschka, K.C. Reddy, Eugene Huebschman, Maurice Wright and Ron Kohl.

Dr. Jimmy Wu, right, explains a wind tunnel model to Col. Joseph Regan, AEDC vice commander. From left are Dr. Jim Maus, Dr. Goethert, Dr. Walter Frost, Bob Kamm, Wu and Dr. Bob Young. (Years before, Dr. Wu with Dr. Art Mager published the first analytical paper on the secondary injection thrust vector control on solid propellant rockets. Aerojet General then used the side force formula derived by Wu in designing the Polaris missile.)

Bob Kamm receives a citation in 1967 from President Johnson while Kamm was chairman of the Los Angeles Federal Executive Board. John Macy, chairman of the Civil Service Commission, watches.

New Dean Charlie Weaver visits on campus Aug. 17, 1976. From left, front, are Weaver, Bob Young, Prasad R. Mikkilineni, Dieter Kurt Nowak, Chung-I Wu; back row, William B. Baker, Jr., Balbir S. Arora, Morris Perlmutter, Marvis Kelly White, and Ahmad Vakili.

Dr. Goethert, left, listens while Senator Reagor Motlow, right, makes a point with Senator Ernest Crouch.

Dr. Goethert, right, encourages Dr. Andy Holt as the UT president breaks ground for an Industry-Student Center at UTSI.

Chapter Fourteen

UTSI's Darkest Hour

IN JULY, 1975, AEDC was converting its large MHD generator to do special research in the technique of extracting power from a hot plasma, using an exotic fuel that would simulate the temperature of coal. UTSI, which for many years had had the only coal- fired generator in the world, had a new $8.1 million contract from the U.S. Office of Coal Research and was looking for up to $35 million in federal funds to step up its research program, still hoping to be able to construct a pilot MHD plant. **Senator Bill Brock,** during a visit to UTSI and AEDC, pledged his support of the plant. It would, however, be a long time before the new MHD facility was operating.

Rosalynn Carter, guest at a reception in Manchester, hosted by State Rep. **J. Stanley Rogers** and his wife Pat, told Coffee County Democrats that "Jimmy can win."

Colonel Webster English announced plans to retire on September 1 as Arnold Center commander, and **Colonel Oliver H. Tallman II,** a graduate of the U.S. Naval Academy and a command pilot, would succeed him.

Morris L. Simon was elected president of Tennessee Press Association. (Three months later, Governor Blanton would name him to the State Board of Education.)

America joined Russia in a spaceship "link-up" (Apollo-Soyuz) one hundred and forty miles above the earth.

The U.S. Soil Conservation Service contracted with UTSI to use the Institute's expertise for gathering data on land and water resources throughout the state. **Dr. Firouz Shahrokhi,** assisted by **Dr. Jack Hansen,** associate professor of civil engineering, was in charge of the project. **Dr. Konrad Dannenberg,** associate professor of aerospace systems, was expecting scientists and engineers from across the nation to attend his short course on solar power in August.

Dr. Weaver announced that a domed observatory with a 12.5-inch (diameter) telescope would be built at UTSI and that **Dr. Conley Powell,** assistant professor of mechanical engineering, would coordinate its use. Later, Weaver accepted a recommendation from Powell and **Dr. Terry Feagin** that the observatory be located on the east side of the road, near the AEDC guard shack.

On July 28, 1975, **Albert J. Moore,** a native of Indiana who moved to Tullahoma in 1950 as ARO's ninth employee at AEDC, died in Clearwater, Florida, at the age of seventy-three. Moore had headed the materials division at AEDC until 1955 when he was named chairman of employee relations. In 1957, he was promoted to director of maintenance and administration. His title later was changed to chief of the general service division. After his retirement on September 30, 1967, Moore served for five years as executive director of the Coffee County Industrial Board. Among his survivors was his daughter, **Mrs. Robert L. Young. Martha** was the widow of **Richard C. Robertson** (once Bob Young's lawyer), who died in 1968. **Phyllis Young** had died on August 11, 1968, after a long illness. Bob and Martha were married on November 15, 1969. They were to have nine happy years before tragedy would strike again and for the second time Bob Young would become a widower.

Dr. Walter Frost edited a book—"Heat Transfer at Low Temperatures"—that he co-authored with **Dr. Kenneth Tempelmeyer,** professor of engineering science at UTSI, **Water Hayes,** a research assistant at UTSI, and **J.A. Roux,** project engineer in the von Karman facility at Arnold Center.

In August, **Frank Glass,** admissions and records dean at MTSU for four years, was named dean of instruction at Motlow College, succeeding **Don England,** who had resigned. Eleven years later, Dr. Glass would succeed **Dr. Harry Wagner** as Motlow president.

A Tullahoma High School graduate, attending a dance sponsored by Tullahoma Sub-Deb Club, went to the hospital after falling from the patio at the Space Institute.

Nine members of the state's new joint senate-house task force on energy, headed by Senator Crouch, visited the Space Institute on August 5 for a briefing on MHD by Dr. Dicks.

The Air Force at Arnold Center awarded contracts totaling $161,000 to Schmiede Machine and Tool Corporation, Normandy, and Universal Machine Company of Tullahoma, to provide parts to modify the MHD generators at AEDC.

In late August, **Morris L. Simon** and **Harry M. Hill** (who had followed **Ralph Harris** as Simon's partner in H&S Publishing Company) sold their newspapers, including The Tullahoma News, the Manchester Times, and the Elk Valley Times in Fayetteville, as well as part ownership of the Herald Chronicle in Winchester, to Lakeway Publishing Company, Inc. of Morristown for "more than two million dollars." **R. Jack Fishman** was president of Lakeway and publisher of the Citizen-Tribune, a daily newspaper published in Morristown. The newspaper that some folks in 1946 had told Simon would never make it had survived for nearly thirty years and in the process become the "mother ship" for a group of newspapers that enjoyed a splendid reputation throughout the state.

On September 11, 1975, the Support Council took action on a matter that had been simmering for some time and one that was not popular in Knoxville; members adopted a resolution asking that UTSI be listed as a **line item** in the budget for the University system. At that time, all state funds for the Institute were included in money allocated to UTK, so the Space Institute did not have its own separate allocation of state funds. The council members saw the line item as the first step toward increased state financial support for UTSI. They noted that this had been a "key recommendation" made a few years earlier by a visiting committee of the Southern Association of Colleges and Schools. Proponents wanted to assure that the Institute had a foundation of state funds to undergird its fiscal structure. It was an ambitious move that would require a lot of work by **J. Stanley Rogers,** assisted by **Ernest Crouch** and **Ed Murray,** before it became a reality. The new dean was in an awkward spot to push for the change since he also was still a vice president at UTK, but it was clear that he supported it. Years later, **Bob Young** said that **Charlie Weaver's** role in this matter was never fully recognized. **Morris Simon,** who was a vocal proponent of the line-item proposal, also said that Dr. Weaver wanted it. Weaver himself, in a "report" on the Space Institute published in the bicentennial edition of The Tullahoma News on July 2, 1976, wrote: "The action by the Legislature to establish UTSI as a line item within the total educational budget of the University formalizes a mode of operation that has been the case since the beginning of the institute and in the long run should clear the way for a much more meaningful and satisfactory relationship between UTSI and its parent, the Knoxville campus."

Also at the September 11th meeting, the Support Council authorized the appointment of a committee to see if private financing of the proposed student apartment complex could be arranged. **William H. Crabtree** was chairman of the committee and **W.R. Carter** was assistant chairman. Both lived in Fayetteville.

Dr. Fletcher Donaldson, having forced MUMPS (Massachusetts General Hospital Utility Medical Programming System) into the public domain, began teaching this system as a required course at UTSI in 1975—a major victory for the Institute professor. The struggle had begun in 1969 when developers of the system refused Donaldson and his class access to the system because, he later told a reporter, "They were afraid we would be calling on them all the time for help. They didn't know that in Middle Tennessee we had some of the best computer people in the United States—and that they were in my class."

Since the government had spent millions of public dollars in developing the system—originally for medical applications—Donaldson contended that educational institutions should be able to use it. In the early 1970s Donaldson had fought to bring the system into the public domain under the Freedom of Information Act, working through members of Congress and the General Accounting Office (GAO). GAO subsequently investigated the way the U.S. Department of Health, Education and Welfare (HEW) evaluated proposals, issued and monitored grants and contracts. As a result, Donaldson said, drastic changes were made at HEW. He had played a major role in instigating the investigation.

In a proposal submitted to HEW, seeking funding for a regional medical system, Donaldson had sought the use of **Dr. Lawrence Weeds's** thirty thousand medical displays and called for his programs for manipulation of these displays to be reprogrammed in MUMPS, which was developed by **Dr. Octo Barnett.** HEW turned down the request on the grounds that the work of the two men could not be forced into the public domain even though public funds had been spent on their work. Donaldson said another factor in the denial of funding was that the evaluators of his proposal "could not believe that we had the talent available in the middle of Tennessee to do the job when in fact some of the best talent in the country is here." He tried to get help from his Tennessee congressmen and on September 30, 1974, Donaldson wrote to **U.S. Senator William Proxmire** of Wisconsin. On October 9, Proxmire wrote to GAO's **Elmer Staats,** and GAO immediately sent a representative to visit Donaldson, and HEW's investigation followed. Among the "drastic changes," Donaldson said, was that a number of grants and contracts were cancelled, "saving the government millions of dollars" and Weeds and Barnett were taken off grants and put on contracts "and required to have their work ready for public use within one year." The UTSI professor would persist in promoting the MUMPS system, specifically pushing to make it available in Veterans' Administration hospitals across the land. One breakthrough would come in August, 1982, when **VA Administrator Robert P. Nimmo** would announce a switch to a decentralization system. Some day, Donaldson would see the system expand to Brazil and other South American countries, Japan, and Western Europe, as well as across the United States. But that would take more speaking, presentation of papers, and behind-the-scenes work by the soft-spoken professor, including letters to the Pentagon, **President Reagan, Senator Barry Goldwater** and others.

Dr. K.C. Reddy was promoted from associate professor to professor of mathematics in 1975, and **Dr. Kenneth Kimble** was promoted from assistant professor to associate professor of mathematics. Shortly afterwards, Kimble headed up a project, funded by a NASA grant, to study pressure on aircraft flying near the **speed of sound.** (In Manchester that fall, spectators turned their attention to the **sound of power** as the county seat hosted the first "tractor pull" to be held in Coffee County.) **Dr. Horace Crater,** formerly of Vanderbilt University, joined the Space Institute faculty in October, 1975, as assistant professor of physics. And in October, things began shaping up for the new MHD facility at UTSI as the U.S. Energy Research and Development Administration authorized UTSI to award a five-hundred-thousand-dollar architectural and engineering contract to **Burns & Roe Inc.** of Oradell, New Jersey, to design the facility for which the Institute had set aside thirteen acres. Several other contracts were in the works, too.

Miss Virginia Richardson, registrar, noted that UTSI's fall enrollment, while not a record, was better than it had been in recent years with two hundred and twenty-eight students registered.

The UT Trustees in October voted to ask the Tennessee Higher Education Commission (THEC) for $1,460,000 for new buildings at the Space Institute. The bulk—$1,035,000—would be asked for a fifty-unit student apartment complex. The trustees were asking for another $425,000 to add a two-story, 9,700-square-foot wing to the academic building. Forty students were living in apartments in the Industry-Student Center; proponents of the new apartments, including Dr. Weaver, said the new units would free the existing center to be used as a small conference center, augmenting the Institute's short course program. THEC, in December of that year, ranked the apartment complex request as "first priority," but gave the proposed addition to the academic building a "third priority" ranking.

A group of Southern legislators and power officials, touring UTSI that fall, expressed hope that MHD would help solve the nation's energy problems. Dr. Dicks told them the problem was more "political" than technical.

The U.S. Office of Naval Research approved a $54,000 extension of the light-weight metals research project headed by **Dr. Maurice Wright.**

The start of what was one of Dr. Goethert's proudest achievements and what would become one of the Space Institute's most cherished traditions came at 4:30 P.M. on October 16, 1975. **Dr. Egon Krause,** professor of fluid mechanics at the Technical University of Aachen and director of its aerodynamics institute, gave the first annual "Quick-Goethert Lecture" in the UTSI auditorium. His subject was: "Looking Ahead to Future Problems in Fluid Mechanics." Signing of the agreement for the A.W. Quick-B.H. Goethert Lecture Series had been a high point of the Space Institute's tenth anniversary celebration.

The German Scientific Society for Aeronautics and Space Flight honored **Harvey Cook** and **Dr. Goethert** that fall by naming Cook a fellow in the society and Goethert an "honorary fellow." Goethert also was extended membership in the German Society for Aeronautics and Astronautics. Goethert, as a co-director with **Gifford Bull** and **Carl Birdwell Jr.,** offered a short course in experimental flight mechanics late that year. Bull was principal engineer of the flight research department of Calspan Corporation, Buffalo, New York, and Birdwell was a retired Navy captain from Patuxent, Maryland. The Tennessee Section of American Institute of Aeronautics and Astronautics also honored Goethert with a "Festchrift"—a volume of technical writing.

Dr. Young congratulates Dr. K.C. Reddy on his promotion to professor. Years later, Reddy would become Academic Dean.

Loyd L. Lester, who shared UTSI's September 24 birthday, joined the Institute's staff on November 1, 1975, as supervisor of the new machine shop (maintenance and storage building). He would stay in this position until his death on December 1, 1990, just before he would have received a fifteen-year service certificate. The sixty-year-old Lester had built a reputation for making whatever was needed. **Joel Muehlhauser** said Lester, who worked for him in the early days, was "a master at taking surplus equipment, refurbishing it, and making good use of it," and of swapping parts from one machine to another to make it work. While Lester was the first supervisor of the machine shop, he was not UTSI's first machinist. **Gary Payne** was a machinist with the MHD bunch before Lester arrived, and he worked with him in the shop for the next fifteen years. During this time, **Bobby Quick** and **Wayne Whittington** joined Payne and Lester in the machine shop.

It was in 1975, too, that the first seeds of a Graphics Center were sown. In November of that year, ECP hired **Artist/Photographer Larry L. Reynolds,** a Winchester native and graduate of Middle Tennessee State University. Reynolds first worked for Dr. Dicks, out of the Energy Conversion Facility (old MHD) lab building. Over the next few years, he would have several bosses, and for awhile was assigned to the Project Control Office, and he would be

Dean Weaver congratulates Dr. Ken Kimble on his promotion. Later, Kimble would head the Computer Center.

stationed in a half-dozen different locations, including the print shop. By UTSI's Silver Anniversary, the staff included two other artists, **Sherry L. Veazey** and **Lisa Blanks** and photographer **Bryan Glasner,** and Graphics was located in upper C Wing as part of the Design and Engineering Department. Several ECP photographers had come and gone since ECP had first hired Reynolds, which came a few months after **Jules Bernard** had hired the **Space Institute's** first official photographer.

Up until that summer, Bernard, using his own cameras, had taken pictures for the Institute and developed them in a tiny darkroom in lower D Wing, next to the Institute's post office. (**ECP's** photo lab later was located across the hall, and ultimately Glasner would use the smaller room as a color lab.) Before Bernard's tenure, the Institute depended on Air Force and Arnold Center photographers.

On June 23, 1975, Bernard hired **Elizabeth Motlow,** whom he later remembered "had a heart for photography." She quit on April 21, 1978, to continue her education at the Rochester (New York) Institute of Technology. **Kent Turner** began working for Bernard two days before Motlow left. In May, 1979, Reynolds hired Kent as an ECP photographer, where he worked until February 22, 1980.

Bernard hired **Maurice A. Taylor** on June 25, 1979.

By 1989, Larry Reynolds, left, had assistance from Lisa Blanks, Sherry Veazey and Brian Glasner.

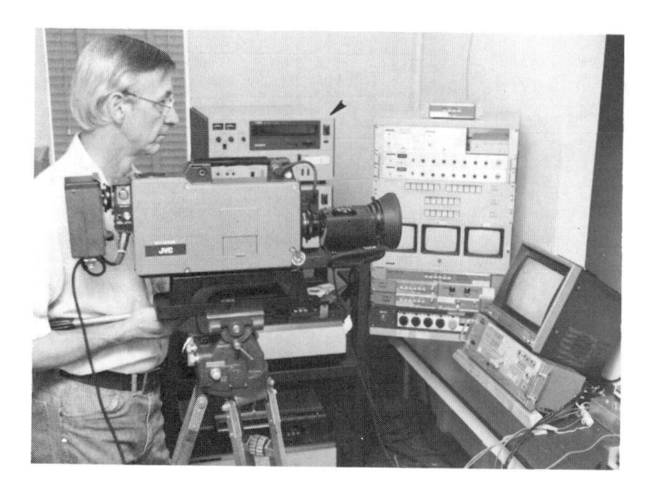

Boyd Stubblefield

"I had other applicants with more experience, but I hired Maurice because he was eager," Bernard would say years later. "I got him from Tennessee State, and I hired him because he reminded me of **me** as a young fellow. He gave it twenty-four hours a day."

But ECP offered more money, and eventually **Taylor,** too, went to work for **Larry Reynolds,** continuing there until February 27, 1987. **Brian Glasner** was hired to replace him.

After Taylor had joined ECP, Bernard hired a Vietnam veteran, **Phillip F. Gatto** of Lynchburg on September 22, 1980, as UTSI photographer. Gatto stayed with him until December 27, 1981, and the next year, Bernard would hire **Boyd Stubblefield** of Tullahoma, a retired Air Force sergeant. Stubblefield also would help set up a video program and would still be on hand to take pictures for the Institute's Silver Anniversary. After Stubblefield came, another photographic lab and dark room was eventually established upstairs in D wing.

In Graphics, **Eddie Macon** was hired on February 27, 1978, as photographic technician—Larry's first photographer. Macon left on May 11, 1979. **Robert C. Nichols** was Reynold's first "artist helper." Nichols began work February 6, 1978, as a draftsman, working for **Jim Martin.** In March, 1979, as an illustrator, he worked in Graphics for awhile, but was again classified as a draftsman when he left February 25, 1980. Other ECP photographers who worked for Reynolds included **Steve Graham** of Tullahoma, from March 24, 1980, till October 2, 1981, and **Jan Riddle** of Manchester. She was first employed in September, 1979, as a clerk typist but the next year started work in Graphics as a photographer and later was a lab technician. Her last work experience at the Institute ended on January 31, 1987.

✧ ✧ ✧

A **major reorganization,** including creation of five new divisions, went into effect at UTSI on December 1, 1975. In announcing the new divisions and the "newly defined duties" of seven administration and faculty members, Weaver said it was designed to "tie together with the Institute's mission in teaching and research and to strengthen its relationship" both with UTK and AEDC. Dr. Young would continue as associate dean, but he also would be director of the academic program. **Bob Kamm,** executive assistant to the dean, would become director of administrative services. Young was to be responsible for all credit and non-credit academic activities, recruitment of faculty and students, as well as various facilities including the library and registrar's office. Kamm's responsibilities encompassed all administrative areas not related to academics.

Five professors were put in charge of the new divisions: **Dr. Firouz Shahrokhi** headed the Remote Sensing Division; **Dr. James Wu** the Gas Dynamics Division; **Dr. Walter Frost,** Atmospheric Science Division; **Dr. Maurice Wright,** Materials Division, and **Dr. Goethert,** Aeroacoustics Division. **Dr. John B. Dicks** remained in charge of the Energy Conversion Division. Weaver said the division heads would be members of the dean's advisory council and responsible for preparing contract proposals as well as administering the contracts.

Goethert had never favored huge research projects, preferring small groups with graduate research assistants and a minimum of technicians. He felt that having many small groups, each with several contracts from different funding agencies, led to a productive and stable operation. Dicks, on the other hand, leaned toward big operations with many technicians, special engineers, a few faculty members and as many research assistants as could be accommodated.

Weaver knew about the conflicting views. His philosophy was to free the faculty and let them have projects, large or small. While he established five other research divisions in addition to the one headed by Dicks, and named directors for them, he preferred to call the directors "lead faculty."

As the year drew to a close, UTSI signed a $1.6 million contract with Babcock and Wilcox Corporation of New York for the design, engineering and manufacture of components of the MHD facility. The firm was to build the main pieces that would make up the one thousand kilowatt generator for the pilot plant.

Dr. Tempelmeyer, who was on the Energy Conversion staff, said the Space Institute also had asked the government to approve a contract for $150,000 with Magnetic Corporation of America of New York to furnish a 32,000-gauss lowfield strength magnet—one of two sought by the Institute. The other, a more powerful, 60,000-gauss super-conducting magnet valued at $1.3 million, would be furnished by the government as part of the $8.1 million contract, Tempelmeyer said. (UTSI never got this magnet.) At this time, the Institute was using a one-hundred kilowatt MHD generator with a 20,000-gauss magnet. That "small" magnet, however, was powerful enough to snatch a tool from the hand of anyone standing near it.

In his "Bicentennial" article the next July, Dr. Weaver would note that the MHD program "is burgeoning, and it looks as though the spin-off from the program at the Institute might be very, very significant to our entire state and area. At the same time, work within the Institute on MHD continues to expand in a most gratifying manner." Actually, the MHD program was to give "growth" a new meaning in the days ahead.

On Friday, the second day of 1976, the man who had been persuaded to form the company of ARO, Inc., and a true friend of the Space Institute, died eight days short of his seventy-eighth birthday. **General Leif J. Sverdrup** collapsed while having lunch at a duck hunting club west of St. Louis and was "dead on arrival" at St. Francis Hospital in Washington, Missouri. A native of Norway, he had come to the United States at the age of sixteen, and he became an American citizen four years later. A letter from **King Olav** of Norway and a telegram from **President Gerald Ford** were among tributes to the general after his death. He had last visited AEDC on the previous August 28 for the "change of command" ceremony, and more recently had visited his old friend, **Bob Williams,** at his home.

A governmental survey, commissioned by NASA and done by General Electric Company and Westinghouse Corporation, looked into ten proposed alternate ways for generating power, and concluded that MHD was the "best bet" for the future. **Dr. Dicks** cited the importance of the study, saying that it would have an impact on which projects were funded by the U.S. Energy Research and Development Administration. He said he had learned that the 1976–77 budget would include a request for thirty-six million dollars for MHD work nationwide, compared to thirteen million for the current fiscal year. Dicks expected that it would take another strong fight to get a "meaningful share" of the money for Tennessee.

Two of the first part-time teachers at UTSI and one of its early graduate students were named in January as new directors of ARO's main test facilities, freeing **Dr. Jack Whitfield, Mike Pindzola** and **Joel Ferrell** to devote full time to their duties as vice presidents. (Whitfield was corporate vice president and vice president for technology, Pindzola was vice president for propulsion, and Ferrell was vice president for support operations.) **Dr. Wendell S. Norman** was the new director of the von Karman Gas Dynamics facility, **Dr. Leon E. Ring** became director of the Propulsion Wind Tunnel, and **R.E. (Bob) Smith Jr.,** who had earned his master's degree at UTSI while working at ARO, headed up the Engine Test Facility.

Dr. Basil N. Antar joined the Institute's faculty early in 1976 as assistant professor in Engineering Science and Mechanics. It was the start of a long association. Antar had received his Ph.D. in Aerospace Engineering and Engineering Mechanics a couple of years earlier at the University of Texas. He earned his bachelor's degree in Mechanical Engineering in 1966 from Leeds Polytechnic, Leeds, England, and his master's in Mechanical and Aerospace Engineering from Rice University, Houston, Texas. He had been a Resident Research Associate in the Space Sciences Laboratory at Marshall Space Flight Center for two years prior to joining UTSI.

The foundation for the Space Institute's new observatory was completed by **R.E. Wells,** Tullahoma contractor, for $1,900. Ash Dome Company of Plainfield, Illinois, was asked to send a representative down to supervise installation of the dome and shell.

John Dicks would testify in a public hearing in Nashville regarding U.S. government forms and paperwork that UTSI had experienced a two-year delay in getting started on the MHD plant because of unnecessary paperwork. He said red tape delays would cost taxpayers between $112 billion and $449 billion in the long run.

NASA contracted with UTSI to research better ways to forecast major storms, using satellite pictures. In charge of the $43,800 project was **Dr. Walter**

Frost, director of the Atmospheric Division, assisted by **Dr. James Connell,** assistant physics professor.

In late January, 1976, President Ford asked Congress to appropriate $437 million for construction of an aeropropulsion test facility at Arnold Center—the largest single addition ever made to the center. It was to have a staff of four hundred and thirty-six persons. An overall budget of $100.6 million also was being asked for Arnold Center. Ford himself vetoed the bill when it came up in the summer, but in the fall he signed a bill that included $439.6 million for the AEDC facility.

Motlow College and Middle Tennessee State University entered an agreement to offer a course leading to a bachelor of science degree in elementary education at the Motlow campus in the fall; a few months later, the starting date was moved ahead a year.

Franklin County filed a request for the state highway department to build a 2.5-mile access road from Estill Springs to the site of the proposed UTSI research park. The county also applied for two hundred thousand dollars in federal funds to pay for a water tower for the park. **Clayton Dekle,** UT architect, visited UTSI to talk about preliminary plans for the park. UT expected in a few weeks to ask the State Building Commission to approve the plans.

In the first few minutes of the last day of January, 1976, a tragedy that in the words of one UTSI faculty member "flattened us" occurred about a quarter of a mile from the entrance to the Space Institute. Four UTSI students were killed in a one-car accident early that Saturday morning, and a fifth one was injured.

Tennessee Highway Patrol Sergeant **James Webb** said the 1968 Camaro, driven by **Marshall Hutto,** was traveling at a high rate of speed on the UTSI access road in Franklin County when it left the road, plunged down an embankment on the left side, sheared one tree and wrapped around another, several feet above ground. **Enice Hutto,** 26 of Oklahoma City, brother of the 25-year-old driver, and **Sharon Fields,** 21 of Stephens, Kentucky, were thrown from the vehicle while the driver and two other passengers—**Mark Anthony Stow,** 22 of England, and **Geraldine Ison,** 21 of Sandy Hook, Kentucky,—were trapped inside. The only survivor was **Marshall Hutto,** who was seriously injured.

Friends said the students were returning to the campus from Tullahoma. All were residents of the Industry-Student Center. Both women were honor graduates in math from Morehead (Kentucky) State University, seeking master's degrees in computer science. Both also did special work in remote sensing. Stow, a native of Kuala Lympur, Malaya, resided in Buckinghamshire, England, and had graduated with honors in aeronautical engineering from the University of Manchester in 1975. Preparing for service in helicopter aviation with the British Royal Army, he was seeking a master's degree in aviation systems at UTSI. The Hutto brothers had both graduated from the University of Oklahoma. Both were seeking master's degrees in mechanical engineering and were assisting with research for NASA.

Roy Gabriel from Trivandrum, India, penned a poem in memory of his classmates and circulated it among grieving students at the Space Institute. A memorial service was held at 3 P.M. February 5 in the UTSI auditorium. **Mrs. Kenneth Kimble** played the piano for the service, and remarks were made by **Dr. Weaver, Don Ey,** president of the Student Government Association, and the **Rev. A. Richard Smith,** pastor of Trinity Lutheran Church.

It was the darkest hour the Space Institute would experience in its first twenty-five years.

(Two years later, on May 27, 1978, **Timothy R. Anglin,** a 28-year-old student from Westlake Ohio, was killed in a two-car collision in Decherd.)

Late in February, 1976, Senators Ernest Crouch and Representatives Rogers and Murray introduced two bills with great significance to the Space Institute. One measure authorized establishing a State Energy Institute at UTSI primarily to develop a $70 million pilot MHD plant. The other set up a separate, line-item appropriation for UTSI. This measure also provided that UT's vice president for continuing education (who happened to be **Dr. Charles Weaver,** also UTSI dean) be responsible for administering the line-item budget. The bill specified that the Space Institute would remain a part of the UTK academic structure. Rogers, who took a special interest in getting the legislation approved, said Governor Blanton's 1976–77 budget would be amended to accommodate the change.

The other measure, establishing a State Energy Institute, represented a victory of sorts for Dr. Dicks. The non-profit organization, according to a story written by **Joe T. Gilliland,** associate editor of The Tullahoma News, was to build and operate a pilot MHD plant. Gilliland quoted Dr. Dicks as saying the special legal entity was being created because of a decision by high University officials that such an operation would be beyond the role and scope of the University. (While this separate entity is what Dicks had asked for earlier and what UT had once opposed, the University would have a strong representative on the energy institute's board—**Dr. Joseph E. Johnson,** executive vice president at UT.) Weaver told Tullahoma Rotarians that the two bills held the **key** to UTSI's "becoming a more effective force."

The Space Institute would continue with the $8.1 million contract, Dicks said, adding that it might be doubled, and UTSI would remain closely involved in the MHD work. Dr. Weaver said UT would be interested in being an "active partner" in the undertaking. Gilliland wrote that it would cost an estimated $70 million to build the facility and another $70 million to operate it for five years. He quoted Dicks as saying it would require fifty or a hundred acres with a good water supply, and about two hundred employees and professional staff members.

The five million dollars voted by the state could be used with about fifteen million from industry and utilities to get about fifty million in federal funds, said Dicks, adding that it was "time to get started." It would take three years to build the plant, he said, and Montana had already let a contract for the design of such a facility.

Gilliland reported that Dr. Weaver, a few days earlier, had written to Dicks, summarizing decisions reached by top officials in Knoxville. First, he said the University was "very proud of the accomplishments in this area" and was "anxious to give all possible support to the present effort and a possible doubling of that effort."

To the extent that the funds were available, Weaver said in the letter, the University strongly supported obtaining federal funds and utilizing a bond issue set up the previous year. The University was willing and anxious to pursue federal funds, using bond money from matching purposes.

"The University feels that it should not be a majority leader in putting together a package involving borrowed income plus similar contracts for several other non-federal sources in order to obtain a $60 to $80 million MHD project, " Weaver wrote. It was felt that this was "truly beyond the role and scope" of the University. On the other hand, UT would be interested in "being an active partner in such a venture, utilizing both technical and administrative resources."

Weaver had closed the letter on a personal note:

"John, speaking as dean, I am anxious to do everything possible to continue to bring to our state all possible benefits of the work you and your group have done in MHD."

The dean promised to continue to "work closely with you in every possible way to secure for the state the largest possible share of the MHD effort. As I've said, we certainly deserve it."

Thus, UTSI was expanding its MHD work with an $8.1 million contract from the U.S. Energy Research and Development Administration and had hopes of perhaps doubling this contract, and a **Tennessee Energy Institute had been established to run the MHD pilot plant.** The University had scotched earlier ideas that UTSI would manage the pilot plant, saying that this was beyond the University's scope. However, the state institute would be on the Space Institute's campus and relationships would be close.

Meanwhile, an official of the higher education commission was anticipating no problems in obtaining state revenue bonds to build a $1,035,000 student apartment complex at the Space Institute—**a dream that would eventually fade.**

The state was appropriating $750,000 for UTSI for the 1976–77 fiscal year, an increase of $100,000 in state funds. For the first time, this money was listed as an individual item in the state budget after both houses in March approved the line-item plan. This appropriation represented about a third of UTSI's revenue since slightly more than two million dollars came from research contracts.

Joe L. Evins, who had returned from World War II with his sights set on Washington, was retiring after thirty years as a U.S. congressman. **J. Stanley Rogers** and **Albert Gore Jr.** butted heads in a heated race to succeed Evins. Gore won.

Ernest Crouch, who had been a member of the state legislature since 1955, was seeking re-election as a senator, facing opposition from **James Austin,** McMinnville realtor.

NASA approved $115,000 to continue Dr. Frost's research into wind disturbance.

C.O. Prince Sr. of Winchester, who had served as both Franklin County judge and county court clerk, died at the age of ninety. He had been a friend of the Space Institute.

On March 10, both the House and Senate unanimously approved the bill establishing a Tennessee Energy Institute, and two days later the State Building Commission approved preliminary plans for the multi-million-dollar complex. With Senator Crouch, Representative Rogers and Dr. Dicks on hand the next week, Governor Blanton signed the bill.

In Knoxville, **Dr. Hilton Smith,** vice chancellor for graduate studies and research, was retiring after thirty-five years with the UT administration.

UTSI was applying for a $26,000 grant from the U.S. Appalachian Regional Commission, to help finance a water tank—a step designed to clear the way for the research park.

Dr. Dicks was preparing to tell a House subcommittee in Washington that the United States could save between $120 billion and $274 billion in the next twenty-four years if MHD technology was developed.

Bob Kamm attended a dedication of NASA's **Hugh L. Dryden** Research Center in California, named in honor of his long-time friend, who had died in the 1960s. Dryden had served as director of the National Advisory Committee for Aeronautics (NACA). Kamm had gotten acquainted with him in 1948 while working on the Department of Defense's Research and Development Board. Dryden had proposed a national unitary wind tunnel plan that had some influence on what ultimately became Arnold Center. Kamm said that Dryden's plan was two-fold: The first part was for a totally separate NACA facility, and the second part became AEDC. **Before Arnold Center was established,** Kamm said, Dryden eyed Camp Forrest as a possible site for a NACA hydrosonic center. At some point, some of his plans were married with others that were in the works for what became AEDC.

In April, 1976, **Marion Wright Hickerson,** former Manchester mayor and former president and chairman of the board of First National Bank in that city, died at the age of ninety-two. His father, **W.P. Hickerson,** one of the founders of the bank in 1900, had married **Heloise Ramsey.**

Lieutenant Governor John Wilder appointed Senator Crouch, and House **Speaker Ned McWherter** named **Ed Murray** as directors of the new energy institute, and Crouch soon was named chairman of the board for the facility, and Murray was secretary. **Dr. Joe Johnson,** executive vice president of UT, was vice chairman. Dr. Dicks headed an executive committee of the board, which included Crouch and Murray as members. The board's first meeting was on May 10, 1976.

John Dicks, back from Washington, said that **Senator Mike Mansfield** had spoken highly of UTSI's MHD program.

UT's College of Engineering gave it's top engineering award to Dr. Weaver. The **Nathan W. Dougherty** Award—named for UT's engineering dean for forty years—was presented to Weaver in Knoxville. At the same time, **William R. Carter** of Fayetteville, a member of UTSI's Support Council, was chosen as one of UT's "outstanding engineering alumni."

An alumnus of Bob Young's early class, **Dr. Jack Whitfield,** was promoted to executive vice president of ARO, Inc. UTSI hosted members of the Tennessee Higher Education Commission with a dinner May 23. On this same Sunday, **Gordon Browning,** former Tennessee governor and U.S. congressman, died in Carroll County Hospital at the age of eighty-six.

In June, 1976, UT trustees gave the go-ahead on developing a ten-acre research park and approved installation of a water tower. Weaver said this cleared the way for developing the park, which would be leased to Greater Franklin County Inc., with headquarters in Decherd. GFC then would sub-lease it to the non-profit "UTSI Research Park Inc."

A summer employee, young **John Stubbs** (who had just finished Tullahoma High School where he was an outstanding distance runner) discovered that $285 in cash and checks had been stolen in a rare break-in at the Industry-Student Center. But it was a much quieter campus than Weaver had known during the student demonstrations while he was UT chancellor. Some thought that Weaver might move to the Tullahoma area, but he had strong ties in Knoxville and continued driving back and forth while wearing two hats. **Bob Young** and **Bob Kamm** took up the slack when the dean was out of town, and in August, Weaver named **Art Mason** as assistant dean, which proved to be especially fortunate when Young, in late 1978, asked to be relieved of his administrative duties and to concentrate on being a professor.

Weaver's views as dean did not always fit snugly with his views as vice president, and during his stay at UTSI he entertained the idea that the Institute should eventually have a vice president similar to the Institute for Agriculture, as well as keeping its dean. It was an idea whose time had not yet come.

✦

Chapter Fifteen

MAGNET . . . WHO'S GOT THE MAGNET?

Dr. Goethert frequently cited the contrast in the tranquil environment of the Space Institute and the turmoil of campuses across the nation in the era of student riots and demonstrations. Dr. Weaver also observed, in an article published July 2, 1976, in the bicentennial edition of The Tullahoma News, that as UT chancellor for three years (starting in 1968), he "viewed the Institute as a sort of haven from the troubles of the main campus." It was, he said, "one of the few places in the university that at that time seemed to be making any sense" with "deer on the lawn rather than protestors at the door!" This contributed, he said, to his spending a disproportionate amount of time as chancellor with matters relating to the Space Institute.

Indeed, the seclusion and relaxed atmosphere seemed to tempt students to stay on even after completing their work. In 1989, Retired Air Force Colonel **Peter J. Butkewicz** of Lebanon, Ohio, recalled hunting and fishing with **Roger Crawford** and **Pat Madiera** and shooting pool with **Hrishi Saha** and **Dick Nenninger** (later killed in the line of duty) while earning a Ph.D. in Aerospace Engineering in 1970. He also remembered 16 hours of preliminary exams for Ph.D. candidacy as "among the most gruelling I have ever spent," and recalled that the guidance and counsel of **Jimmy Wu** and **Gerhard Braun** "carried me through," along with help from **Bob Young, Jim Maus** and **Marcel Newman.** When he retired in 1983, he was Director of Flight Systems Engineering at Headquarters, Aeronautical Systems Division Wright-Patterson and as a civilian joined General Electric Aircraft Engines in Cincinnati, and claimed his UTSI education as a "dominant forcing function of my career and of my life."

Dr. Peter J. Butkewicz

Despite its relative tranquility, however, the Space Institute was the scene of at least one brief demonstration against the Vietnam War. This occurred in the fall of 1968 in sympathy with a nationwide, student-led "moratorium." A few protestors, wearing arm-bands, assembled in the Institute's lobby and displayed an open casket. Apparently the demonstration, which stirred some lively debate with students from the military such as Air Force **Captain Roger Crawford,** was short-lived—perhaps dampened somewhat by Dr. Goethert's arrival that morning.

✧ ✧ ✧

Dr. Arthur Mason was one of five full-time faculty members when the Institute opened. By the summer of 1976, when Weaver named him assistant dean, Mason had directed thesis and dissertation research of more than twenty Master of Science and Ph.D. graduates. He had established a chapter of the Society of Physics Students and of Sigma Pi Sigma physics honor society. In 1973, he had been instrumental in chartering a UTSI Club of the Society of Sigma Xi, and honorary scientific group, which he served as its first president. He would serve as assistant dean and/or associate dean for nearly ten years and then continue as a professor of physics and be on hand as a member of the Silver Anniversary Committee.

Arnold Center celebrated its twenty-fifth year in June of 1976, with Air Force Secretary **Thomas C. Reed** stressing in his speech that AEDC was a "key defense unit."

Dr. Fletcher W. Donaldson, computer science professor, was asked by the University to be co-director of a conference in Key Biscayne in December, dealing with advances in medical systems. He was constantly pushing for adoption of the MUMPS computer system. He also sponsored a seminar at UTSI on Advances in Medical Systems in 1976 with high level personnel from such government agencies as the Department of Defense, the Indian Health Service and the Veterans' Administration attending. During his fourteen years at the Space Institute, Donaldson would conduct twenty-one short courses. Among speakers invited by the professor were **Evmenios Damon,** an executive with Goddard Space Flight Center, **Dr. Robert Frosch,** assistant secretary of the Navy for Research and Development, **Admiral Grace Hopper** from the Pentagon, and **Brigadier General Gustav Lundquist,** retired AEDC commander and head of the Federal Aviation Agency.

The Coffee County School Board banned Walter D. Edmonds' classic "Drums Along the Mohawk" from the high school library after one member, the **Rev. Jesse Garner,** complained that the book was obscene.

UTSI was planning twenty-six short courses and two special lecture series for the next academic year. **Grant Hansen,** past AIAA president and a former assistant Air Force secretary, presented Dr. Goethert with AIAA's first simulation and ground-testing award in San Diego.

State **Senator Reagor Motlow** announced that after thirty years in the state legislature, he would not seek re-election. But politics were plenty hot with Nashville's **Jim Sasser** winning the Democratic nomination to run against U.S. Senator **Bill Brock** from Chattanooga, **Jimmy Carter** slugging it out with **Gerald Ford,** and **Ernest Crouch** holding on to his state Senate post.

Brock visited UTSI in July, expressing concern that the U.S. Department of Energy Research and Development Administration was spending seven million dollars to help Russia's MHD research program. **Dr. John Dicks** shared his concern, saying this was one way the government was "hamstringing" MHD research at UTSI. He said the U.S. government, through the "scientific information exchange," was putting Russia ahead of MHD research in Tennessee and Montana. Dicks got a "beautiful response" when he spoke to state energy officials in Oak Ridge about the problem. He said the energy agency had budgeted about two million dollars for developing a super-conducting magnet—a key part of the MHD generator—which would be sent to Moscow. Furthermore, he said, they were planning in the next year to spend another five million on the Russian project and to build them a generator channel based on UTSI's design, while paying for personnel and scientific testing in Moscow. (Several Russians had visited UTSI to observe MHD research being done there.)

Dicks learned that the magnet promised to the Space Institute was not to be delivered until a year after it was promised, which he said meant that it would not be available when it was needed. He was especially incensed that in spite of this, the U.S. was spending money to send a magnet to the Russians.

The agency was pressuring UTSI to cut back on research with eastern coal, Dicks said, while they also resisted authorizing more funds to UTSI's contract to cover inflation. Dicks said escalation of costs meant that the original $8.1 million should be increased to $12 million. The "bureaucracy," Dicks charged, was throwing cold water on plans to get public and private funding for the MHD pilot plant that was to be constructed by the Tennessee Energy Institute. An official from Washington countered that the Space Institute was only one of several MHD projects supported by the federal agency, including the one at AEDC where the existing generator was being revamped to permit research into more efficient ways for extracting power production from high temperature fuels.

Dicks also spoke in Wheeling, West Virginia, at a hearing on coal mining held by a committee of that state's legislature. Since West Virginia was rich in high-sulphur coal, they were interested in MHD and planned to introduce a bill to create a state MHD power authority. The Tennessee Energy Institute might build a demonstration plant for them, Dicks said. Bidding was delayed in August on the first construction phase of the new MHD energy research complex at UTSI. When proposals did come in October, Perfection Builders and General Contractors of Ringgold, Georgia, was low with an offer of $264,600, but this bid later was rejected as being $50,000 too high, and Architect Dekle began redesigning specifications. The next month, Gillespie Construction Company of Brentwood and Taylor Ironworks Supply Company of Macon, Georgia, submitted low bids totaling $46,865 for constructing a water tower to serve the research park, but this exceeded existing funds, and new bids were requested.

Dr. George E. Mueller, head of Systems Development Corporation in Santa Monica, and the man who had directed America's first manned flights to the moon, gave the Quick-Goethert lecture at 4:30 P.M. on September 30.

In the fall of **1976,** the UT trustees recommended as top priority, building projects for UTSI in 1977–78 totaling $3,105,000. Included were proposals to add three wings to the academic building, totaling 30,000 square feet at a cost of $1,750,000; to erect a twenty-five unit apartment building for married students, covering 32,000 square feet and costing $795,000; to add twenty units to the Industry-Student Center apartments, 16,000 square feet at a cost of $400,000, and to spend $160,000 to add to the pumping plant. Tennessee Higher Education Commission responded by recommending that the state issue $1,195,000 in revenue bonds to finance both the new housing for married students and the addition to the center. The bonds were to be paid off by rental payments.

THEC rejected the addition of three wings and the expansion of the water plant as well as turning down a request for $130,000 to construct an electronic research building. However, these projects would be resurrected in 1977.

Jimmy Carter and **Walter Mondale** won trips to the White House in November, and **Jim Sasser** retired **Bill Brock** from the U.S. Senate.

The Space Institute welcomed **Dr. John Thomason,** a native of Dyersburg, Tennessee, as assistant professor of electrical engineering. And in October, 1976, a man who had been born in a log cabin in Alabama and who would become dean of the Space Institute, signed on as director of the new gas diagnostics research division at UTSI and professor of aerospace engineering. **Dr. Kenneth E. Harwell,** a former faculty member at Auburn University and a consultant to industrial and military organizations, had graduated first in his aeronautical engineering class at the University of Alabama and second in the entire University class of 1959 before going to California Institute of Technology for his master's and Ph.D. degrees. He had worked for General Dynamics at Fort Worth and in the Jet Propulsion Lab at Pasadena and had spent a year on special assignment at the Army Research Development and Engineering Lab at Redstone Arsenal in Hunstsville.

Early the next year, Harwell would be named principal investigator on a fifty-thousand-dollar contract awarded by the Army Missile Command to analyze exhaust plumes of jets and rockets.

Dr. Goethert received the "Senior U.S. Scientist Humboldt Award" on behalf of the government of West Germany in early 1977 because of his international standing as a scientist. West Germany, with the Alexander von Humboldt Foundation, had created the award to memorialize the twenty-fifth anniversary of the American Marshall Plan, which had helped to rehabilitate Europe after World War II.

The Support Council added the following five members: **Dr. W. B. Bigbee** of McMinnville, State Representatives **Walter Bussart** of Lewisburg and **Ray Johnson** of Manchester, **James M. Malone,** president of S&M Manufacturing Company and Serbin Fashions Inc. of Fayetteville, and **David Segroves,** assistant publisher of the Times-Gazette in Shelbyville. **Dr. John B. Dicks** and **Dr. Wayne Brown,** executive director of the Tennessee Higher Education Commission (THEC), were named ex-officio members.

UTSI signed three new NASA research contracts totaling $127,692. **Dr. Maurice Wright** was in charge of two dealing with materials, and **Dr. Kenneth Kimble,** with **Dr. J. M. Wu** as consultant, was in charge of one dealing with studies in aerodynamics. **Dr. Firouz Shahrokhi** was designated as scientific adviser to the United Nations, and Manchester voters chose **Clyde Vernon Myers** as mayor for a fifth term. Alderman **Robert Huskey** had challenged him after **Buster Bush** did not seek re-election.

In March, the Support Council asked the legislature to appropriate $925,000 in UTSI operating funds for the next fiscal year, as recommended by THEC. **Governor Ray Blanton** had trimmed the budget to $843,000.

Tennessee's new congressman, **Albert Gore Jr.,** said one of his top priorities was to get more money for UTSI's energy research. Testifying before a subcommittee, he recommended that the federal energy budget be increased by fifteen million dollars, mostly for stepping up research at the Space Institute. He later was joined in this request by the newest senator from the Volunteer State, **Jim Sasser.** (In April, Gore repeated the request before a subcommittee of the House Interior Appropriations Committee.) Dr. Dicks was quoted as saying UTSI would need a "rock bottom amount" of $6.2 million the next year, including the two-million-dollar magnet that had been promised. The U.S. government was still planning to send a magnet to the Russians the next year despite protests of Dicks and others involved in the MHD program at UTSI. Gore took this protest to **Dr. James Schlesinger,** national energy chief, and reported that Schlesinger, in Gore's presence, instructed his chief of staff to re-evaluate priorities assigned to the MHD program.

Following a briefing by **John Dicks,** officials of the new Tennessee Energy Institute pledged to support an effort to get a forty-million-dollar MHD pilot plant in Tennessee. **Senator Crouch** and **Representative Ed Murray** were pleading with **Governor Blanton** to help.

Crouch introduced an amendment to the general budget bill, raising state funding for the Space Institute to $925,000 for the next year, an increase of $82,000 over Blanton's figure, and a healthy increase over the $750,000 in state funds for the current fiscal year. A few weeks later, the legislature approved the additional funds.

In late April, Dr. Dicks told a House subcommittee in Washington that MHD could be commercially feasible in ten years with proper government support. Two breakthroughs in MHD research took place in the spring of **1977**—one on the UTSI campus and one in the Soviet Union—and both due to research that had been done at the Space Institute. In Moscow, scientists ran a generator for 250 hours, producing 10,000 kilowatts of power, which Dicks said was "by far the largest power run" ever made in MHD research. On May 10, scientists at UTSI scored their first success in removing close to ninety-five percent of sulphur from high-sulphur coal.

Dr. Nathan W. Dougherty, engineering dean and professor at UT for forty years, died on May 18, 1977, at the age of ninety-one.

In Washington, the House approved an amendment sponsored by Congressman Gore to raise the MHD appropriation from $50 million to $67.6 million.

Smith & Bryant Construction Company of Tullahoma, with a $13,412 bid, was chosen to prepare the site and foundation for a water tower at UTSI, and Brown Steel Contractors of Newnan, Georgia, was awarded the job of constructing a two hundred thousand-gallon water tank for $38,565.

Joe Gilliland wrote in a Tullahoma News Story that the U.S. government would ship the big $2.5 million magnet sought by UTSI to Russia on June 18. **Dr. William D. Jackson,** an official of the Energy Research and Development Administration, said they still intended to provide a magnet for UTSI, but Dr. Dicks insisted: "We need it now."

The magnet was flown non-stop on an Air Force C-5 transport loaded at Chicago's O'Hare Airport, near the Argonne Lab where the magnet had been built by the University of Chicago. Loss of that magnet, which would have greatly increased the capacity of the MHD generator at UTSI, was a bitter pill for Dicks and UTSI to swallow.

UT trustees in June approved five construction projects totaling $887,683 for the Space Institute. They were: An operating and lab building, $354,856 (bid by Southeastern Engineering and Development Company of McMinnville), and an electric substation, $290,850 (bid by Dillard Smith Contracting Company, Chattanooga), both for the new MHD energy research complex; a $160,000 addition to the water treatment plant designed and supervised by Sverdrup & Parcel and Associates of Nashville; $51,977 for a water storage tank, and $20,000 for sewer lines from the main administration building to the Industry-Student Center. At that meeting, UT President **Edward J. Boling** reported that overall, the Space Institute "continues to grow at a rapid pace, fulfilling the covenant made with the people of the state. . . ."

Contracts were awarded in July, and UTSI was planning to seek bids on the second phase of the work on the MHD complex. These construction costs were part of the $8.1 million contract; much of the money had already been obligated for equipment.

Aubrey J. (Red) Wagner, chairman of Tennessee Valley Authority, and several other TVA officials accepted **John Dicks'** invitation that summer of 1977 to visit UTSI for an MHD demonstration, which impressed Wagner with the program's "promise and potential." (Actually, Dicks had difficulty getting a response from Wagner until Wagner's old friend, **Morris L. Simon,** called him and convinced him that it was important for him to visit the Institute and hear the presentation by Dicks.) Wagner directed that TVA form a panel to work with the Space Institute, which Dicks said would be valuable help for him and his staff with the design and specifications of the proposed power plant.

Meanwhile work began on the MHD complex. **Roger Hankins,** UTSI project architect, said he would ask for bids on August 25 for work on the final phase of the project: Building a three-thousand-foot control room, and a six-thousand-square-foot test building. In a conference with contractors, Hankins set a target date of December, 1978, for having the complex in shape to begin its work.

James Franklin (Jim) Martin of Estill Springs was named manger of UTSI's Design Lab effective July 1, 1977, marking the start of what would become the Design and Engineering Department. Martin's office, for awhile, would be in B-101. For the previous year, Martin had served as MHD manager. He had been associated with the Space Institute since starting work on a master's degree as a graduate research assistant in September, 1968. He got the degree two years later and remained on the staff as an engineer. **Dr. Robert A. Frosch** was appointed administrator of NASA. In December, 1970, Frosch, then assistant secretary of the U.S. Navy for research and development, had accepted an invitation from **Dr. Fletcher Donaldson** and spoke at UTSI on "The Simulation of Large Systems."

Twelve scientists were at UTSI in the summer of 1977, studying the potential for computer use with wind tunnels. Dr. Young was project manger, and **Dr. Bernard (Barney) Marschner,** technical director. Marschner, who had been the first project engineer for the Propulsion Wind Tunnel at AEDC where he worked from 1950 to 1957, now was chairman of the computer science department at Colorado State University. Among the scientists were **Dr. Frank Collins** of Tullahoma, professor of aerospace engineering at UTSI, **Dr. J. L. Jacocks** of Manchester, an employee of ARO, Inc., and **Carlos Tirres** of Tullahoma, assistant professor at Motlow College.

But on August 16, 1977, world attention focused on West Tennessee when **"The King,"** Elvis Presley, died at age forty-two.

Eddie Shaw, state transportation commissioner, wrote to Franklin County **Judge Roy Tipps** that the state would build an access road to serve the research and industrial park at the Space Institute, climaxing efforts that had stretched over several years.

Then **Red Wagner** created considerable excitement by proposing to President Carter that TVA be allowed to build a two-hundred-million-dollar MHD pilot plant. In a sixty-four-page document from TVA to the White House, the proposal was touted as a way that TVA could "fulfill the President's request" to be a national energy agency. Dicks was "very pleased" and thought it was obvious that the TVA official's recent visit to UTSI "had something to do with it." Congressman **Al Gore** called on President Carter and his top aides to support TVA's proposal.

Southeastern Engineering and Development Company was low bidder with an offer of $404,831 to construct a 120-foot tall coal-processing superstructure for the MHD complex. Two other bids were received, the highest being $547,425.

Stephen S. Strom, who had been with Babcock & Wilcox Research Center in Ohio for eleven years, joined UTSI in the fall of 1977 as section manager of chemical engineering for the MHD research project. (He would be a participant in an exciting experiment in December, 1979, when a UTSI team produced a cupful of gasoline from coal and generated a flurry of interest in building a synthetic fuel plant in East Tennessee.)

Another B&W veteran, **Richard C. Attig,** also signed on as the new manager of the fuels and combustion section of UTSI's MHD research project that fall. He was to work with **Dr. Susan Wu,** manager of the basic and applied laboratory. Attig, a graduate of Pennsylvania State University in fuels technology, had engaged in research and development related to air pollution, corrosion, and other problems resulting from using coal and oil as fuel. At Babcock and Wilcox's research and development facility in Alliance, Ohio, Attig had been research and development coordinator.

The UT trustees in September recommended spending $2.2 million for academic buildings—to add the three wings previously rejected—and gave approval for selling $860,000 in revenue bonds to pay for a family apartment complex. They also okayed a thirty-five million-dollar fund-raising campaign for UT—"Tennessee Tomorrow"—with a goal of raising $555,000 for UTSI. **William R. Carter** and **Bill Crabtree** of Fayetteville headed the campaign. Dr. Weaver said UTSI would use the money to build a low-speed wind tunnel for the Institute's research work and instruction at a cost of $250,000, to buy a new research aircraft for $150,000, and the rest to bring in new faculty members.

On October 27, at 4:30 P.M., **Dr. Theodor Benecke,** president and board member of the federal German Aerospace Industry Association, gave the annual Quick-Goethert lecture on "The European Aerospace Industry—Present Situation and Future Goals." Dr. Young introduced the speaker.

Senator Crouch, Support Council chairman, also chose this occasion to recognize Dr. Goethert's 70th birthday, which had been a week before, on the twentieth. **Dr. Quick** was present along with **Chancellor Reese, Hilton Smith,** and several others from Knoxville. Goethert had spent a week in St. Thomas Hospital earlier in October after suffering a respiratory ailment upon his return from a visit to Germany.

CFW Construction Company of Fayetteville bid $197,143 to complete the last phase of the MHD complex work—putting in water and sewer lines and a sewer treatment plant for ten acres of the fifty-acre track that had been set aside for a research park. This bid exceeded the $159,000 federal grant that had been obtained for this purpose.

The higher education commission in November, called for allocating $1,153,000 to UTSI for the next fiscal year—an increase of $228,000—specifying that

$76,000 be used to add two professors and necessary staff members. The commission recommended spending $2.2 million for a new academic and research building at UTSI as part of a total $52 million capital outlay budget for 1978–79. **Bob Kamm** said the sooner the better because UTSI was running out of office space. The Institute had acquired a 720 square-foot trailer behind one wing of the main building to use for overflow office personnel. But it would be a long, drawn-out affair before the Institute would get the needed expansion. And Dr. Goethert would have to defend the design of the academic and administration building.

President Carter vetoed the energy research administrations' budget because it contained eighty million dollars in development funds for the Clinch River Breeder Reactor at Oak Ridge, which he opposed. Dicks said the veto would likely have no effect on the MHD program.

The UT trustees on October 29 tentatively approved a bid of $997,481 from Southeastern Engineering and Development Company to construct an electric system. The bid ran $220,000 over estimates, and it was sent on to the U.S. Department of Energy.

Just before Christmas, 1977, the energy department extended for three months the MHD contract; **John Dicks** was expecting to get an answer soon on his request for $4.5 million more.

Bing Crosby's recorded rendering of ''White Christmas'' was particularly poignant that Christmas; the old Crooner had died on October 14 of a heart attack after a round of golf at the La Moraleja Club outside Madrid, Spain.

✦

Chapter Sixteen

A Snag in Expansion Plans

TWO HUNDRED and forty-one students registered for the winter quarter in January, **1978**—an increase of twenty over the previous winter's enrollment. Sixty-three were full-time students. **Hubert Humphrey,** LBJ's vice president and an unsuccessful presidential candidate in 1968, died in January. Members of the Unification Church—commonly called "Moonies"—were planning to sue the City of Manchester for denying them a permit to solicit money. **Senator Jim Sasser** recommended that former Tullahoman **Tom Wiseman** be appointed as a federal judge in Nashville. (On August 2, President Carter named Wiseman to the district judgeship.)

In Huntsville, dates were being set for the public to see the first space shuttle, the "Enterprise," which was scheduled to undergo testing at Marshall Space Flight Center in March.

President Carter's budget contained $2.7 million for new construction at Arnold Center and $115,903,000 in operating funds at the center.

The Department of Energy had extended UTSI's $8.1 million MHD contract and, a few weeks later released $2,553,000 for the research and promised to provide another $1.7 million soon. The ECP team reported drastically reducing nitrogen oxide from the exhaust of its coal-burning plant—a major cause of air pollution. A few months earlier, the team had, for the first time, succeeded in removing more than ninety percent of sulfur dioxide in the coal-burning process. These two achievements meant, Dr. Dicks said, that "a MHD power generator will be able to deal with the two most troublesome air pollution problems associated with a coal-burning power plant." TVA proposed to begin studies into the feasibility of a 500,000 to one million kilowatt MHD-steam combined cycle power plant. **Dr. Richard V. Shanklin III,** one of the first full-time students at UTSI and a former Space Institute professor, assumed duties as head of the MHD division of the Department of Energy in Washington. (Shanklin had lived for ten years in Tullahoma.)

A spokesman for the Blanton administration said plans were to get a $2.2 million expansion project at UTSI approved in 1978 and to seek funding for it in 1979. But things already were expanding with MHD research on the move.

Wesley T. McMinn came in January as a senior engineer in design, working for **Jim Martin**—first in B Wing, later in upper C Wing and, for awhile, in the UTSI Research Park as the Design Engineering group went through various stages of independence (lab, section, department). **James W. (Jim) Knight,** who had come to the Space Institute in 1975 from Planning Research Corporation, also was working for **Jim Martin** in Design Engineering and later would work for **Wes McMinn.**

McMinn had worked with Combustion Engineering Inc. in Chattanooga since 1970, the year he got his bachelor's degree in engineering science from Tennessee Tech. While in Chattanooga, he acquired his master's degree in engineering administration from UT. At the Space Institute, McMinn would ride out a topsy-turvey period for the design engineering group—including being shifted as a research lab to the UTSI Research Park before eventually returning to upper C Wing—and he ultimately would succeed Martin as head of the group. He also would complete work at UTSI in 1988 for another master's degree—this one in mechanical engineering. From 1980 to 1981, McMinn was group supervisor and for three years he was section manager before being named manager of Design and Engineering in 1984.

The design unit had moved into upper C Wing, but during the peak of the MHD build-up, it was shuttled to the research park. **Walter C. Hamilton,** a senior drafter, joined the group during this period. So did **Betty Wilkinson,** as a print clerk, and later she would become principal secretary for the Department.

Terry Jobe, specialist, and **Perry Waller,** designer, were added to the staff after the group moved back to upper C Wing. **Cliff Wurst,** who got his master's degree at UTSI in 1970, after two years as a research assistant, came back to the Institute in 1979 as a design engineer, having worked for five years with Fluor Engineers and Constructors, Inc. Later he was named as a supervisor in Design and Engineering. A surf-board enthusiast and former water skier, Wurst would later serve as commodore of the Yacht Club. **David Patrick** joined McMinn's crew on January 3, 1989, as a drafter.

Jim Martin would leave UTSI April 30, 1982. After McMinn was named manager in 1984, Graphics was moved to upper C Wing and made a part of the Design and Engineering Department. **Emmett W. Cox** would come from Cobra Manufacturing and Engineering Company in 1980 as a senior design engineer.

In February, 1978, **Dr. Mitsuru Kurosaka** joined the UTSI faculty as associate professor of mechanical and aerospace engineering. Born in Mukden, China, of Japanese parents, Kurosaka received his bachelor and master's degrees from the University of Tokyo, and his Ph.D. in mechanical engineering from California Institute of Technology at Pasadena. He came to UTSI from Schenectady, New York, where he had been a research scientist at General Electric's research and development center. Later in 1978, he was put in charge of a two-year Air Force contract to study special vibration problems in jet engines.

Dr. Conley Powell, UTSI professor, told Manchester Kiwanians that man had the capability to "reach the stars," perhaps within the next fifty years. **Dr. James Connell,** assistant professor of physics, was conducting research on fog for NASA. **Dewey Vincent** went to Boston in late February to pick up a newly acquired "lab on wheels"—a van measuring twenty-seven feet long. **Dr. Kenneth Harwell,** head of the Gas Diagnostics Division at UTSI, said the lab would be used for research work by his group and by **Dr. Walter Frost's** Atmospheric Science Division.

Former state **Senator Reagor Motlow** of Lynchburg, who had served more than thirty years in the legislature, died on March 12 after collapsing at his home.

In April, the Energy Department granted UTSI another $1.2 million for MHD research, which Dicks said would keep things moving for about another month. This appropriation brought to almost twelve million dollars the amount of money spent on the research contract.

Charlotte Melton Campbell of Tullahoma, who had worked as a secretary for the Air Force at AEDC, became the first full-time secretary at the physical plant in the spring of 1978. A mainstay—especially in later years when illness sidelined **Dewey Vincent**—she would become senior administrative services assistant and work closely with **Robert Parson,** who was destined to fill Vincent's leadership position. Robert and Charlotte would play a major role in coordinating efforts for the Institute's Silver Anniversary Celebration.

Two construction projects, totaling $1.1 million, were started in April of 1978. UT awarded a $977,481 contract to Southeastern Electric and Development Company of McMinnville to build a main control room and a $135,000 contract to Dillard Smith Construction Company of Chattanooga for an electrical system, both part of the ECP plant. **P. D. Grissom,** head of Southeastern, said he would complete the control room in eight months rather than the twelve months specified, which would get the project back on schedule. Southeastern was winding up construction of an office lab and was building a structural steel tower to be used in the MHD facility.

At this point, DOE was still planning another twenty million dollar research facility at UTSI; Shanklin said formal requests for the proposal would be submitted soon. In June, **Bill Zechman,** 29-year-old McMinnville newspaperman, joined the staff as administrative assistant to Dr. Dicks.

On Saturday night, May 27, 1978, a UTSI student died in a two-car accident at Decherd. **Timothy R. Anglin,** son of Mr. and Mrs. Albert E. Anglin Jr. of Westlake, Ohio, was killed in the head-on collision with a car driven by **Mollie Lucas** of Winchester. Timothy had entered UTSI in the fall of 1977 as a candidate for a master's degree in physics. He also was a student employee in the Energy Conversion Program.

By the summer of 1978, tensions with the Soviets had increased, threatening to halt the United States' exchange of technology. Court decisions regarding two Jewish dissidents and the charging of two American newsmen with libel had put a strain on relations between the U.S. and Russia. **Dr. Susan Wu,** senior member of the ECP staff who had visited Russia in June, said the increased tension was obvious. Earlier that year, the UT Singers, heading for a Soviet tour, upon their arrival were told that the concerts had been cancelled, and their instruments and passports were seized.

In August, Dicks was expecting a superconducting magnet and a forty-million-dollar grant to be approved soon to keep the energy research alive. The U.S. Senate appropriated $100 million for MHD research nationwide and also called for the Space Institute to get an eight-million-dollar magnet. Dean Weaver and Dr. Dicks went to Washington to accept a $1.9 million check from the Department of Defense for the MHD program. Employment was starting to rise at the MHD complex with 100 employed there in late August and twice that many expected by the time the facility was operating.

Pope Paul VI suffered a heart attack on August 10 and died at the age of eighty.

Dick Fulton, Bob Clement and **Jake Butcher** were fighting that summer for the Democratic nomination to face **Lamar Alexander** in the race for governor. Butcher won, and in November, Alexander, popular for his plaid shirts and walk across the state, was elected by a large margin. **Howard Baker** won a third term in the Senate; **Albert Gore Jr.** was sent back to Washington for a second term as congressman, and on the state level, **Representative Ray Johnson** of Manchester was elected to a second term.

Six UTSI faculty members were granted tenure, and four of them were promoted to associate professors. Promoted were **Dr. Horace Crater,** physics, **Dr. Terry Feagin,** aerospace and computer science, **Dr. Ronald Kohl,** physics, and **Dr. Trevor Moulden,** aerospace engineering. **Dr. Jack Hansen,** associate professor of civil engineering, and **Dr. Virgil Smith,** assistant professor of aerospace and mechanical engineering, were granted tenure.

Plans for the expansion of the academic building hit a snag in the fall of 1978 when the state building commission deferred appropriating $60,000 in pre-planning funds requested by UTK for getting the design under way so that bids could be taken by July, 1979. The commission acted on advice of **Bill Jones,** state finance and administration commissioner, who said no pre-planning money should be appropriated until new projects were recommended by the Tennessee Higher Education Commission (THEC). The UT Trustees were scheduled to meet on October 20 to adopt construction priorities. Architect **Clayton Dekle** was quoted as saying his staff would recommend a high enough priority to assure that the expansion proposal would go out for bids the following July.

On August 31, the UT Trustees had approved the Nashville architectural firm of Thomas & Miller to design the addition. In October, the trustees ranked the expansion as No. 1 priority, and THEC recommended $2.7 million for the expansion. However, THEC lowered the priority to second place.

Dr. John Caruthers joined the Space Institute's faculty that September as assistant professor of engineering science and mechanics. He had received his bachelor and master's degrees from Auburn University and his doctorate from Georgia Tech. He was scheduled to teach aerospace and engineering science courses and conduct fluid dynamics research. Caruthers came to UTSI from the Detroit Diesel Allison Division of General Motors Corporation at Indianapolis. As a scientist in the analytical mechanics research section of the research department, Caruthers had developed advanced methods for aerodynamic designing of jet engines and for prediction of aerodynamically induced vibration in jet engines.

Two hundred and sixty-four students enrolled for the fall quarter with the following eleven foreign countries represented: Greece, Libya, India, Vietnam, Germany, Republic of China, Iran, Mexico, Peru, Canada and Saudi Arabia.

The "voice" of the world's most famous dummy, **Charlie McCarthy,** was silenced in October, 1978, when the famous ventriloquist, **Edgar Bergen,** died at age seventy-five.

On October 26, 1978, **Dr. John B. Dicks Jr.** presented the fourth Quick-Goethert Lecture at the Space Institute. His topic was "Power From Coal by MHD Through Space Technology." **Dr. Egon Kraus** of Germany had given the first lecture; **Dr. George Mueller** of NASA, the second, and **Dr. Theodor Benecke** of Bonn, Germany, the third.

In addition to the $2.7 million expansion funds, UTSI also was asking for $405,000 in 1979 for construction of a new laboratory building and some paving work. Estimated cost of the eight thousand square-foot lab building was $305,000. **Bob Kamm** said this was the first of several that UTSI hoped to build during the next five years. It would house facilities for calibrating and repairing electronic equipment and provide storage space for sensitive electronic equipment. Another $100,000 was being requested to pave a section of road that intercepted with the access to Estill Springs. (Four owners of land condemned by Franklin County for rights of way for the access road filed for an injunction, saying the 21.5 acres was not necessary or desirable for a road.)

During the first sixteen months of the "Tennessee Tomorrow" fund drive, UTSI had received $111,375 by November, 1978.

Engineers and technicians, in November, began assembling a fifty-ton "Iron Core" electromagnet built by the Magnet Corporation of America at Boston. **Jim Martin,** design engineering manager for the MHD group, said the magnet would probably be ready for experimental use by late 1979. Eleven years later, the magnet was still in use at the MHD facility.

In December, 1978, it was announced that **Dr. Leon Ring,** who had been a part-time faculty member at UTSI, would become TVA's new general manager in January.

Bill Holt, Student Government Association president, was named to the student affairs committee of UT Trustees.

Dr. Lewis Pinson, a member of the faculty at Auburn University since 1973, joined UTSI as associate professor of electrical engineering late in 1978. A graduate of the University of Alabama, he had received his master's and Ph.D. degrees from the University of Florida.

A few days before Christmas in 1978, Dr. Young, who had been associate dean for seven years, was granted his request to return to his teaching and research duties effective January 1, 1979.

It was a difficult decision, precipitated by the death of his wife, **Martha Ann,** on November 9, 1978, six days short of her ninth anniversary as **Mrs. Bob Young.**

Martha, a native of Aurora, Illinois, was a tireless civic leader and active member of the Republican Party. She had served as Coffee County finance chairman for **Lamar Alexander's** successful campaign for governor. From 1956 to 1961, while working for ARO Inc. in the office of the managing director, Martha had been editor of **High Mach,** a monthly publication for employees of Arnold Center.

Although he relished administrative duties and missed them greatly, **Bob Young** made the decision that he felt he had to make. He had spent a month away from the Institute while Martha was in intensive care. **Art Mason** and **Bob Kamm** "carried on well," but when Young returned to work in December, he was apprehensive about his role. Weaver had been grooming him for dean, sometimes admonishing him to "Act more like a dean." Young felt that he was "still somewhat tainted by the Goethert brush," but in good times, the prospect interested him. Now, for the second time having to pick up the pieces of his personal life, Young "on the spur of the moment asked Weaver to consider making me a professor. He was very sensitive . . . he always said a professor was the best job at a university. He was very kind and accepted my wish with no problem."

Both Weaver and faculty made nice statements about Young, and a petition was circulated, asking him to continue as associate dean. In a memorandum, the dean expressed his regrets at losing the associate dean, adding: "Not that Bob doesn't deserve the change. No one could deserve it more. He has served the Institute unselfishly, tirelessly, and with distinction since its very beginning, and in many ways the Space Institute **is** Bob Young. It is obvious that Dr. Young is irreplaceable, serving the Institute as he does in so many ways."

Young was especially grateful when the faculty, staff and students put up a "magnificent" photograph of him in the corridor between B and C wings. An inscription informed that it was for "his pioneering and tremendous efforts in developing the academic and research programs" at UTSI. Later, Dr. Goethert had a plaque mounted next to the picture, in recognition of Young's "dedicated service and his many significant contributions to the UT Space Institute, and of his loyal cooperation and friendship."

Nevertheless, it was a difficult adjustment for Young, who had been involved in the graduate program since 1957, to walk away from the administrative duties. Years later, Young felt that he had made the right decision. He also remembered that he had received exceptional support from the UTSI family during that trying time.

✦

Chapter Seventeen

A CUPFUL OF HOPE

IN 1979, the administration and academic building at the Space Institute extended only as far as E Wing. The library was located in C Wing, in space later occupied by the book store, short course and business offices. For three years, officials had been trying to add on to the Institute. In 1978, the UT trustees had assigned the expansion a third-priority rating, but **Governor Ray Blanton** put nothing in the budget for the work. In January, 1979, with **Lamar Alexander** as Tennessee's new governor, State **Representative Ray Johnson** of Manchester said he was optimistic that the $2.7 million construction program would be approved to pay for adding 47,000 square feet to the academic building and 10,000 square feet for the physical plant and storage of maintenance equipment. Two months later, Governor **Alexander** did indeed recommend the $2.7 million for the expansion and also called for a 25 percent increase in UTSI's operating budget, to raise it to $1,171,300. Legislators in May approved the building funds and upped the operating budget to $1,731,000. But the addition of F, G and H Wings to the academic building did **not** quickly follow. The UT architect, **Clayton Dekle** objected to following the original semi-circular design because, he said, it was not cost-efficient. Specifically, he was against continuing with the glass exterior. Dr. Goethert, who was still very much a part of the life at UTSI, insisted that the wings would be constructed to conform with the original structure. Dr. Weaver agreed, but in mid-January, 1980, final specifications and plans for the $2.7 million construction were still being prepared.

Late in May, 1980, representatives of thirteen contractors attended a pre-bid conference to indicate their interest in the expansion project. It was August of that year, however, before contractors were selected. **Hardcastle Construction Company** of Nashville was low among eleven firms bidding with an offer of just over $2.1 million on the entire project, but this contractor asked that this bid not be considered because of a miscalculation. The next lowest bids were by two companies bidding on separate parts of the project. **Sharondale Construction Company** of Nashville was selected for the main building with a bid of $1.9 million. **Wiley Reed Construction** Company of Woodbury was chosen to build the new physical plant for $295,000, for a total cost of about $2.2 million.

On February 1, 1979, the Department of Energy released $1.317 million in additional contract funds for energy research at UTSI. Approval of the funds cleared the way for the Institute to award a contract for the **last** construction phase of its new test facility. This phase—installation of a cooling tower, utility lines, a fire-protection system and coal and seed material shelters—had been held up since the previous September. The money was part of forty million dollars that DOE had said it would spend on UTSI's MHD program over the next five years. DOE authorized the Institute to award a $797,743 contract for this final phase of CFFF to Southeastern Engineering Development Company of McMinnville. **Henry Markant,** program manager for UTSI's energy conversion program, said the five-month delay in awarding the contract had resulted in a $90,000 increase in Southeastern's original bid. (Mr. Markant died on May 10 at Vanderbilt Hospital, following an apparent stroke suffered a few days earlier. It was a pressurized time, wrangling with architects and contractors, and Markant left a meeting complaining of a headache and asked secretary **Pat Esslinger** for aspirin. Associates, sensing the seriousness of his "headache," took him to Drs. Snoddy and Galbraith in Tullahoma, and soon afterwards he was transferred to Nashville. Markant had moved to the United States from Germany as a child, and he came to the Space Institute from the Babcock & Wilcox Company in 1977.)

Another Babcock & Wilcox veteran, **Norman R. Johanson,** who as a private consultant in Johanson and Associates had been involved in the CFFF project, joined the UTSI staff in July, 1979, and would become program manager of ECP. Johanson, whose experience with Babcock & Wilcox dated back to 1948, brought nine years of experience in nuclear and fossil steam generation manufacturing. He had served as plant manager of a thirteen-hundred-employee manufacturing operation in West Point, Mississippi. He had taken both bachelor's and master's degrees from the University of Akron as well as advanced management courses at the University of Virginia and Harvard University graduate schools.

Members of Manchester Chamber of Commerce in January had accepted an invitation from Drs. Weaver and Dicks to tour the MHD facilities and were enthused over the promising positive effects on the local economy. Dicks discussed an even larger proposal—a $1.5 billion **commercialization** project that would have funded development of three successively larger electric power plants over the next decade, the first being the Coal Fired Flow Facility (CFFF). However, this was not to be, and it would be fall of 1980 before the CFFF would become operational.

Meanwhile, a proposal by the Energy Department that UTSI should share an eight million dollar super-conducting magnet with a facility being developed in Montana caused a furor in Tennessee and in Washington. U.S. **Senator Jim Sasser** said this would be "illegal and unconstitutional" because language that he had sponsored in an appropriations bill in 1978 specifically earmarked the magnet for the Space Institute's energy research program. State **Senator Ernest Crouch** of McMinnville, chairman of the Energy Institute board, said the suggestion could open the door for "the sabotaging of the Tennessee MHD program." State Representative **Ed Murray** of Winchester said the plan could mean "the loss of technical progress" that could adversely affect "hundreds of millions of dollars worth of new manufacturing, marketing and jobs in Tennessee." Protests also came from various directors of the **Energy** Institute, including State Representative **I. V. Hillis** of Sparta, and State Senator **Ray Albright** of Chattanooga. **John Dicks,** testifying before the interior appropriations subcommittee in Washington, said he feared the plan might mean another diversion of a critical component for the Institute's program.

"The giant magnet is scheduled for delivery (from Argonne National Laboratory near Chicago) in about a year and a half," Dicks said. "It will require two special trains to move it to UTSI from Chicago. We are unable to find any reason for this (sharing the magnet) because there is a Montana magnet being constructed by GE."

U.S. Representative **Albert Gore Jr.** met with **John Deutsch,** assistant energy secretary, who pledged to investigate the proposal. As of UTSI's twenty-fifth anniversary in 1989, the magnet still had not been delivered and remained in Chicago.

In March, 1979, the UT Trustees authorized UTSI to buy water from Estill Springs and authorized spending $55,875 in capital outlay funds to build necessary water lines. Until this time, the Space Institute had operated its own filtration plant, drawing water from Woods Reservoir.

Colonel Michael H. Alexander was tapped to succeed **Colonel Oliver H. Tallman II,** who was retiring in May as AEDC commander. **Robert E. (Bob) Smith Jr.,** a part-time associate professor of mechanical engineering at UTSI, director of the von Karman Gas Dynamics Facility at AEDC, and ARO's assistant vice president for technology, was put in charge of a new office for management of research and technology for ARO, assisted by **Dr. J. Leith Potter,** deputy director for technology. **Ralph Kimberlin,** a former test pilot and a graduate of the Space Institute, joined UTSI's faculty in April as associate professor of aviation systems. A major in the Air Force Reserves, he had been test pilot and flight test engineer for Piper, Rockwell International and Beech Aircraft companies. Kimberlin received his bachelor's degree from the U.S. Naval Academy and his master's degree from UTSI. After being commissioned to the Air Force, he was one of three officers responsible for the development and Vietnam combat evaluation of the AC-47 side-firing gunship.

The Space Institute also was getting a single-engine, short takeoff and landing (STOL), aircraft from the Ball Corporation from Boulder, Colorado, which Bob Kamm said would be housed at the Tullahoma Municipal Airport and used in research.

Jim Stephens reported to the Space Institute in June, 1979, as its first Personnel Manager. Stephens developed a stand-alone Institute personnel function where none had existed before. Prior to his arrival, personnel support and services had been provided by the UT main campus at Knoxville.

"It was reported that the relationship with Knoxville personnel was constructive," Stephens remembered years later, "but the problems of Knoxville being two hundred miles away finally became significant."

Stephens, who had formerly worked for a public agency in Atlanta, was first interviewed by the Energy Conversion Division (ECD), predecessor of the Energy Conversion Programs (ECP). Because of the rapid build-up of the coal research program, EDC management had decided to hire their own personnel director. In the interim between the first interview and the recommendation to offer the position, it was determined instead that a **UTSI** Personnel Office would be established to support not only ECD but the other offices, programs and divisions at the Institute as well. Stephens, approached about assuming the post, eagerly accepted the challenge.

The first problem was office space. The Institute's physical space appears to have been crowded throughout its existence. After a brief stay in the Dean's Office in Lower "A" wing, a suite of offices was prepared on the upper balcony of the Industry-Student Center. The walls, door, and windows on the southwest corner of the balcony are remnants of that first office.

"I had a spectacular view of Woods Reservoir from my office," said Stephens, "but the frequent travel over to the main academic building through **rain, snow,** and **sleet** made me wish for an office in that building." It eventually would come about.

When the new library was occupied, the location of the old library on the ground floor of "C" wing in the main building, was vacated. Walls were installed to partition the old library into office space, and in 1983, the Personnel Office was moved from the Industry-Student Center into the main academic building.

In the beginning, the Personnel Office staff consisted of **Stephens, Freeman D. Binkley, Delia Burchfield, Diane Chellstorp, Glenda Dodson,** and **Barbara Sartain.**

"Freeman was assigned to the Personnel Office and did benefits and insurance work," Stephens said, "but he wore an important hat outside the personnel function, that of Manager of Student Services."

Diane Chellstorp and Stephens were the only members of this original personnel staff still at UTSI in 1989, and at that time, Diane was an administrative assistant in the ECP. She began her career at the Institute in 1975 as a secretary for Dr. Goethert, and she later was one of those who handled personnel duties in ECP before UTSI had a personnel office. **Glenda Dodson** and **Delia Burchfield** processed applications for UTSI, and Diane screened these and interviewed applicants for ECP. When **John B. Melton** of Woodbury joined ECP on June 18, 1979, as administrative assistant, Diane joined Stephens' staff in the new personnel office. Melton was named ECP's assistant business manager on July 1, 1980, and continued in that post until leaving on March 12, 1982. **Howard Parker** had started with ECP on February 4, 1974, as assistant business manager. He became ECP's business manager the next year, and on July 1, 1977, he was named manager of staff services, a job he held until October 21, 1978. **Joe Baron** was assistant business manager, ECP, from January 3, 1977, through August 19, 1978, but his duties were basically those of a personnel manager for the MHD people. During this general period, **Roy Dennis Farmer** worked in the Space Institute's business office under the supervision of **Paul West.** Farmer started on March 1, 1975, as assistant business manager. Five years later, his title was changed to administrative assistant, and on October 1, 1980, he became manager of sponsored research budgets and accounts until leaving UTSI on May 20, 1983.

Stephens remembered as "most satisfying" the experience of starting a new function. "We developed new policies unique to UTSI, transitioned away from Knoxville support, developed processes and the capability to help our people in ways they needed help, and made it through a land-office growth mode. UTSI's maximum headcount of around four hundred and forty was reached in the 1980's. We're down to three hundred and fifty gross headcount now (1989). The Institute is maturing and is not presently seeing the former steadily increasing numbers of employees.

"On the failure side, I regret the internecine conflict between a former Dean and a former Administrator of the coal research program. The politics of that situation inevitably colored and affected the way in which we provided support and services to some folks at UTSI. I am unhappy with those memories. However, our relationship with our offices and divisions is outstanding these days—amicable and professional all up and down the line."

(The "conflict" referred to by Stephens obviously was the small war that may already have been brewing between **Charlie Weaver** and **John B. Dicks** and that would reach its peak about two years later when Dicks would be bumped out of his role as ECP administrator, and Weaver would head back to the classroom.)

In the Space Institute's 25th year, the personnel staff consisted of four persons—Stephens, **Patricia Burks,** employment and benefits, **Betsy Hastings,** and **Barbara Ervin,** secretarial and database support. (Burks, who began work at UTSI in April, 1988, later (1991) would be named Grievance/Affirmative Action Officer.)

The "land office growth" to which Stephens referred was indeed about to take place soon after he joined the Institute.

✧ ✧ ✧

A delegation of Chinese scientists came to visit UTSI's MHD facilities in June, 1979, and **Dr. Egon Krause,** who had given the first Quick-Goethert lecture, returned to lecture on June 8. Governor Alexander named **Charlotte Parish** of Tullahoma to the UT Board of Trustees in July, replacing **Don O. Shadow** of Winchester.

The Department of Energy awarded a twenty million dollar contract to Babcock and Wilcox of New Orleans for a new energy research unit at MHD—a project that had been pending for months, but alas one that would later be **cancelled.** This contract was to provide a heat and seed recovery system facility for an MHD plant and would have been located at the site of the Coal Fired Flow Facility, which would open the next year. **The new unit never materialized, however.** After **Ronald Reagan** swamped **Jimmy Carter** in the presidential election in November, 1980, the emphasis on energy programs changed. The new president talked of doing away with the Department of Energy, and **he proposed abandoning MHD research altogether.** The pilot MHD plant at UTSI, which from the start has been leased by the Department of Energy, became more of a "proof of concept" facility. Despite the lack of support by the Reagan Administration, MHD research continued at UTSI with significant break-throughs in the 1980's. The mood had been different in the summer of 1979. **President Carter,** noting that the United States had large reservoirs of coal, outlined a new National Institute on Energy, prompting Dr. Dicks to comment: "They're finally realizing in Washington that we have to depend on coal in a large part, and I think it is a new emphasis on technology . . ." Dicks saw Carter's crash program on energy as a sign that the government was turning in a direction that would be of "substantial help" to the MHD program.

In late July, 1979, **George Fumich,** fossil energy project director for the Department of Energy, visited UTSI to look at is energy research program. His group supported MHD and had been involved in federal-funded coal research since the administration of **President Kennedy.** Fumich would return in about fifteen months to speak at a dedication of the new ECP facility.

On August 1, UT **President Edward J. Boling** signed the largest research contract ever received by UT—$37.4 million for energy research. This five-year agreement with the Department of Energy called for UT to complete the pilot plant at UTSI and to test the feasibility of using MHD to increase the amount of electricity generated by coal-fired steam plants. Next to nuclear work at Oak Ridge, this was said to be the largest energy research contract ever issued in Tennessee's history. Among those at the signing with Boling were **Dr. Charles H. Weaver,** UTSI dean and vice president for continuing education at UT, **Dr. John B. Dicks,** director of the Energy Conversion Division of the Space Institute, and **Dr. Evans Roth,** UT Knoxville vice chancellor for graduate studies and research.

Wilson Pritchett, alternate energy specialist for the National Rural Electric Cooperatives, said, after visiting UTSI, that MHD was "one of the two most promising technologies for increasing the effective use of our coal resources for electric power production." The other, he said, was fluidized bed coal combustion, and he added, "I would say that MHD is probably the closest technology at hand."

Dr. Bruce A. Reese, head of Purdue University's aeronautics and astronautics department, joined AEDC on September 1, 1979, as chief scientist, filling a position vacated a year before by **Dr. William H. Heiser.**

T. L. (Pete) Fletcher, a former Tullahoma mayor and alderman, died on September 18.

UTSI's initial enrollment of 283 was an increase of 19 over the previous fall quarter.

Construction began in late November on a five thousand square-foot lab building in the ten-acre research park on the UTSI campus. The building was being raised by the Research Park Development Inc. of Fayetteville, that had been founded by CFW Construction Company. **Paul Bagett,** executive vice president of CFW, said there were "strong prospects" for occupying the first building. (Eventually, it became the first headquarters for the Center for Advanced Space Propulsion.)

Tentative plans were to conduct testing in the new MHD facility by the following March. However, **John Dicks** and his team made the news during the Christmas holidays by producing about a cupful of gasoline from coal, using a process that they thought could be used to create new synthetic fuel. **John Lanier,** staff chemist, said the "break-through" came in the early hours of Friday, December 21, 1979, culminating several months of work. Dr. Dicks said not only had they demonstrated that their approach worked, but they had realized two other major achievements at the same time. First, he said, cancer-causing agents usually found in gasoline were destroyed, and secondly, an eastern, high-sulphur coal was used—a fuel that had not previously worked in most coal liquefaction efforts.

Senator Ernest Crouch, who also was chairman of the Tennessee Energy Institute, was excited about the possibilities of the spin-off program, saying he hoped the legislature would provide "seed money" to expand the synthetic fuels program.

"I envision a plant being built in Tennessee that will have the capacity to produce 50,000 gallons of gasoline a day or more,'' Crouch said. Two other Energy Institute board members—**State Senator Albright** and **Representative Ed Murray**—were present for some of the experimental runs.

The process made gasoline in two steps: First, it extracted carbon monoxide and hydrogen in the gas that formed after burning coal at a very high temperature with pure oxygen and inserting water into the burner. Next, the gas passed through a catalyst of ''mill scale'' (a low-cost by-product of steel production, which Dicks called ''a column of rust'') to produce gasoline.

Dicks said a big accomplishment was getting a yield of two ''synthesis gases''—carbon monoxide and hydrogen. A large amount of these gases could be produced, he said, by burning half a ton of coal an hour, which was enough to yield one and a half barrels of synthetic gasoline. Because of the existing small apparatus to produce liquid fuel, the team used commercially bottled carbon monoxide and hydrogen to make the gasoline. A major hurdle was overcome, Dicks said, when the synthesis gases were produced from coal.

''A big thing about this is that it uses existing technology,'' Dicks said. ''We came upon it through our experience in burning coal at extremely high temperatures, and even then we were rather slow to realize what we could do.''

For the past year, Dicks had been urging area leaders to try to promote the commercialization of MHD and its spin-off technologies. He said economic analyses showed that synthetic gasoline could compete with natural fuels. Circumstances at the time were encouraging for development of alternative fuel supplies with much concern expressed over an ''energy crisis'' and ''gasoline shortage.'' Dicks reminded that during World War II, Germany produced most of its gasoline and diesel fuel from coal, using various synthetic methods. The Organization of Petroleum Export Countries (OPEC) was meeting to agree on world oil prices. Before that conference began, Saudi Arabia boosted prices to twenty-four dollars a barrel for contract deliveries, and prices on the ''spot market'' were already forty-five dollars a barrel.

The synthetic fuel idea generated considerable excitement that carried into 1980. State Representative **Ray Johnson** of Manchester, also a member of the Tennessee Energy Institute, joined **Murray, Crouch, Hillis, Albright, Senator John Rucker of Murfreesboro, Lieutenant Governor John Wilder, Senator Milton Hamilton** of Union City and **Tom Garland** of Greenville, and others, including **Dr. James M. Cothran,** commissioner of Economic and Community Development, in pushing for funds for a $700 million plant. Area legislators sponsored a bill authorizing up to $600 million in state bonds to finance a demonstration synthetic fuels plant, contingent on Congress providing federal loans to guarantee payment. Senator Crouch offered a back-up amendment to provide other funding. At least two private firms—**Koppers Company** and **Pullman-Kellog Inc.**—were interested in cooperating with UTSI in building a plant in Campbell County in East Tennessee, the state's leader in coal production.

Plans drew fire from environmentalist groups, and state legislators only went so far, enabling legislation poised, waiting to see what Congress would do; the Alexander administration also was taking a wait-see position. **In May, 1980, the governor vetoed a hundred-thousand dollar item contained in a budget amendment** sponsored by Senator Crouch; a spokesman said the governor felt money was available from other sources, such as Koppers Company. The veto prevented the budgeting of a hundred thousand dollars to the Tennessee Energy Institute (TEI).

Rosemary Bates, state reporter for the Knoxville Journal, quoted Alexander as saying the veto would have ''no effect on the preliminary application from Tennessee for a coal liquefaction plant.'' But, Bates reported, Dr. Dicks, chairman of TEI's executive committee, thought the veto had seriously jeopardized the state's chances to become a national leader in synthetic fuel production. Dicks was in Washington, trying to promote the proposed plant and, according to Bates, had submitted a four-hundred-page proposal to the Department of Energy. Tennessee was competing with the largest private industries in the nation for a share of tax-free bonds expected from **President Carter's** National Alternative Fuels program, Dicks said. He thought coal production in Tennessee might entitle the state to twenty such (liquefaction) plants as outlined under the program. Bates wrote that Dicks told her: ''Frankly, it looks like the Energy Institute is working out of somebody's garage and that won't impress the federal people.''

The Tullahoma News on May 16 quoted Dicks as saying the veto would ''seriously undermine our credibility'' with those federal officials who were studying the proposal for the Campbell County plant. On May 27, the Nashville Banner quoted Dicks as saying ''It seems illogical not to be willing to spend $100,000 to enhance the chances of a breakthrough on a project like this one which could bring $700 million into Tennessee.'' Eventually his criticism of the veto would come back to haunt him. Or so Dicks suspected.

On May 28, the Tullahoma newspaper reported that Governor Alexander had done an about-face and said that if necessary his administration would come up with $100,000 in state money to help get federal funds for the program. This, of course, is what had been proposed in the bill that Alexander vetoed a few days earlier. The governor had gotten flak from persons other than Dicks, not the least being **Ernest Crouch,** prime sponsor of the energy bills, and, reportedly, from some East Tennessee coal executives. (The county judge of Campbell County had announced that he was willing to hold a **doughnut sale** to raise the money.) In his latest statement, Alexander said if necessary he would call a special session of the legislature to enact synthetic fuels legislation. He vowed continuing "top priority" to efforts to get the plant—a statement welcomed by Dicks, who said it was "extremely important that Tennessee be politically unified at the highest levels." By August, Dicks was recommending that the Energy Institute pull out of plans for the synthetic fuels plant, saying the private firms would have a better chance of getting Department of Energy money without the Energy Institute's involvement. On September 12, The Tullahoma News quoted **Dr. James Chapman,** vice president of the Tennessee Energy Institute (TEI), as saying that TEI was out of the coal conversion business—at least for the time being. The Energy Institute had agreed to withdraw after the Campbell County proposal was rejected in the first round of grant awards made by the U.S. Department of Energy. Chapman, program director for UTSI's energy conversion division, said good prospects still existed for another plant in Tennessee. Interest for a Tennessee plant remained high into the next year, but plans for the plant eventually folded, as the price of gasoline dropped, and talk of a "shortage" died down, and internal problems bubbled at UTSI.

Robert O. Dietz of Manchester, who for six years was director of the von Karman Institute in Brussels, Belgium, and an ally of UTSI, retired at the end of 1979 as the senior level Air Force civilian employee at Arnold Center after more than thirty-one years of federal service. He had served in various executive capacities with ARO Inc., and after returning from Brussels, became director of technology at AEDC, helping to plan for future test facilities, including a new aeropropuslion system test facility. Dietz, a 1944 graduate of the Missouri School of Mining and Metallurgy, had taken his master's degree in March, 1962, under Bob Young in the UT-AEDC graduate program.

Dr. Young, on December 1, married **Mrs. Betty Sullivan Delk** of Shelbyville.

Dr. Firouz Shahrokhi, on January 31, 1980, spoke before the United Nations in New York, reporting on a UN-sponsored meeting he had chaired in December in Damascus, Syria. The meeting concerned peaceful applications of outer space activities in the Middle East, including surveys of water and mineral resources, land-use surveys and anti-pollution measures. A native of Iran, Shahrokhi had served for two years as technical assistant on Middle East affairs, in the U.S. State Department. (Probably no one noticed, but on a dare from a friend, Shahrokhi wore dinghy boots in his UN appearance.)

Bob Kamm, who had been on the Institute staff for twelve years, was named assistant dean in January, 1980. Dr. Weaver said Kamm would also continue serving as director of administrative services. Weaver said UT Chancellor **Jack Reese** and **Dr. Evan Roth,** vice chancellor for graduate study, research, had both recommended Kamm for the new position.

"The unique character of the Institute as a campus-like entity of UT makes it imperative that there be strong administrative support and control over a very diversified set of activities," the dean said, adding that "**Bob Kamm,** as chief administrative officer of the Institute, operates in this capacity in a very superior fashion."

At the same time, Weaver announced that **Dr. Arthur Mason** was the new director of academic programs, succeeding Dr. Young who, in 1979, had returned to teaching and research duties. Mason continued to be an assistant dean.

Noting that UTSI carried on a "wide variety of graduate teaching and research," Weaver said this required that "every ramification of the activities be guided and supervised with . . . care and boldness," adding, "I am pleased to announce that Art will be responsible for all of this."

William H. (Bill) Benson was named Quality Assurance manager at the start of 1980. He had joined the Design Engineering staff in April, 1979. When Benson left on December 5, 1984, **Emmett Cox** would succeed him as Quality Assurance manager until retiring in 1990. **Billy W. Davis,** a tournament bass fisherman, worked for both Benson and Cox. After Cox retired, **Pat Lynch,** Safety Officer since June, 1980, was made manager of Quality Assurance and Environmental Compliance. Lynch had worked both for UTSI and for ECP. In 1986, he took his master's degree at the Institute in Engineering Science and Mechanics. He served several offices, including chairman, of the Highland Rim ASME chapter.

Another alumnus of the Space Institute, **Dr. Samuel R. Pate,** former director of the Propulsion Wind Tunnel, was elected vice president in charge of ARO's technology group in the spring of 1980. **Dr.**

John William Davis, research director of NASA's Ames Research Center, Moffett Field, California, succeeded Pate at PWT.

In Manchester, **David W. Shields Jr.,** former Coffee County judge, school superintendent, highway commission chairman, state representative and attorney, died on May 18, 1980.

In June, the MHD team was pleased with the successful testing of a 120-foot tall coal-pulverizing apparatus in the power plant. **Dr. Joel Muehlhauser,** who from 1978 to 1981 was manager of the Facility Lab, said initial operation of the equipment was "apparently completely successful," and that further operation was planned to verify the satisfactory performance. There would be no "hot test" in the facility until fall, he said.

Dr. Jim Chapman, ECP program manager, said the success was particularly gratifying since it was the first attempt to check out faults.

"It is a very elaborate piece of equipment," he added, "and I don't think we would have been surprised if we had encountered some minor problems. Fortunately, everything seemed to be working very satisfactorily."

Dean Weaver was pleased with this "additional evidence of the progress being made in the MHD program," and Dr. Dicks called it an "important milestone in our program." Dicks had invented the "diagonal wall MHD generator" that was acclaimed as a major improvement in the energy-extraction technique. The design was adopted by the Soviets for use in its MHD program, according to published reports at that time.

Several men who would assume prominent roles in the energy conversion programs came to the Institute in 1980. **Marvin Sanders,** who would become manager of the Facilities Department, had been at Arnold Center (Propulsion Wind Tunnel) for fourteen years. Sanders had received his master's degree in 1971 from UT. **Dr. Harold J. Schmidt,** with three degrees from the University of Illinois, had been a research engineer at AEDC since 1973. He would become manager of ECP's Modeling and Analysis Section. **Marvis K. White,** a senior project engineer in the Materials & Component Development Section, had been at the Space Institute since 1973, having earned his master's in engineering science in 1976. Before starting his tenure at UTSI, White had been a student engineer with NASA.

✧ ✧ ✧

Dr. Ed Boling appointed **K. Suryanarayanan,** president of the Space Institute Student Government Association, to the Student Affairs Committee of the UT Board of Trustees. In the summer of 1980, UTSI got a $109,000 NASA contract to continue studies on the effect of wind on aircraft and space vehicles with **Dr. Walter Frost** as principal investigator of the three-year project.

Dr. Kenneth Harwell in July was elected chairman of the Graduate Studies Division of the American Society for Engineering Education at its annual conference at the University of Massachusetts.

UTSI was expecting to get a jet that could take off and land at speeds of sixty miles per hour. The Ball Corporation of Muncie, Indiana, was giving the Jet Wing aircraft to UTSI, and the Navy also was providing a $99,000 contract for UTSI to continue testing it with **Ralph Kimberlin,** associate professor of aviation science, as the pilot.

In September, 1980, Tullahoma Paving Company was low with its $189,527 bid to pave a 2.876-mile road that had been planned for about four years to provide direct access between UTSI and AEDC and other points in Franklin County. The road would extend from northwest of Woods Dam to Highway 41-A, a mile north of Estill Springs. Contractor **Cecil Jamison** of McEwen was nearing completion of preliminary work on the road under a $529,318 Department of Transportation contract let in October, 1978.

October 21, 1980, was a big day at the Space Institute. About 250 people, including state and national figures, crowded in for the dedication of the first pilot electricity-producing plant designed to burn coal—a $12 million facility that gave the energy conversion program a major place to continue its MHD work. **George Fumich,** assistant secretary for fossil energy with the Energy Department, was keynote speaker. He lauded the Tennessee congressional delegation for getting approval and funding of the facility and its forty-million-dollar, five-year demonstration program. Fumich praised the efforts of **Dr. John B. Dicks,** saying that it was his "foresight and determination" that brought it to Tennessee.

"J. B. was involved with coal research when coal was a dirty word," Fumich said, "and when energy was coupled with the word abundant. He saw the problem of the eighties and worked on them."

Senator Jim Sasser said UTSI would "demonstrate to the world that we can get fifty percent more energy for a ton of coal," and U.S. **Representative Albert Gore Jr.** called the plant's opening a "success story" and praised Dicks and his team.

Among others identified in press reports as attending the ceremonies were Tullahoma **Mayor George Vibbert, Manchester Mayor Roy L. Worthington, Gentry Crowell, secretary of state, State**

Representative Ed Murray, State Senator Ernest Crouch and several Energy Department officials.

It seemed like a happy time, but Dicks' quarterbacking of the dedication ceremony drew fire from **Dean Weaver.** Their relationship by then was on a downhill slide, and as it turned out the last week of that October would play a pivotal role in its deterioration. Certainly, the dedication ceremony did nothing to ease the conflict.

It looked a little like a miniature Democratic convention. Conspicuously absent were **U.S. Senator Howard Baker, Governor Lamar Alexander, UT President Ed Boling,** and others. **Charlie Weaver** was there, on the platform, but he may well have been fuming inside.

In a memo dated October 24, **Weaver** complained to **Dicks** that "Several of your activities related to this event were below minimum expectations for a senior administrator of the University . . ." He itemized four transgressions dealing with invitations to the dedication and lack of consultation with the vice president and dean of the Institute (Weaver) in planning the ceremonies. He complained that neither President Boling nor Governor Alexander was invited to the event and that Dicks' actions could only be construed as "a deliberate insult to both the Dean's Office and the University."

Among the partisan employees attending the affair were some who obviously savored the feeling that Dicks had "put one over on" Weaver. Dicks himself replied on October 29 with a three-page response to Weaver in which he said his actions "were all directed and authorized by DOE, which had come to the conclusion that the dedication would be a monumental disaster otherwise."

Calling Weaver's characterization of him as a senior administrator incorrect, Dicks said, "I have no state money at my disposal; I am on a nine-month appointment as a professor of physics . . ." Except for the magnitude of the contract, he said, "I am not unlike any other university professor conducting sponsored research." This effort by Dicks to portray himself as "only a professor" seems to have been particularly irritating to Weaver at various times during their conflicts.

Dicks concluded his memo to Weaver by expressing regrets that Weaver's actions may have prevented **Energy Secretary Charles W. Duncan Jr.** and Boling from appearing on the same platform "to extol the accomplishments of the university in the field of energy research and development."

Exactly what Weaver's objectionable "actions" had been was not made clear. But the quarreling was taking its toll on and off campus, and eventually DOE's representative at UTSI, **Jay Hunze,** would warn Knoxville that the fuss had to end or DOE would pull out. **Ed Boling** and **Joe Johnson** wanted the conflict resolved "quietly and decisively," and in November, 1980, at least one UT trustee thought the whole board would be in favor of "reassigning or firing both" Weaver and Dicks. But things would continue festering for yet awhile.

Joel Muehlhauser, who had grown up with the MHD program, was deeply involved in construction of the CFFF. As lab manager, he had been responsible for fabrication and installation of the facility. In 1981, he would become manager of the Research and Development Lab, responsible for project engineering functions on all ECP programs. He would become ECP's deputy program manager in 1986 and, eventually, would be administrator of the program. (Ten years later, **Muehlhauser** also would become UTSI's first Dean of Research and Development.)

George W. Charboneau, who had come to the Space Institute in 1975 after twelve years with Lockheed Aircraft in Burbank, California, had designed and built the data acquisition system for the Coal Fired Flow Facility. He earned his master's degree at UTSI in engineering science with **Dr. Frank Collins** as his major professor and had first designed a data acquisition system for UTSI's gas dynamics laboratory. Charboneau would serve as group supervisor and instrumentation and controls manager.

Dicks said the plant would go through a six-month "shakedown" period before becoming fully operational and generating power. He anticipated collecting data from the facility in about three years and then working "toward a full-scale plant."

A lot of things would change before those three years would pass.

✦

Chapter Eighteen

ROCKING THE MHD BOAT

WINDS of change blew hard through the Space Institute in 1981, and MHD was in the eye of the storm. In mid-January, a memorandum went out from **Dr. Charlie Weaver** informing **Dr. John B. Dicks** that a team from UT Knoxville would visit the Institute on January 15 to determine whether UTSI would continue its present efforts in MHD research. The decision, according to a story by **Elton Manzione** in The Tullahoma News, was whether to extend or to give up the $40 million DOE contract. Dicks confessed to being "astounded that the University would consider withdrawing" from the project, which he said had received "tremendous response" from DOE officials. The program was two-fold: Conducting research and development and operating the $13 million demonstration plant.

In his memorandum, Weaver had written that it appeared reasonable that "future activities by the University would be between two boundaries." The minimum boundary, the dean said, would be a high level of research and development, using UTSI's current **small** MHD facility, and engineering and support would not be included. The maximum boundary would include operation of the DOE facility and support of a large engineering effort.

"We can't say we want one piece and not the other," Dicks complained, warning that with the prevailing competition, "We could lose all of it." Most of ECP's $6.2 million payroll and most of the $40 million contract was on spent engineering, design and construction, rather than for operation, he said, adding, "The dollars have been going to our drawing boards, computers, and engineering efforts."

Weaver contended that the minimum effort would "fit into the existing operational framework of the Space Institute."

The DOE contract was expiring in September of 1981, but there had been promises to extend it for two years with ten million dollars appropriated each year. Manzione wrote that sources had speculated that the action announced by Weaver was prompted by Dicks' criticism the previous year of **Governor Lamar Alexander's** veto of $100,000 in funds for the proposed liquefaction plant, but Weaver said this had nothing to do with it. "None of this is a criticism of Dr. Dicks or his activities," Weaver said. "That has been no problem. We just need to look at where we are and where we want to go." He had had **"absolutely no interaction"** with the governor on the matter, the dean said. There had been "some friction, bumps or whatever in the program between Dr. Dicks and myself," he said, "but this is just a matter of trying to do the right thing at the right time." He thought UTSI would continue playing a central role in the MHD program, but UT saw a need to "study the options."

Adding to the nervousness at UTSI was the fact that word had leaked back to the campus of at least one meeting between Weaver and ARO officials in which the possibility of ARO assuming responsibility for the MHD program was discussed.

Dicks said two hundred and thirty-three people could be affected since any contract to run the small facility would only require about fifty people, and these would be of a lower skill and a lower pay level than many of those involved.

Senator Ernest Crouch wasted no time in sending a letter to Knoxville, expressing his opposition to the proposed study. He told a reporter that "It doesn't make sense to go in and tinker with a federal program that is working well."

Weaver insisted that it was "just the first step in an orderly review procedure that will take several months." He and Dicks agreed, he said, that "we're at the point where we need to make some decisions."

As a matter of fact, on December 16, 1980, Dicks had advised Weaver that "the university must arrive at a decision whether to participate energetically and enthusiastically, and to expand according to national needs, or to inform DOE that it wishes to withdraw from the program. In my view, these are the alternatives. Any attempt at an intermediate solution will almost surely lead to frustration and disappointment, as well as possible erosion of public confidence in the university.''

Dicks expressed these feelings in a memorandum in response to Weaver' request for Dicks' thoughts on the future of the MHD program after September 30, 1981. Dicks first noted that any changes in the management structure would have to be approved by the energy department.

''The plan on which we are now operating represents my concept of the most efficient and effective approach to the tasks assigned to us,'' Dicks wrote, adding that this concept had won the ''strong endorsement of DOE.'' His opinion regarding university involvement over the next years was that the ''university should, as a matter of clearly-stated policy, either wholeheartedly pursue MHD, along with its spinoffs and other programs in which DOE is interested, or it should abandon the activity and retreat to very basic research.''

A mainstream program such as MHD demanded ''responsive, aggressive, participation,'' Dicks maintained, and ''Neither the government or the taxpayers it represents will tolerate a lukewarm, unresponsive and unproductive incumbency in this vital area of energy research and development.''

Some of the ''friction'' and ''bumps'' that had cropped up in the Weaver and Dicks relationship certainly had to do with Dicks' private consulting firm, J. B. Dicks & Associates, which he had founded in the summer of 1966. The UT professor's activities as president of the Tennessee Energy Institute also had rankled Dr. Weaver. He was especially disturbed about recommendations, which Dicks presented to Representative **Rabon (Ray) C. Johnson** in November, 1980, for possible legislation in support of the development of energy and technological industry in Tennessee. In the fall of 1980, Weaver had asked—and directed—Dicks to spend **all** of his ''Space Institute time'' on the MHD program.

On January 7, 1981, The News had reported that tests were scheduled in the spring on a method of disposing of polychlorinated biphenyls (PCB's)—cancer-causing industrial chemicals—using high-technology techniques developed by UTSI. **Holley Industries,** a Florida electrical firm, and **J. B. Dicks Associates** were working together to form a demonstration program for the Environmental Protection Agency. Dicks said it could start in the spring if Holley worked out arrangements to use some of the Department of Energy equipment.

''EPA is really excited about it, and they are using their influence with the Department of Energy to help get the equipment,'' Dicks said. Holley was providing the money; Dicks' firm was providing the technological expertise. There was talk of floating incinerators, placed on barges that would move from harbor to harbor in the war on PCB's, using technology developed at the Space Institute.

Weaver apparently was becoming increasingly concerned that the line between Dicks' private pursuits and UTSI's business might be blurring. As early as November, 1979, Weaver had circulated a memo to faculty and staff outlining policies designed to avoid conflicts of interest governing outside employment and the use of Institute facilities. This included this statement: ''With the exception of facilities for the use of which there are established procedures and fee schedules, University facilities, equipment or services may not be used in the performance of outside consulting work without written agreements and authorizations . . .'' And even earlier—in May, 1979—UT Attorney **Ronald C. Leadbetter** had advised against the University's entering into an agreement between UT and J. B. Dicks & Associates. Dicks, on May 1, 1979, had informed Weaver that the University was being invited ''to participate in proposing work to the Department of Energy'' regarding hot gas cleanup. UT's part in the proposal—$428,000 over about two years—would go to support the construction and operation at UTSI of a gas-cleaning experiment.

UTSI's Energy Conversion Division had ''developed expertise in operation which would certainly be an asset for obtaining such an effort in Tennessee,'' Dicks wrote to Weaver. Locating the apparatus near the MHD experiment would mean that it could be used later for seed regeneration studies and for cleanup of gases in the MHD system for preheater firing, Dicks said. He said UT's participation was desirable but not essential since another site could be found. While he could not see any ''conflict of interest'' in the proposal, Dicks said that to avoid any such appearance, he should not participate directly in UT's decision whether to participate. The next day, Weaver informed Dicks that ''in general, I give enthusiastic support to the actions proposed. . . .'' He said the Space Institute could be ''either the leader, with spinoff to industry, or the strong base for industry to take the lead.'' He asked Dicks to go ahead with preparation of a proposal to submit to UT.

Leadbetter's response, dated May 24, expressed his opinion that ''state law prohibits the University from entering into such an agreement.'' The work proposed to be done by J. B. Dicks & Associates was ''directly related to the University's MHD project which is conducted under the supervision of

Dr. Dicks,'' Leadbetter noted. University policy, he said, based on state law and ''a desire to avoid even the appearance of conflict of interest,'' would prohibit execution of the contract by the University.

Weaver may have felt that the Institute—and its dean—were being by-passed on the 1981 PCB test plans because apparently Dicks had given Weaver no notice of the planned tests. This already was a sensitive topic with Weaver, having come up during the previous October, and the PCB tests and Dicks' consulting firm ultimately became a major issue, focusing specifically on whether Dicks had attempted, improperly, to use facilities during a PCB test in **October** of 1980. And plans for the 1981 tests never materialized.

Meanwhile, an aroused **Ernest Crouch** suggested that the Tennessee Energy Institute might be willing to take over MHD if the University saw fit to curtail the program at the Space Institute.

One option being bandied about was to split the forty-million-dollar DOE contract and allow such functions as equipment testing to be made available for competitive bidding. Some prognosticators saw the whole show being moved to Montana. But Crouch, for one, did not intend to let that happen. He estimated that the DOE contract with UTSI indirectly put about fifteen million dollars into Franklin and Coffee Counties alone each year. Dr. Dicks told his employees that the legislators had indicated a willingness to intervene in any efforts to dismantle the program.

Crouch, who was chairman of the Energy Institute's board, said he would strongly support initiation of discussions on the possibility of transferring the MHD program to the Energy Institute. Risks were so great otherwise, he said, that such discussions were ''imperative.'' The senator noted that the Tennessee Energy Institute had been formed by the legislature with the possibility of work being transferred when the DOE contract reached forty million dollars. Now, in light of the proposed UT review, he saw the creation of the Energy Institute as ''prophetic inspiration or at least exceptional wisdom on the part of the legislature.''

Dr. L. Evans Roth, chief research officer of the University and vice president for graduate studies, headed up the review committee, which met January 16 at UTSI with Dicks and other MHD researchers. Other members of the committee included **Dr. John W. Prados,** vice president of academic affairs, **Dr. William Bugg,** head of the UT Physics Department, **Dr. Reese Roth,** and **Dr. Igor Alexeff,** professors of electrical engineering. Crouch requested that the committee's findings be submitted to him and to speakers of both legislative houses.

Rumors had it that prior to the committee's meeting, several UTSI officials had flown to Knoxville. And despite Dr. Weaver's earlier statement that the review process would take months, there were whispers of a ''forty-eight-hour deadline.'' As it turned out, the whispers were not that far wrong.

On January 21, The Tullahoma News reported that the committee had given the ''green light'' to the MHD program. In a memo to **John Dicks, Charlie Weaver** was ''pleased to announce'' that the committee had recommended the continuation and expansion of the program. **Evans Roth** had found the program ''in large part very strong and greatly appreciated'' by the Department of Energy. ''We were very impressed with what we saw,'' Roth said.

Dicks was pleased of course, but another black cloud was on the horizon: America's new president wanted to scuttle the whole MHD program.

Ronald Reagan was sworn in as America's fortieth president on January 20, 1981, and minutes later, Iran released fifty-two American hostages. The timing represented a final slap in the face of **Jimmy Carter,** whose administration had been mortally wounded by the four hundred and forty-four-day nightmare.

At this point, prospects were still looking bright for a synthetic fuels plant to be built somewhere in Tennessee. The Department of Energy advised **Governor Alexander** that a $1.1 billion coal liquefaction facility project in Tennessee proposed by Synfuels Associates was one of three selected for ''fast track development.'' The Tennessee proposal was a joint venture by the Koppers Company, Cities Services Inc. and the Continental Group, with Koppers as the lead company. Alexander said ground might be broken in 1981 for construction of the facility, which would use two million tons of coal a year to produce about ten thousand gallons of synthetic gasoline per day. Prospective sites were in Campbell, Marion and Roane counties, and it promised to be the largest private capital investment in the history of the state. DOE had awarded a $5.75 million contract to Koppers and companions for a feasibility study.

It was a spring that would see **President Reagan** shot in an assassination attempt (March 30). **Edward M. Dougherty,** president of Sverdrup-ARO Inc. since 1974, would resign, and **Dr. Jack D. Whitfield** would succeed him as president of the company, which was renamed Sverdrup Technology Inc. Manchester would be invaded by the Ku Klux Klan led by **Bill Wilkinson,** and before the Saturday night rally ended, twenty-two of the robed Klansmen would be whisked off to jail.

Ernest Crouch had won re-election to the state senate by a 131-vote margin over his Republican opponent, **Pedro Paz,** who filed a complaint about "irregularities" in the election—Grundy County polls stayed open an hour late, he said. A bi-partisan committee failed to substantiate the claims. The AEDC commander (**Colonel Alexander**) got a general's star. **Dr. Max Kinslow** of Manchester, who got one of the first two master's degrees awarded in UT's graduate program at AEDC and later earned his doctorate at UTSI, died on February 1 after a long illness. **Dr. Basil Antar,** associate professor of engineering science at UTSI, joined four other scientists in planning an experiment for a spacelab mission aboard NASA's space shuttle. The experiment to simulate the upper level wind circulation in Earth's atmosphere later was scrapped. Dr. Weaver named **Dr. Kenneth E. Harwell** as his "special assistant" in February, saying he would be involved in all aspects of the "operations of my office"—a statement far more prophetic perhaps than either suspected.

Captain Carmen A. Lucci of Youngstown, Ohio, who got her master's degree in aerospace engineering at UTSI in March, 1977, died in an airplane crash near Edwards Air Force Base, California, on March 3, 1981.

Preparations were under way in March to start demonstration runs at the MHD plant. State Representative **Ray Johnson** talked about how improved technological advances such as those being made in MHD helped promote Tennessee's "image for technological sophistication and competence," helping to attract industry and other business investors. Then the bomb dropped.

President Reagan, in his budget recommendations, called for **total abandonment of the nation's MHD program.** Dicks spent a week in Washington with Energy Department officials. **Senator Jim Sasser** and **Representative Al Gore** pledged to help save the program. Dicks issued a directive, ordering that work on short-term objectives be accelerated in order to get "all the data we can" so as to not lose twenty years' worth of work. Coffee County Executive **Don Darden,** Manchester Mayor **Roy L. Worthington,** and Tullahoma's Mayor **George Vibbert** fired off letters to Washington; the Manchester and Tullahoma city boards passed resolutions. **Ray Johnson** said he was in contact with Washington and confident that Congress would not do away with the MHD program. (Nevertheless, he noted that the Clinch River Breeder Reactor was the only energy program in Tennessee not being cut; Memphis had lost its $700 million gasification plant.)

Reporter **John Talley** wrote that Gore, calling Reagan's recommendation "a bad mistake," pledged to save the MHD appropriation and while anticipating a "tough fight," was confident that "the facts are on our side." The administration, Gore said, did not realize that the energy crisis was not over or that MHD was a real attempt to lessen the United States' dependence on foreign oil. Late in March, Dicks was confident that Congress would reinstate funding. At a subcommittee meeting on science and technology, Gore got no objections when he announced his intentions to request that sixty million dollars be put back into the budget. A couple of weeks later, Gore succeeded in getting the House Appropriations Committee to reject the president's proposal to eliminate six million dollars from MHD spending in the current year. Gore introduced the proposal to reinstate the funds. A pattern was established that spring that would be followed for the next eight years: The administration would not put a dime in the budget for MHD research; Congress, always prodded by Tennessee's delegation, would add the funding.

Dr. Philip A. Kessel, who had earned his Ph.D. at UTSI, joined FWG Associates in the spring of 1981 as director of research and development. NASA had awarded a $140,000 contract to the company to develop and test a fog dissipating device for use at airports, on bridges, and along coastal highways. **Dr. Walter Frost** had founded the company in 1976, operating from a building in the old lab complex. Eventually, the firm was moved into a building in the research park.

In April, the State Senate's finance, ways and means committee agreed to appropriate $100,000 for synthetic fuels research at UTSI. Sponsors included Senators **Ernest Crouch** of McMinnville, **John Rucker** of Murfreesboro, and **Ed Gillock** of Memphis.

Bob Kamm presented a plaque to Aachen University in Germany in recognition of the academic cooperation between Aachen and UTSI and in honor of **Dr. B. H. Goethert,** UTSI dean emeritus, and **Dr. A. W. Quick,** dean emeritus of Aachen University.

The Tennessee Section of the American Institute of Aeronautics and Astronautics (AIAA) gave its highest award—the H. H. Arnold Award—to **Dr. Kenneth Harwell** that spring. It was in recognition of Harwell's contributions to the "plume and smoke technology, in particular to the understanding of the infrared radiation characteristics of aircraft exhaust gases and leadership in the development of sophisticated gas diagnostics." In honor of AIAA's fiftieth anniversary, four members were given awards for their work in the Tennessee Section. These were **Dr. Susan Wu, Dr. J. Leith Potter, Robert O. Dietz** and **L. E. Rittenhouse.** Among those getting twenty-five-year membership awards were **Dr. Goethert, Dr. Marion L. Laster,** and **Robert E. Smith Jr.**

The Space Institute in June hosted the nineteenth annual symposium on the engineering aspects of MHD, which drew scientists from five foreign countries. **Dr. John B. Dicks** told participants that he had concluded early in his research that coal was the logical choice for fueling commercial scale MHD power plants. **Dr. Heinz Pfeiffer** of the Pennsylvania Power and Light System, was keynote speaker. **Dr. S. T. Wang,** head of the superconducting magnet construction division at the Argonne National Laboratory, said his staff was in the process of writing shipping specifications for a magnet under construction for UTSI. It should be shipped by September, he said. Nine Septembers later, the magnet had not yet arrived.

Dr. Goethert attended ceremonies in Germany marking the one hundredth birthday of **Theodore von Karman,** who immigrated to the United States in 1934. In remarks at the ceremony in Aachen, Goethert said both AEDC and UTSI owed their existence to von Karman. They reflected "the ideas of von Karman," Goethert said. "These facilities in Tennessee are visible witnesses to the visionary genius of von Karman looking far ahead into the future. All of us who have associated with von Karman and the Arnold Center and the UT Space Institute recall particularly today, with gratitude and admiration, the genius of this man who has opened our eyes to the tremendous challenges of the future." Von Karman was appointed as science and technology advisor to the U.S. Air Force. At the end of World War II, he had written a report calling for development in the United States of a center for aeronautics testing that would include wind tunnels, and propulsion, control and electronic research testing facilities—and Arnold Center was the result.

In a memorandum dated March 24, 1981, **Dr. Charlie Weaver** announced that he was taking "direct charge" of a project to conduct gasifier tests for the Koppers Company. This action was to "make certain" there were no delays and to eliminate "troubles that are being reported to me about interference through outside consulting and disappearance of data. . . ."

In a memo the next day, **Dr. Jim Chapman,** ECD program manager, told Weaver that Dicks had asked Chapman to "inform you that he does not object to the management of this program from your office. . . ."

"We are anxious to see Koppers get the support it needs in this program and will cooperate in any way possible to assure its success," Chapman wrote. He said he knew of no "disappearance of data." He also sought to spike any impression that **Dr. William Holt** and **John Lanier** had been taken off the Koppers project because they no longer worked for J. B. Dicks & Associates. Rather than being removed from the project, Chapman said, Lanier was "assigned the key responsibility of providing the sampling and analysis support." Holt **had** been relieved as project engineer on March 19. While Holt was "a very talented individual," doing detailed planning work in a timely manner was not one of his demonstrated talents, Chapman warned, adding, "I am sure you will soon find this out if you restore him to the project engineer position with this program."

Weaver overruled the decision involving Holt and Lanier. On May 11, he advised Dicks and other Energy Conversion Division (ECD) personnel that **Holt and Lanier** would "continue to serve as project engineers for the Koppers Test Program" with **responsibility for all technical and operational aspects** of the tests. In a memo, Weaver spelled out a procedure to be followed ("Now that UT has a contract with the Koppers/KBW companies") in the operation of the Energy Conversion Facility (ECF) for the Koppers tests. The small ECF (the original MHD building) would be dedicated to the tests. On these tests, the ECF deputy manager, **Marshall Seymour,** whom Weaver designated as "operations manager," would be in charge of the facility, reporting directly to the dean's office and to **Dr. Kenneth Harwell,** the dean's "coordinator" for the tests. Otherwise, Seymour was to continue reporting to **Dr. Joel Muehlhauser,** manager of the Facilities Lab. While involved with the Koppers tests, Holt and Lanier would report directly to Dr. Harwell; otherwise, they would continue reporting to their supervisor, **Dr. Susan Wu,** who was manager of ECD's Research and Development Lab. All data and information produced during the tests was to be delivered to Harwell "as soon as possible after the test completion" and the dean's office would transmit the data to Koppers. This May 11 memorandum, addressed to **Dicks, Wu, Muehlhauser, Seymour, Nick Munn,** Data Documentation Lab, and **Ed Moulder,** ECD technologist, effectively relieved Dr. Dicks of responsibilities for the Koppers project. Weaver's decision also left Muehlhauser out of the loop and, although Seymour was **supposed** to be in charge of the facility, he apparently had no authority over Holt and Lanier on these tests.

Four mornings later, at 10:30 A.M., an explosion rocked the facility during a Koppers test, causing an estimated fifteen thousand dollars worth of damage. No one was injured.

"There had been eight successful tests before," Dr. Weaver told reporters, "one only the day before, and they had seemed very pleased with the results so far."

Dr. Harwell said the explosion was relatively minor and drew attention only because it occurred at about the same time the U.S. House science and technology authorizations committee was voting to approve reinstatement of $29 million to the 1982 MHD research. Actually, the Koppers test was **not** related to MHD, as Harwell explained, but "the crew running the test was the same crew that had previously run MHD."

The explosion may indeed have been minor, but it also was symptomatic of the confusion and tug-of-war tension that prevailed during this period of feuding between Dicks and Weaver.

Dr. Dennis Keefer headed up a committee appointed to investigate the accident. Other members were **Dr. Lloyd Crawford, Pat Lynch, Dr. Joel Muehlhauser, Dr. Carroll Peters, and Stephen Strom.**

A few days before the explosion, **Strom,** as section manager, had insisted to Weaver that he would not allow his people to participate in the tests under prevailing conditions, and Weaver in turn had taken Strom off the project.

Pat Lynch, UTSI's safety officer, also had called Weaver in Knoxville a day or two before May 14, to complain about aspects of the test procedure, insisting that no one be allowed near the test apparatus during start-up or shut-down. He recalled that during the conversation, Weaver asked, "Do you mean that things are being run in a herky-jerky manner?" This was exactly what Lynch meant.

"This is the one time that I feel that I may actually have saved a life," Lynch later said. "Otherwise, **Bob Douglas**—and possibly others—would have been standing outside right beside the downstream (stack) when the gases exploded inside it."

Lynch, in a peculiar position during the feuding time since he worked half-time for ECD and half for General Administration, had been standing in the control room at the time of the blast and saw "the ceiling raise up just a little."

The committee, in an undated report, attributed the direct cause of the accident to "failure of the coal to ignite when it was first admitted to the test cell." Although instrumentation was available to indicate occurrence of ignition, the test procedure did not require that these instruments be monitored to verify ignition, the committee concluded, so test participants were not aware of the ignition failure and continued to operate the cell using normal test procedure.

"Unburned pulverized coal and oxygen accumulated in the test cell for some 45 seconds," the committee reported, "at which time the explosion occurred."

Two heat exchangers were damaged beyond repair, according to the committee, numerous steel wall panels along the rear wall and roof were loosened, and the cover cap on the roof "running along the ridge line was also lifted during the explosion, which occurred inside the building."

All procedures, including safety procedures, were being followed, but "it has become clear during the course of this investigation that those operational and safety procedures were wholly inadequate to insure the safety of personnel and the integrity of test hardware during coal gasification tests," the committee concluded. The test plan was completed in final form only a couple of days before the explosion and the plan was not formally reviewed by the project coordinator, facilities manager, or safety officer although each of these held "informal discussions" with **Holt** and **Lanier** before the plan was completed. The committee zoomed in on the lack of a check to insure that ignition had occurred, emphasizing that such a check "is fundamental in the safe operation of any combustion system." Recommendations included a proposal that each potentially hazardous combustion facility should be equipped with "a device to verify ignition each time fuel is admitted to the system."

Dr. Chapman informed members of the committee on June 26 that "Dr. Dicks wants to document the (explosion) more fully than the Committee Report" and offered a marked-up copy. This version made minor, though perhaps significant, changes. For instance, it noted specifically that the reason there was no review of the test plan or its execution by the normal ECD managers who had planned and executed tests in the facility in the past was that **Weaver had put Seymour, Holt and Lanier** in charge. Also, in stating that there was no formal review of the test plan by the project coordinator, it noted that he **(Harwell)** had been in California. Where a committee member had written in an earlier draft that test participants had "continued to operate the cell using the normal test procedure," was this added comment: **"except that Dr. Holt, the project engineer, was taking on the telephone to Dean C. H. Weaver during the time coal was on and during the time of the explosion."** Some who were in the area later said that **Holt** abruptly left the phone after the blow-up and that **Seymour,** seeing the telephone receiver dangling, spoke into it, discovering that **Dr. Weaver** was on the other end, asking what had happened.

"The test train just blew up!" **Seymour** informed the dean, who was in Knoxville.

More fireworks lay just ahead.

✦

Chapter Nineteen

WHO'S IN CHARGE?

IT was, perhaps, inevitable that a power struggle would arise. Things had happened very quickly. A city within a city had grown up almost overnight on the Space Institute campus. The Energy Conversion Division (later called Energy Conversion Research and Development Programs) had become a gigantic entity—part of UTSI, yet separate. Three letters, **MHD,** told the story, and it was one of stupendous growth, growth that very likely had surprised even **John Dicks,** who had secured record research contracts. ECD had emerged as sort of a research institute within a research institute. It had its own business office, its own shipping and receiving, its own purchasing and bookkeeping, even its own photographer . . . it was a rather independent set-up.

While MHD may have altered or even cramped the lifestyle of the Space Institute, it also was the "golden egg" that enabled UTSI—even in its twenty-fifth year—to exist with only about thirty percent of its budget consisting of state money. It was the single largest outside source for paying the bills.

Dr. C. H. Weaver was the designated leader of the Space Institute; **Dr. John B. Dicks,** as ECD administrator, was leader of the "inner city." Conflicts over leadership roles and cross-purposes had intensified by 1981, so that UTSI's family members felt pressured to "choose sides," and tension mounted, and morale waned. Within ECD there was a fierce loyalty to Dicks, even from individuals who did not especially like him personally or agree with some of his actions; Dicks had, afterall, given them good opportunities to advance in their profession, and they had been in on the start of the exciting times of getting the "C Triple F" complex established and operating.

Some resentment lingered toward Weaver—or toward Knoxville for "forcing him" on the Space Institute without following a standard search procedure. It was commonly believed that Weaver's strong-arm handling of student unrest and Vietnam "moratoriums" had cost him his job as UT chancellor. (He **had** been severely criticized for calling in Knoxville police to break up protests on the UT campus.) That he was a UT vice president and that he spent more time in Knoxville than at the Space Institute was another sore spot. This "non-resident" role probably made Weaver's job more difficult for **him,** but **Bob Kamm** filled the gap on a day-to-day basis. Kamm was a strong "company" man—and the "company" was the UT Space Institute. Kamm advocated unity and opposed division or fragmentation of authority or purpose.

Of course, **"Doc" Goethert** was still active on campus, too. There is some evidence that he had been uncomfortable with the phenomenal expansion brought on by the MHD research. Later, he would comment that it had been "very impressive growth, but it was also dangerous." Goethert differed with Dicks as far as their general philosophy or approach to research was concerned with Goethert favoring diversity over largeness.

Weaver was generally regarded as "likeable"—a man who always had a joke available, and a gifted after-dinner speaker. **Pat Lynch,** who had earned his B. S. degree at Knoxville while Weaver was chancellor, remembered a brief conversation he had with the dean one morning in a Men's Room at UTSI. As the two "faced the wall," Pat casually mentioned that he had been on the Knoxville campus at the time of the demonstrations. With a sideways glance and a grin, Weaver responded: "You weren't one of those little bastards throwing rocks, were you?"

Weaver hated "surprises," and he was somewhat bugged by the fact that Dicks had his own "press agent" in **Bill Zeckman** (whose stringing for the Nashville Tennessean while he was employed at UTSI drew complaints from Nashville Banner State Editor **Larry Brinton**—complaints that found sympathetic ears in the Dean's Office). But the problem between Weaver and Dicks went beyond personal egos, personality clashes, and turf struggles. Weaver was concerned about reports he was getting that "every third person" in ECD was working for J. B. Dicks Associates. He was especially upset when Dicks proposed legislation on October 28, 1980, regarding the Tennessee Energy Institute (TEI) that included the premise that UTSI facilities would be available for tests. Under "special capabilities," Dicks' proposal to **Representative Ray Johnson** included the provision that "TEI through its relationship with the Energy Conversion Division of the University of Tennessee Space Institute, is able to call upon highly experienced and qualified personnel in a variety of technical areas."

An even bolder part of the proposal stated that the Energy Institute **"must have the necessary and appropriate degree of administrative flexibility and must have unimpeded access to the personnel and facilities of the Energy Conversion Division of the University of Tennessee Space Institute."** According to Dicks, this provision would be necessary for TEI to carry out its functions and responsibilities and also "avoids the creation and maintenance of two organizations with essentially similar capabilities and eliminates the costs associated with duplication of resources."

The purpose of the proposal was to "outline the capabilities of the Tennessee Energy Institute in the area of stimulating, promoting and facilitating the growth and development of high-technology, energy-related commercial enterprises in the state of Tennessee," and the document focused on the "specialized services" which the Energy Institute could offer that were not "generally available from any other existing state agency." These services were, in essence, the expertise that the **Space Institute** had accumulated through years of MHD research.

In a letter to Dicks dated December 2, 1980, Weaver expressed his distress about parts of the proposal including a complaint that Weaver, as chief administrative officer of the Space Institute, had not been consulted about a document that so significantly involved the Space Institute. Weaver also was concerned that significant changes were being proposed in the administrative structure of the Space Institute, and that the proposal called for the utilization of UTSI resources that were either non-existent or not available, and that it proposed activities that were "outside the current role and scope of the Space Institute, and suggest **removal of Space Institute accountability for some aspects of one of its major divisions—the Energy Conversion Division.**"

Weaver wrote that UTSI had a "truly major MHD program," and that all of the time, personnel, facilities, and resources of ECD were devoted to MHD, which was fully funded by federal dollars. Under federal laws, rules, and regulations, the dean said, the Energy Conversion Division was tied up with MHD and could not divert resources into other ventures. Weaver had been concerned that ECD employees might be writing proposals (ie: synthetic fuel projects) for the Tennessee Energy Institute and charging their time to the Department of Energy contract. In his letter to Dicks, he wrote that "Synthetic fuels are important and have a lot of national attention. However, synthetic fuels are not an area of interest for the Space Institute at this time, and we do not have the resources for this area of activity. **As your Dean, I have requested and directed that you spend your Space Institute time on MHD.**"

Weaver asked that Dicks meet with the TEI board and some UT officials to "examine the implications of your proposal for the Space Institute." UTSI had, he said, fiscal, legal, and federal regulatory constraints on what it and its personnel could do, and all parties needed to fully understand those constraints.

"The University and its Space Institute cooperate with a wide range of activities, but this cooperation is within a framework that fits the academic institution and its purposes," Weaver concluded. Copies went to other officers and members of the TEI board including Chairman **Ernest Crouch, Vice Chairman Joe Johnson, Secretary I. V. Hillis, Jr., Assistant Secretary Ed Murray, Treasurer Ray C. Albright, Assistant Treasurer Ray Johnson, and James C. Cotham and John Rucker,** members, as well as to various UTSI officials.

Weaver also met with Dicks on that same day (December 2). In a "Memo of Record," Weaver wrote that the most important aspect of this talk was Dicks' "insistence that he is not in any sense a 'Key UT Administrator,' and thus is not subject to any of the customs, processes, and so on in which I had instructed him in writing so explicitly. I told him this was intolerable considering the large number of people and the enormous amounts of money under his supervision." The dean also wrote that he had again "made it clear to him that we wanted him working only in MHD—that he had no strength in or capabilities for synthetic fuel work, and that if requests for such work came to UT, they would not go to him or UTSI."

The problem had been brewing for a long time. In a lengthy Memorandum of Record dated February 6, 1981, Weaver wrote that during the summer of 1979 he and his immediate staff "began to have serious concerns about the possible improper use of UTSI facilities." He said the rapid growth of contract work, specially in the MHD area, "and the steady growth of activities in companies established by Institute professors seemed to call for a much stronger position concerning facilities use.'' So a memorandum setting forth policies and procedures dealing with the subject had been issued on November 15, 1979.

Beginning early in 1980 (the February 6 memorandum continued), ''personnel in the MHD group became increasingly involved in the synthetic fuels efforts of the Tennessee Energy Institute.'' The dean said that he had given TEI all possible support, ''giving the most liberal interpretation possible to the statement in the TEI law that UTSI should help it.'' But, he continued, in the spring and summer of 1980, ''we in the Dean's Office became increasingly concerned about the activities of the MHD group in the **synthetics fuels** area. We knew that a very large amount of work was being done on proposals and reports of various kinds, and we were concerned that some of this work might be occurring on MHD time, using University facilities improperly. From time to time we heard comments such as 'yesterday everybody was pulled off MHD and put on syn-fuels'. . . .'' They also became concerned about ''how time was being charged,'' Weaver noted. A brochure advertising J. B. Dicks Associates added fuel to the fire because it featured a picture of what appeared to be an incinerator, owned by the Defense Department, in the Space Institute's MHD facility.

The issue of use of University facilities by an outside firm had come to a head in October, 1980. A ''Memo for the Record'' dated October 27, 1980, lists the following sequence of events: On Monday evening, October 13, **Nick Munn,** data documentation manager for the CFFF, called **Bob Kamm** to ask if the Dean's Office knew that J. B. Dicks & Associates planned to run tests the following Saturday under contract to another company in the UT-owned MHD research facility. Kamm said no. Munn said he had been offered part-time employment by the Dicks firm to run the computer for the tests, but that he would tell them that he felt the tests were illegal, and that he would not participate. Two days later, according to the memo, Munn told Kamm that **Marshall Seymour,** who was to be Munn's supervisor on the tests, asked Munn why he would not run the computer for the tests and if some of Munn's people could run it. As reasons why he would not participate, Munn referred to the dean's memos of November 15, 1979, and September 9, 1980, dealing with use of the facilities. He said he would lock the computer to prevent anyone from using it. On October 16, Munn told Kamm that **Al Bart,** deputy program manager of the ECD, tried to get him to change his mind, but he told Bart he would lock the computer. The next day, Munn told Kamm that **Dr. Joel Muehlhauser,** chief of the Facilities Lab, had talked with him about participating. Munn stood fast. **Dr. Weaver** then asked **Dr. Mason** to talk with **Muehlhauser** and order him in Weaver's name to not participate in the tests. This was done on October 17, and on the same day, **Kamm** and **Mason** called in **Marshall Seymour** and gave him the same order. (At a meeting weeks before, Muehlhauser asked Dicks if he planned to notify the University of the proposed test, and Dicks answered that this was his business, and that he would take care of it. Apparently Muehlhauser, Chapman and others on this level thought it had been taken care of.)

On October 20, 1980, Weaver issued a memorandum to Dicks on the ''Proper and Improper Use'' of the small MHD facility. He opened the memo by saying that key members of the ECD had ''very serious misconceptions about what is proper operation and what is improper operation'' of the facility. He said the facility was staffed by University personnel assigned one hundred percent to the MHD contract, and all work done by them had to be approved in accord with contract objectives and scheduling.

Determined to prevent the test, Weaver considered cutting off all power to the facility over the weekend. Instead, on October 21, 1980, he sent Dr. Dicks another memo regarding operation of the facility. It was brief:

''Effective this date no activities are to be conducted in the above facility outside of regularly scheduled University working hours except with prior written permission from the Dean's Office.''

The test did not come off, but the question of whether Dr. Dicks had tried to hold a test in the facility without proper authority became a major issue and finally was the subject of an investigation by the comptroller of the state treasury.

The issue was still alive at the Institute early in 1981 when **Dr. Harwell** informed Weaver that he had seen an agreement dated September 26, 1980, between J. B. Dicks and Associates and Holley Electric Corporation, which purportedly was an agreement under which PCBs would have been burned in the small MHD facility.

''Mr. (William) Holt again emphasized that JBDA definitely planned to run a PCB burn test in the UTSI small MHD facility if Dr. Weaver had not stopped the test,'' according to Harwell's memorandum. Harwell said that in preparation for the test burn, a motor belonging to Holley Electric reportedly was installed in a pumping system located at the small MHD facility.

On April 1, 1981, Weaver, in a ''Memo of Record,'' noted that ''Several weeks ago we learned from DOE personnel that a request from the Environmental Protection Agency to conduct burn tests for PCB in the small University-owned facility might come to DOE and eventually to the Institute.'' **Jay**

Hunze had then asked the DOE safety group in Chicago to respond, and this response was, in Weaver's words, that "this could be a very hazardous test and that it would have to be handled very carefully." Weaver thought this was "relevant to the attempted burn by J. B. Dicks and Associates, without any notification to the University or the UTSI safety group or anyone else in an official capacity in the University about what was to be done." Whether Weaver referred here to plans for tests announced in January, 1981, or to the alleged 1980 plans is not clear, but at the beginning of this memorandum, Weaver stated that he had just finished a telephone conversation with **Jay Hunze** that was "relevant to any judgments concerning the attempted PCB tests of last fall (Saturday, October 25, 1980)."

The culmination of the long-smoldering feud was spelled out in newspaper headlines in July, 1981. The Nashville Banner, on July 9, reported: **"UT Strips John Dicks of his MHD Administrative Duties."** UT had removed Dicks from administrative responsibility for the University's energy conversion division, wrote **Rick Locker,** but (officials said) "the noted energy researcher will continue as the division's chief engineer and researcher". UTSI staff members "were told this morning" that **Dr. John W. Prados,** UT vice president for academic affairs, would assume administrative duties for the MHD research program. Locker quoted UT President **Edward J. Boling** and UT Knoxville Chancellor **Jack Reese** as saying that the changes reflect "UT's concern and support for MHD." The changes, they said, would free Dicks for full-time work on the project. In the days that followed, Dicks charged it was "politics," and insisted he had been under "political attack" since criticizing Governor Alexander the year before for vetoing a bill that would have provided $100,000 to the state Energy Institute.

On July 10, a UPI story in The Tennessean quoted Dicks as saying "I think the move in putting Dr. Prados in charge might be a good one, but what concerns me is there may be some hint of political involvement because of my activities in the Tennessee Energy Institute." This story quoted a UT news release as saying Dicks would work closely with Prados in all engineering and scientific aspects of the MHD project. It also noted Dicks' prior criticism of Alexander. The reporter wrote that Dicks spoke of talking with an attorney about investigating the release of a "university study that was critical of him while legislators were debating the $100,000 appropriation." The UPI article quoted Dicks as saying, "It was an amazing coincidence that this study that was adverse to me was released at that time."

John Talley wrote in The Tullahoma News that Dicks blamed the action on "internal problems" in the University and state, quoting him as saying: "Everyone knows we have problems within the state and university. I don't believe these problems led directly to Dr. Prados being named, but they did add to the atmosphere of uncertainty that convinced university officials that something needed to be done."

Talley said Dicks had asked the state comptroller's office to investigate the circulation of a "document" in the State House and Senate charging Dicks with "improprieties" in the operation of the MHD project at UTSI. Dicks told Talley he had been refused access to the document. According to Talley's story, Dicks, a month before, had charged that there was a "deliberate attempt by someone in the state to jeopardize the MHD program by bringing uncertainty about the project's future to the minds of many state legislators." Talley noted that Representative **Ray Johnson** (a member of the Energy Institute) had "confirmed that those uncertainties led directly to a delay in the State Senate's approving a $100,000 appropriation for MHD and coal liquefaction research at UTSI."

Apparently the "document" to which Dicks referred was a UT "audit," or report of Ron Leadbetter's investigation into the suspected use of university equipment and employees for work involving Dicks' company—the information that was turned over to the comptroller's office and in turn investigated. It reportedly had been shown to some legislators in May, just before the General Assembly was to vote on the $100,000 for the Tennessee Energy Institute.

Terry McWilliams, wrote in the Knoxville Journal on July 15 that Dicks thought politics caused his demotion, quoting him: "They say what they'd like me to do is operate without reins, but I have no power to direct anything. They'd like me to continue to bring in large programs somehow but without power" to oversee them.

But **Dr. Prados** told McWilliams it was "not a political matter at all", and in classic understatement added: **"Weaver and Dicks were not getting along together. They created a problem, and it is my job to straighten it out."**

Dicks agreed that the actions were caused by a conflict between Weaver and him, McWilliams wrote, but Dicks also maintained that the decision was "made high in the university—a political decision," and he suspected that if he had been untenured, he would have been fired. The state comptroller's office did conduct an investigation of sorts, dealing with the allegations of possible abuses by Dicks and his company of facilities owned or controlled by UTSI. In a letter dated November 23, 1981, **Comptroller W. R.**

Snodgrass wrote to President Boling (with copies to others including Dr. Dicks) that UT had conducted a review of the "activities" of Dr. Dicks and his company, "concerning their use of the facilities" of UTSI. UT had turned over these results to the comptroller for "consideration and further investigation." Snodgrass attached a lengthy report from **Frank L. Greathouse,** CPA and director of the state audit division.

Greathouse's report concluded that Dicks and others acting on behalf of him or his firm "may have violated at least the spirit and possibly the letter of UT policies and procedures relating to the use of UTSI and DOE equipment, employees and facilities."

Essentially, the state tossed the matter back to UT, recommending among other things that the University review issues raised by Dicks "to determine whether the institute's administration acted reasonably in this matter."

Early in the report, Greathouse wrote: **"Although there was not sufficient evidence to warrant further investigation by this office of these possible abuses, it appears that some university policies and procedures may have been violated."** However, he added that it "would appear that the university's policies and procedures may need clarification and revision" and recommended that UT review the issues raised in the report and "take appropriate administrative action."

Greathouse said the state's "limited review" showed that JBDA had a written contract with Holley, calling for testing and development of a machine that would effectively burn and destroy PCB's. The contract called for three stages, the first being a test burn for which JBDA was to be paid $25,000. Greathouse raised three questions: Whether JBDA was required to obtain UTSI approval prior to attempting such a burn, whether Dicks made appropriate Department of Energy officials aware of plans for a burn and whether JBDA attempted to conduct the burn. Dicks felt no prior permission from UTSI was necessary if DOE approved the test. His position was that the facility in effect belonged to DOE since "practically all of the equipment was bought with DOE funds." (Greathouse said who owned and was responsible for operations in the small facility appeared "ambiguous.") Dicks told the investigators that suggestions that a PCB test burn had been planned for a specific date were "inaccurate," uttered by people who had not been in on his "high-level" discussion. Dicks raised several issues with the investigators, saying that UT Attorney **Ron Leadbetter's** report was unwarranted and an outgrowth of a personality clash between him and Weaver, and complained that Weaver and UTSI had treated his firm more harshly than other companies controlled by UTSI staffers. He accused Weaver of interfering with the MHD program and damaging morale. It appears that UT chiefs were content to let the matter die quietly (though the "other shoe" was yet to fall). And there were rumbles as late as the fall of 1982 of an impending investigation by the U.S. Department of Energy, but no evidence of such ever surfaced.

Dr. T. Charles Helvey, the famous "owl man" who had taught cybernetics at UTSI from 1968 to 1973, was in the news, having developed a "Pax system" designed to help elderly and disabled persons care for themselves at home. The state Health Department was testing the system. Helvey, in his late seventies, was chairman of a cybernetic development firm, the Kappa Institute, and president of Citizens Emergency Systems, Inc., which helped in the development of the Pax system.

Dr. John S. Steinhoff, staff scientist at Grumman Aerospace Corporation, Bethpage, Long Island, New York, was appointed associate professor of Engineering Science on September 1. He held master's and Ph.D. degrees from the University of Chicago and had done post-doctoral studies as a research associate at McGill University. A graduate of Brooklyn Technical High School, Steinhoff had completed undergraduate work at Rennsselaer Polytechnic Institute, Troy, N.Y.

Henry W. (Hank) Hartsfield, a UTSI graduate, was being groomed for his first flight into space on a shuttle mission scheduled for the summer of 1982. He and **Ken Mattingly** formed the backup crew for the second launch of the shuttle Columbia in late 1981. **Dr. Ron Kohl,** one of the professors on Hartsfield's master's committee, remembered him as being very cooperative and well known for his high quality of work. Dr. Young remembered as "excellent" Hank's work on his thesis, "Noise in a Helium Neon Laser Used as a Quantam Amplifier."

Dr. Heribert Flosdorff, director of design and engineering for the Hamburg Commercial Aircraft Division of the Messerschmitt aircraft firm in Germany, presented the seventh annual Quick-Goethert lecture at UTSI on October 22. He said rapid development of a high technology commercial aircraft manufacturing industry in Europe was cutting into the United States' domination of the market. Flosdorff was instrumental in developing the European Airbus program—a joint project using funds and experts from Germany, France, Spain and Great Britain to develop commercial passenger and transport aircraft.

The UTSI dean was not present to introduce the Quick-Goethert lecturer. Weaver had been called to a special meeting of the UT trustees in Jackson, Tennessee. During the month that followed, rumors circulated that Dr. Weaver was "cleaning out his office" at the Space Institute, and the secret came out on Friday, November 20, 1981, at a press conference held by **Ed Boling** in Knoxville: Weaver was going back to the classroom.

On November 22, The Tennessean reported that Dr. Weaver was stepping down as UTSI dean to become "university professor" of engineering at Knoxville. President Boling named **Dr. B. H. Goethert** as interim dean and said UT had launched a nationwide search for a new full-time dean. The sixty-one-year-old Weaver was quoted as saying "I've been asking to be returned to teaching for the past year and a half, even though I've had a very delightful 12 years as a member of Ed Boling's team." **Robert S. Hutchinson,** UT's vice president for public service, was being assigned administrative duties for the Division of Continuing Education. Boling also had announced (on the previous Friday) that **Dr. Susan Wu** had been named director of UTSI's energy conversion division, which had been administered by Prados since July. He was to become UT's vice president for academic affairs and research.

Dr. Susan Wu

Bill Morgan, UT spokesman, said the changes came after a special study on "administrative reporting channels" and "internal relationships" at the Space Institute.

Weaver thought, according to The Tennessean, the problems of administering UTSI had been "solved in a classy, superior way." There were "a great deal of growing pains during such an enormous project at an academic institution," he said. "All of us learned a great deal, and one of those lessons was that some academic flexibility doesn't carry over very well in a big project."

Weaver was the first "university professor" chosen. This position was created the previous June for former UT presidents, chancellors or vice chancellors with a decade of service but short of retirement age. Weaver said these professorships were created "when it became clear that UT had some outstanding people, but a question arose as to what was going to happen when they had completed ten successful years in a pressure cooker and decided that ten years was enough."

On November 21, the Nashville Banner had quoted Boling as saying, "While I don't think we've done anything basically wrong, I'm not sure we were properly structured to deal with such a large amount of (research) money."

David Lyons, Nashville correspondent for the Knoxville News-Sentinel, in a story published November 18, had reported that the administrative shuffle was pending and quoted Dicks as saying much of the criticism against him was caused by a personality clash between Weaver and him and that the conflict would end if Weaver resumed full-time duties with UT in Knoxville.

The comptroller's report was made public a day or two after the administrative changes were announced, and perhaps because of this was robbed of some of its sting. Lyons reported that the audit cleared Dicks of "serious wrongdoing and instead criticizes UT administrators' handling of the situation."

John Dicks continued at UTSI as a full-time professor of physics until his death on September 11, 1990, after an apparent heart attack. A memorial service two months later paid tribute to him, and in February of the next year the state legislature adopted a resolution in his honor.

Charlie Weaver, who often had said the classroom was the best place to be, returned to teaching and conducting laboratories for undergraduates. In October, 1989, he was named interim chairman of the University of Alabama at Huntsville's Department of Electrical and Computer Engineering.

Perhaps to some degree, both men were victims of a program that grew too big, too fast.

Meanwhile, things were going forward at the Space Institute with a major expansion of the academic building and construction for a new maintenance and physical plant building both nearing completion. On December 2, the State Funding Board sold $146.3 million in general obligation bond anticipating notes, including close to two million dollars to cover the projects at UTSI, which were being entirely financed in this way. Sharondale Construction Company of Nashville was contractor for the addition to the academic building while Wiley Reed Construction Company of Woodbury was responsible for the $295,000 maintenance building. The State Funding Board on December 2 sold $146.3 million in general obligation bond anticipating notes, including close to two million dollars to cover the projects at UTSI, which were being entirely financed in this manner. Sharondale Construction Company of Nashville was contractor for the addition to the academic building while Wiley Reed Construction Company of Woodbury was responsible for the $295,000 maintenance building.

✦

Chapter Twenty

The Smoke Begins to Clear

JANUARY 1, 1982: Dr. B. H. Goethert was in the saddle again. In an interview with **John Talley,** Tullahoma News reporter, the interim dean discussed what he saw as a need for greater emphasis on academics. This would come about through obtaining more diversified contracts that would allow for an increase in the academic and research levels. It was clear that the phenomenal growth of the MHD program had troubled Goethert.

"There were thirty employees in the MHD project when I left (as dean in 1975)," he said, "and it soon went to two hundred and fifty employees. That is very impressive growth, but it was also **dangerous.** Our efforts to get the big earnings to get equipment funded for other parts of the Institute was somewhat overshadowed by MHD. When I left, there were about thirty-five professors . . . and we still have about that many, so that shows the academic area was a little neglected." The MHD project was more of an industrial research project than an academic one, he said, but he was optimistic that serious efforts to get secondary contracts would provide the needed balance.

Goethert said the Support Council, while not officially disbanded, had suffered from neglect, and he wanted to revitalize not only the Council but also the Industrial Advisory Group. He restated his interest in having a select number of **undergraduate** students at UTSI. He said such an undergraduate studies program had been approved for one year in the early 1970's but was cancelled after a year because of a shortage of engineering students at the UT Knoxville campus. Goethert thought undergrads could help set up experiments, cutting costs of paying faculty and graduate students for these tasks, while getting the practical experience they needed in their education. (This undergraduate program was one of Goethert's unrealized goals and an issue that **Morris L. Simon** would resurrect before the Support Council in 1990.) Goethert saw the main thrust in 1982 to be a "strengthening of our research facilities." He hoped that within a year a wind tunnel would be added as well as research facilities for testing helicopter designs. Goethert knew the importance of community support; he had not forgotten that the support from the community and state leaders had been vital in getting the Institute established, and he emphasized "the close working relationship UTSI has with the Air Force at AEDC."

"Really, the Institute came about at the urging of the (Arnold) center," he said, "and UTSI has been an asset to AEDC by having an academic facility so nearby." He also wanted no one to forget how **Governor Frank Clement** had gone to bat to help make the dream of an Institute come true.

Construction was essentially complete on the new $1.9 million (west wing) addition to the academic building, and late in January students and faculty began moving in. **Bob Kamm** said once professors were moved out of their temporary trailer facilities, the final stage of construction—hooking up sewer lines—would be completed. An "open house" for the addition was held on April 9. **Joyce Moore** was general chairman of the affair. Other chairmen were **Charlotte Campbell, Patsy Solomon, Sandy Shankle, Roger McCoy** and **Helen Mason.**

Early in March, Dr. Goethert said that preliminary results from a test burn in a new toxic waste incinerator at UTSI indicated the test was a success and emissions from the incinerator were "clean." Polychlorinated biphenyls (PCB)—a highly toxic liquid used as insulation in electric transformers—were burned during the test, which was conducted in UTSI's Energy Conversion Facility (the old MHD lab). Officials of the Environmental Protection Agency were on hand for the test. Pyrotech Systems

Inc. of Tullahoma developed the incinerator, which was believed to be the first portable incinerator of its type in the United States, and contracted with UTSI to do the testing. **Dr. William L. Holt,** president of the firm (and an employee of UTSI's Energy Conversion Division), said that while PCB was banned by Congress in 1978 (after being linked to the formation of cancer), it was estimated that twenty million gallons of the chemical might be stored in various places around the country. The system, mounted on four flat-bed trailers, would allow Pyrotech to travel to hazardous waste dumps to clean them up, Holt said. (Dr. Dicks had been involved in a similar project and had talked about possibly locating PCB-burning incinerators on barges, thus providing easy mobility.)

The test was completed on March 7, 1982. Battelle Combustion Laboratories, which monitored the test, took samples of emissions to ensure that none of the PCB was being released into the atmosphere. The wastes were burned at a high temperature so that their components were broken down into water vapor, oxygen, carbon dioxide and nitrogen compounds. UTSI personnel conducting the test were **William Millard** and **Edwin Moulder.** Millard was new at the Institute, but he had worked five years with Goethert at AEDC. Goethert wanted Millard to run the test; several faculty members had questioned whether the procedure, involving toxic materials, should be conducted on campus.

Millard said the test passed in flying colors, meeting ECP standards and getting the go-ahead for commercial use. Pyrotech had proposed building a fixed-base incinerator near Coalmont in nearby Grundy County for use as a regional PCB disposal site where more than one hundred people would eventually be employed. However, the proposal met heavy opposition from private citizens and county leaders, and the proposed two-million-dollar facility was never constructed.

When Pyrotech was slow about paying its bills, UTSI personnel chained the mobile unit. Pyrotech was bought by an Arkansas firm, Ensco, which paid the bills and moved the equipment to Whitehouse, Tennessee, one Friday. Millard said the equipment was stolen over the weekend but later recovered by Ensco, and the company later built some commercial units.

Holt, who had started working for the Space Institute on March 1, 1977, as a graduate research assistant, left on March 31, 1982. He finished his Ph.D. work at UTSI in the spring of 1981 with **Dr. Susan Wu** as his major professor. The title of his dissertation was: "High Temperature Combustion of Pulverized Coal." He had earned a master's degree in 1964 from the University of Florida.

Two other UTSI staffers—**Jim Martin** and **John H. Lanier**—also were involved with Pyrotech, and they both left with Holt, presumably to devote more time to the new firm. Lanier, who had hired on as a chemist in July, 1977, and was promoted to supervisor a year later, left with Holt at the end of March, 1982. He had been named manager of the Sciences Section, Chemistry & Environment, on December 9, 1979. Martin, who had headed the Design and Engineering program in its formative years, left a month after his partners—on April 30.

A major battle was shaping up in Washington to thwart **President Reagan's** efforts to dismantle the Department of Energy and in particular to save the MHD program. **Senator Jim Sasser** announced that the Reagan administration had decided to immediately scuttle the national research program. He said the Department of Energy had ordered officials to spend half of the $5.2 million 1982 appropriation to close down the program. Senate Majority Leader **Howard Baker** joined Sasser and Fourth District Congressman **Albert Gore Jr.** in fighting the action. Goethert told a reporter that most operations at the Institute would continue normally even if MHD funds were cut off. True, MHD accounted for one hundred and sixty jobs, and those people likely would be furloughed if the funding stopped, but this money was directly responsible for only three of the thirty-five professors. **Dr. Susan Wu** was pleading with congressional subcommittees to save the program. **Dr. John Prados,** UT vice president of academic affairs and research, said major MHD research at UTSI would be mothballed if less than five million dollars was appropriated. By late spring, it appeared that the MHD job had been saved for the time, and attention shifted to the next year's budget. It was a long, month-to-month battle that managed to keep the MHD program afloat, and it was a battle that would be repeated for most of the decade.

Problems of a different sort arose at 9:30 A.M. on February 19 when an explosion occurred outside the new Coal Fired Flow Facility. Dr. Wu said the blast was the result of a "lack of communication" between test workers. Caused by the overheating of some sheet-metal pipes that were part of a venturi scrubber, the explosion caused an estimated twenty thousand dollars in damage, but no one was hurt.

In March, Senator **Ernest Crouch,** chairman, told the Support Council that he expected a dean to be selected within a few months. He said a ten-member search committee would be interviewing between forty and fifty applicants. The following new members were added to the Support Council: **Donald R. Eastman Jr.** and **Robert O. Dietz** of Manchester; **James C. Murray** and **Dr. Jerry L. Kennedy** of Tullahoma; **Frey Drewry** of Winchester; **Repre-**

sentative **Martin Sir** of Fayetteville, and **Dr. Arthur Mason** and **Dr. Susan Wu** of the Institute faculty.

Two former UTSI students—Astronauts **Henry (Hank) Hartsfield** and **Donald (Pete) Peterson**—were tapped for future space shuttle flights. Hartsfield was training for a summer flight aboard the space shuttle Columbia. He and **Ken Mattingly** would fly Columbia's fourth mission. Peterson would be a crew member aboard the United States' second shuttle, Challenger, on its maiden voyage in early 1983. Peterson discussed the upcoming flight with Dr. Goethert during a visit to UTSI in February, 1982. Hartsfield would carry with him a UT flag and, in the fall, after it had traveled 2.9 million miles, he would bring the flag to UTSI.

The Institute got a $432,000 contract from the Army to participate in a field test program to evaluate anti-infrared smokes being developed for armored fighting vehicles operating in Western Europe. Six nations, all members of the North Atlantic Treaty Organization, would be conducting the tests in France during the summer of 1982 and in Norway during the next winter. **Dr. W. Michael Farmer,** associate professor of physics at UTSI, was project manager and principal investigator, under the direction of **Dr. Kenneth E. Harwell,** director of the Institute's gas diagnostics research division. Lundy Electronics Corporation of Glenhead, New York, also contracted with UTSI for development of a software graphics package for Lundy's line of computer graphics terminals. **Dr. Kenneth R. Kimble,** assistant professor of math and director of UTSI's Computer Center, headed this project. NASA gave UTSI's atmospheric science division, headed by **Dr. Walter Frost,** a certificate of recognition for its role in the space shuttle program and awarded another contract to perform the same type services for the upcoming flight. The group was responsible for monitoring upper atmosphere weather conditions during re-entry and landing of the shuttles. NASA also renewed a research contract calling for three UTSI professors—**Drs. Basil Antar, Robert L. Young and Frank Collins**—and graduate students to study the growth of crystals under weightless conditions in space. **Dr. Antar** was one of forty-one faculty members from colleges and universities to participate in a faculty fellowship program at NASA's Marshall Space Flight Center in Huntsville that spring. And one of **Ken Harwell's** aerospace engineering students, **Abdolhossein Nourinejad,** received an award in Atlanta for a paper that he presented at an AIAA regional student conference. **Dr. K. C. Reddy** was heading a two-year study of the mathematical and computer testing of engine compressors to find improved ways of using the testing at Arnold Center. Later in the year, Dr. Antar spent two months as guest professor at the Institute of Aerospace Engineering at the RW-Technical University of Aachen, West Germany. During the summer, he and **Dr. Alfonso Pujol Jr.** participated in NASA's summer fellowship program at Marshall Space Flight Center.

On April 30, 1982, **Dr. A. W. Quick** died in Germany. He was seventy-six. He and his friend Dr. Goethert in 1965 developed an agreement for academic cooperation between the Technical University of Aachen, Germany, and UTSI, and the two universities in 1974 established the Quick-Goethert lecture series in honor of the two scientists.

By mid-May, the list of candidates for UTSI dean had been narrowed to eight, and interviews were being held at UTSI and in Knoxville. Still in the running were **Dr. Harwell,** professor of mechanical and aerospace engineering and director of the Gas Diagnostics Research Division at UTSI; **Dr. Jain-Ming (Jimmy) Wu,** aerospace engineering professor and director of UTSI's gas dynamics division; **Dr. Susan Wu,** professor of aerospace engineering and ECP administrator; **Dr. George R. Inger** of Boulder, Colorado, professor and chairman of the department of aerospace engineering science at the University of Colorado; **Dr. John L. Junkins** of Ohio, visiting scientist in the flight dynamics laboratory at Wright-Patterson Air Force Base; **Dr. David Y. S. Lou** of Arlington, Texas, professor and chairman of the mechanical engineering department at the University of Texas at Arlington; **Dr. John D. G. Rather** of McLean, Virginia, founder of the Pan-Scion research group, and **Dr. Ronald O. Stearman,** professor of aerospace engineering and engineering mechanics and director of the center for aeronautical research at the University of Texas in Austin.

On June 22, Dr. Prados told the UTSI faculty and staff that **Dr. Kenneth Harwell** would become the new dean effective July 1. The forty-five-year-old Wetumpka, Alabama, native was not present for the announcement; he was attending a conference of the American Society of Engineering Educators in Texas. Harwell had been at the Space Institute for five years as a professor of mechanical and aerospace engineering. In 1981, he became director of the Gas Diagnostic Division and also was named an assistant to the former dean, Charlie Weaver. Harwell got his undergraduate degree from the University of Alabama in Tuscaloosa, his master's and doctorate in aeronautical engineering from the California Institute of Technology, and he had been a professor at Auburn University and a Ford Foundation Resident Scholar at Redstone Arsenal in Huntsville. The new dean and his wife, **Sharon** (also a Ph.D.), were honored at a reception on August 6 at the Industry-Student Center, hosted by the assistant dean, **Bob Kamm,** and his wife **Shirley,** and Associate Dean **Arthur Mason** and his wife **Helen.**

Dr. Kenneth E. Harwell

Work was under way on a boathouse, funded by **Dr. Jerry L. Kennedy,** a Tullahoma physician and member of the Support Council, and designed by two UTSI students, **Bryan Alexander** of Tullahoma and **Peter Hoffman** of Sewanee. The boathouse, located between the Industry-Student Center and the beach, featured an observation deck on its roof. Kennedy had designated that his gifts to the UT President's Club be used for the project. The boathouse and dock were dedicated in ceremonies on August 21 in conjunction with the annual summer picnic sponsored by the Student Government Association. Several hundred students, faculty and staff members attended. Soaring Club members showed off their newly acquired Schleicher Ask-13 sailplane. **John J. Sheridan,** UT director of special gifts, presented a plaque to Kennedy.

UTSI's new Soaring Club came about through efforts of **Peter Solies, Pat Lynch,** and **Marshall Hutto.** The constitution for the club was drawn up in February, 1982. The club's new sailplane took its maiden flight (out of Tullahoma) on December 8 of that year.

Linda Hall of Tullahoma began working for the Space Institute on June 16, 1982, as principal secretary for Upper E Wing. About three years later **Linda Williams** of Winchester Springs would succeed her in that position, and Linda **Hall** would go to the Computer Center, where as computer operation coordinator she would be involved during the center's impressive growth. In 1987, **Richard Hopwood** would come from Motlow College to UTSI's Computer Center as system manager, filling a vacancy created when **Marshall Hutto** left. **Edward J. Mirtes,** senior energy research technician with ECP, would divide his time as a computer trouble-shooter between ECP's Computer Services and UTSI's Computer Center, headed by **Dr. Kenneth Kimble.** Mirtes, a retired Air Force sergeant, had started work at UTSI in April, 1979. **Donald L. Huebschman,** who had been at the Institute since June 1, 1974, was manager of **ECP's** Computer Services.

Huebschman's father, **Dr. Eugene C. Huebschman,** had been professor of electrical engineering at the Institute since February 1, 1966. He would retire at the end of 1982. However, he had been on leave since becoming president of the Nathaniel Hawthorne College, Antrim, New Hampshire, on October 10, 1981. Born in Evanston, Indiana, on Halloween, 1919, Dr. Huebschman received his bachelor's degree in 1941 from Concordia University, his master's in 1946 from Purdue University, and his Ph.D. from the University of Texas. He taught at the University of California in 1960 and at Brevard Engineering College from 1961 to 1964.

Actually, UTSI had only had a computer center since 1980. The Institute's first access to major computers had come in its infancy through a cooperative effort with the Lockheed Company. **Bob Young** recalled that **Dr. Roy Smelt,** a vice president with Lockheed and later employed at Arnold Center's Engine Test Facility, "got us into the network after Lockheed got C-5A computers for all Lockheed's offices including London." For years, the Institute's computer program was run by committee, and it was located in the crowded C-102 conference room until the expansion program was completed in the early 1980s.

Kimble said that **Dr. Jim Maus,** as committee chairman, arranged to lease IBM 1130's early in 1970. **Dr. Goethert** was motivated to okay paying the rental cost because of the high travel costs resulting from a lot of UTSI personnel going to Knoxville to use the IBM's there. "People were standing in line to punch the cards," Kimble said. Then **Dr. Fletcher Donaldson** succeeded in getting a link to Knoxville "which allowed us access to the big machines there," Kimble said.

Soon after **Dr. Terry Feagin** joined the faculty in the fall of 1973, Dr. Goethert put him in charge of computers. In July, 1978, Feagin was awarded tenure and shortly afterwards was picked to fill in for **Dr.**

Asa O. Bishop Jr., assistant director of UT's Computer Center, who was taking a year's leave of absence. Dr. Kimble was put in charge of computers while Feagin was gone. Feagin returned briefly, but in October, 1980, UT chose him as head of the Computer Science Department in Knoxville. At this time, **Ken Kimble** became manager of the Computer Center, and he hired **Marshall Hutto** full time as manager of the computer system. Hutto later designed the first computer room in F Wing.

Kimble remembered that **Dr. Charlie Weaver** got cold feet about spending nearly a quarter of a million dollars to install the first Vax system, and Kimble coaxed **Bob Kamm** off his sick bed to persuade the dean. Kamm's support made the difference, and the Vax was delivered in February, 1980, and became the nucleus of UTSI's first Computer Center; however, it was all located in C-102. According to Kimble, a student—**Ron Kolbe**—was responsible for getting larger quarters for the Computer Center in F Wing when the building was expanded. Kolbe, a member of the Building Committee, "presented our case," Kimble said.

In 1985, ECP needed more computer capabilities, and Kimble suggested that UTSI get a second Vax, cluster the systems, and manage it for ECP.

Dr. Kimble remembered several Graduate Research Assistants who made contributions to the computer program, including **Ray Ricco, Linda Binkley, Sean Chi,** and **Mary Jones.** And **Rhonda Dawbarn** came to the center under a six-month CETA program but continued for a while as a computer operator and an assistant to Hutto before finishing her degree at Tennessee Tech and eventually working for TRW.

After **Harry Joseph Ferber II** got his master's in Computer Science at UTSI in December, 1988, he joined Kimble's staff as a systems programmer. (Later, he would marry **Mary-Frances Pomykal,** John **Dace's** sister and Registrar **Bettie Roberts'** assistant.)

In July, the industrial advisory group, dormant since 1975, met and reviewed its academic and research programs. **Dr. Alfred Ritter,** director of technology for Calspan at Arnold Center, was named to the group, which Goethert said was being "revitalized" to strengthen cooperation between industry and the Institute. Ritter had once headed the aerodynamic research department at Calspan's Advanced Technology Center in Buffalo, N.Y. He would leave Calspan and AEDC and move to Huntsville, but he would remain on the advisory board and later become its chairman.

Dr. Roy Schulz was program manager for a twenty-eight month, $342,000 NASA contract with UTSI for research related to design technology of turbo-jet engines. **Dr. Thomas Giel,** one of the principal investigators, and Schulz had prepared the proposal in nationwide competition for the contract. Three other UTSI professors, **Dr. John S. Steinhoff, Dr. Ahmad Vakili,** and **Dr. W. Michael Farmer** also were principal researchers.

The Institute promoted a former student, **Dr. Edward M. Kraft,** to associate professor, part-time, of mechanical and aerospace engineering. In the early 1960s, while studying at the University of Cincinnati, Kraft was a "co-op" student with ARO, Inc. at AEDC. In 1968, he joined ARO as an aerospace engineer in the large rocket facility and later moved to the wind tunnel as a project engineer. In 1968, Kraft was made a supervisor in the engine test facility and when the Air Force split the operating contract in 1981, he joined Calspan. He earned both his master's and Ph.D. in aerospace engineering at the Space Institute, and in 1977, he began teaching courses at UTSI in advanced aerodynamic theory and intermediate fluid mechanics. In 1982, Kraft was assistant manager of the aeromechanics branch and supervisor of the technology applications section in the propulsion wind tunnel facility.

In September, 1982, **William R. Carter,** chairman of the board of CFW Construction Company, Fayetteville, succeeded **Senator Ernest Crouch** as chairman of the Support Council. Crouch, who had headed the council for all of its seventeen years, was elected chairman emeritus for life. State Representative **Ed Murray** of Winchester was elected vice chairman, **Robert W. Jones** of McMinnville, treasurer, **Mrs. Charlotte Parish** of Tullahoma, secretary, **James C. Murray** of Tullahoma, director of publicity, and **Clinton Swafford** of Winchester, legal counsel. **Dr. Thurman Pedigo** of McMinnville, and **Dave King,** Manchester banker, were new members, along with UT's executive vice president, **Dr. Joe Johnson.** (Five years later, on March 26, 1987, the Support Council voted Ed Murray as lifetime vice chairman.)

Dr. Young was elected chairman of the engineering accreditation commission of the Accreditation Board for Engineering and Technology. UTSI enrolled three hundred and twenty-seven students for the fall quarter, an increase of forty-three over the previous fall enrollment. The U.S. Senate passed a bill in late September to continue the existing level of funding for the MHD program until mid-December, which Dr. Wu estimated would total about $450,000 a month.

Boyd Stubblefield joined the UTSI staff that fall as an Institute photographer, filling a vacancy left the previous December when **Phil Gatto** resigned. After retiring as an Air Force sergeant, Boyd had worked for Carrier Corporation, Micro-Craft and Koch manufacturing company. At this time, **Maurice Taylor** was photographer for ECP with a darkroom in lower D Wing, and Stubblefield used the smaller lab across the hall. Larger, upstairs quarters later were provided for Stubblefield.

At 4:30 P.M. on October 20, **Dr. J. J. Cornish III,** vice president of engineering and planning of the Lockheed Georgia Company, gave the eighth Quick-Goethert lecture at UTSI. His lecture dealt with vortex flow. A banquet was held that evening in the Industry-Student Center, and afterwards, Dr. Goethert was honored—it was his **seventy-fifth birthday.** At this affair, Mr. Carter announced that the B. H. Goethert Distinguished Professorship had been established and that more than twenty-five thousand dollars had been pledged toward the honorary professorship. The family of **Dr. Robert L. Young** also founded the B. H. Goethert Graduate Student Fellowship Award for an M.S. degree. **Dr. Hans-Chistoph Skudelny,** vice rector of the Technical University of Aachen, presented Goethert as an honorary citizen of the university. Numerous testimonials were given during the dinner. When the guest of honor was called upon to speak, he said, "I feel that my place is now here. I want to help establish close ties between this country and Germany." A symposium honoring him also was held at the Arnold Center Officers' Open Mess during the afternoon, sponsored by UTSI and the Tennessee section of the American Institute of Aeronautics and Astronautics.

The American Institute of Physics recognized a group of Space Institute researchers for their discovery of an explanation for the "Ranque-Hilsch" effect—a mysterious separation of swirling air into hot and cold streams when injected through tangential holes into a pipe. This phenomenon was discovered in 1933 by **George Ranque,** a French engineer. He found that the temperature of the air near the tube centerline became freezingly cold while the air near the tube wall became very hot, thus temperature separation had taken place without the aid of any external mechanical device. This effect was popularized in a later paper by **Rudolf Hilsch,** a German scientist, whose work was discovered after World War II by an Allied team investigating German technology.

Dr. Mitsuru Kurosaka, Jim Goodman, and two graduate students, **Joe Chu** and **R. D. Fizer** attributed the cause to a whistling sound present in the vortex flow of the device. Kurosaka said he and his fellows were led to the mechanism of sound through chance observations made in the swirling flow of aircraft engines. He said sound served to deform the distribution of swirl velocity in the radial direction through the mechanism of acoustic streaming.

Keith Anspach of Tullahoma, who had enrolled at UTSI in September, was selected as the first "Goethert Scholar" for the 1982–83 academic year, receiving the scholarship established in October by the family of Dr. Young. Keith was seeking a master's degree in computer science.

On December 31, 1982, **Miss Virginia Richardson,** who was the first person from the Tullahoma area on the staff when UT first started a graduate program at AEDC in 1956 and had been with the Space Institute from its inception, retired as the Institute's first registrar. The UTSI family turned out at a reception to honor the longest-serving staff member in continuous service. Dr. Harwell presented her with a gift. **Dr. Joel Bailey** from UT Knoxville conveyed greetings to Miss Richardson. He was director of the graduate program from its opening until Dr. Young took over in 1957. Tributes also came from **Dr. Goethert, Dr. Arthur Mason,** assistant dean, and **Dr. Young,** associate dean. **Miss Richardson,** a graduate of Peabody College in Nashville, was a member of a prominent Tullahoma family. Her grandfather, **Robert H. Richardson,** was a mayor of Tullahoma and founded the Richardson Grocery Company. Her father, **Warren W. Richardson,** was a Tullahoma merchant who was active in political and civic affairs. Miss Richardson had taught at South Jackson and East Lincoln schools in Tullahoma before starting her twenty-six-year career with UT and the Space Institute.

Harwell said she had "been of great assistance to every student who has registered here for the past twenty-six years. . . ." Those students included astronauts Hank Hartsfield and Donald Peterson and . . . Miss Richardson never forgot the thrill of a summer morning in 1972 when **Jules Bernard** introduced her to **Neil Alden Armstrong.** The man who had been first to step on the moon (on July 20, 1969) wanted to register for a class at UTSI. He had come to the right person.

✦

Chapter Twenty-One

A NEW PROGRAM TAKES OFF

IN JANUARY, 1983, **Dr. Merritt A. Williamson** of Nashville joined the Space Institute as director of what **Dr. Kenneth Harwell** called efforts to "establish one of the best **engineering management** programs in the country." As professor of industrial engineering, Williamson would administer at the Institute the statewide program that was offered by the college of engineering of UT, Knoxville. A former dean of engineering at the Pennsylvania State University, Williamson had established the engineering management master's degree program at Vanderbilt University in 1966 and served as director until 1981. He had seen the programs grow from fewer than ten to almost ninety within sixteen years, and he predicted continued growth as Tennessee became more industrialized. Harwell hoped eventually to have "a nationally recognized program which will become to technology management what the Sloan School of Management at MIT is to industrial management." Indeed, the EM program did become the fastest growing engineering program not only at UTSI but in the entire UT system. **Dr. Jerry Westbrook** was named associate director. At that time, Westbrook was professor of industrial engineering at UT, Knoxville, and director of the UT Nashville engineering graduate program. Prior to its merger with Tennessee State University, he had served as dean of engineering at UT Nashville for six years. Westbrook received his bachelor's degree in electrical engineering from Vanderbilt University, a master's in industrial engineering from UT Knoxville, and a Ph.D. in industrial engineering and operations research from Virginia Polytechnic Institute. Westbrook actually was involved in getting the program started before Williamson—whose reputation could only enhance the new program—was named director. Williamson had an apartment in the Industry-Student Center and normally spent a couple of days a week on campus.

Dr. Merritt Williamson

Dr. Walter Frost received the Losey Atmospheric Sciences Award at the twenty-first aerospace sciences meeting of the American Institute of Aeronautics and Astronautics in Reno. Named after **Captain Robert M. Losey,** meteorological officer (the first U.S. officer to die in World War II), the award was given to Dr. Frost for his leadership and contributions to environmental research in the fields of atmospheric turbulence, fluid flow, and aeronautical systems safety, and for his contributions "to the definition of space shuttle environment inputs." The Middle Tennessee Boy Scouts council also had given **Bob Young** its highest honor—the Silver Beaver award.

Bettie L. Roberts of Estill Springs, who had worked for two years as recorder in the registrar's office, was appointed registrar, succeeding **Virginia Richardson,** UTSI's first registrar, who had retired in December, 1982. Mrs. Roberts was a graduate of Thornton College in South Holland, Illinois, and a licensed Tennessee affiliate broker. Several months later, **Mary-Frances Dace Pomykal** of Manchester joined the office as recorder. **Ralph D. Cantrell** joined the Institute's staff as manager of the contract administrative services office. This office was part of the administrative services office headed by **Bob Kamm,** assistant dean. A graduate of UT, the Nashville native got his master's in contract management from the Florida Institute of Technology. Harwell thought that Cantrell's experience with defense contracts would enhance UTSI's plans to increase its activities in defense technology.

Mary Frances Pomykal, standing, later joined Bettie Roberts, seated, in the Registrar's office.

On March 17, the UTSI Support Council called upon the state legislature to establish a "high-technology corridor" in Middle Tennessee with the Space Institute as its hub. This action was a follow up to a meeting at UTSI the previous August when the state's High Technology Task Force had been given a look at the broad spectrum of specialized research facilities available at the Institute to attract and support new industry. Dr. Harwell emphasized the potential cooperation between industries and the Institute, and several UTSI professors including **Drs. Jimmy Wu, Michael Farmer** and **Maurice Wright** discussed specific possibilities.

The Council, which on September 14, 1982, had named **Ernest Crouch** as lifetime chairman emeritus, at the March 17 meeting named **Morris L. Simon** as lifetime vice chairman emeritus, and **Dr. Ewing Threet** as lifetime treasurer emeritus. An effort to raise funds to support the B. H. Goethert Professorship, which the Support Council had created the year before, had produced about fifty-five thousand dollars toward the goal of a hundred thousand dollars.

The really hot item at the meeting, though, was Simon's offering of a resolution requesting UT to recognize the Institute as a primary campus headed by a vice president or a chancellor. Simon first recounted the council's success in getting a line-item budget for UTSI, which he said was the first step toward more autonomy. Now it was time for another step.

Dr. John Prados, UT vice president for academic affairs and research, said flatly that it would never happen. While he praised the council for its support of UTSI and interest in the economic development of the state, he said he would have to oppose the proposal. UT was, he said, an "integral part" of the academics of UTSI and to separate them would hurt both institutions. Dr. Harwell, stressing that academic ties had to be maintained with UT Knoxville, said he supported Prados fully and could not speak for or against Simon's proposal. Dr. Goethert said the proposal represented a very significant change, and being a primary campus should be a goal. While he was not sure about the timing, Goethert was "all for taking these steps in the long run" and felt that it definitely should be a long-range goal. Simon said he recognized that his proposal would not be acceptable to UT at that time, but he thought it was time to start considering such a change. **G. Nelson Forrester** strongly supported Simon's position. **Ernest Crouch** said the proposal required more thought. He suggested that **Chairman William R. Carter** appoint a committee, which would report at the next meeting. Carter agreed to do this, **Ike Grizzell** withdrew his second, and Simon agreed to put his motion on hold. On April 13, Carter sent letters to the five members, informing them of their appointment as a committee to study Simon's proposal. **William Crabtree** of Fayetteville was chairman, and members included **Simon** and **Forrester** of Tullahoma, **Franklin Yates** of Shelbyville and **Clinton Swafford** of Winchester.

Simon frequently was point man for Goethert, and a document prepared by Goethert shortly after the March meeting suggests that Simon's motion on the primary campus issue had enjoyed Goethert's blessings. In response to a request from Simon, Goethert wrote some of his thoughts about the issue, including the following:

"We must not be timid in developing our long-range plan for the future. Only bold ideas can lead to major jumps. Small things must not remain small. **The UTSI can grow to a major primary campus if vigorously and wisely pursued."**

It is altogether possible that Senator Crouch's proposal for Carter to name a committee was scripted before the meeting as a ploy to keep the issue alive. (Simon always contended, too, that Harwell privately supported having a vice president in charge of the Institute even though it might jeopardize his own position as dean.)

Goethert's idea of a primary campus included establishing additional institutes on campus. As soon as the concept was accepted in principle, Goethert said the UTSI should start to expand its faculty, student body, staff and program "to serve as a nuclei for the proposed new institutes. . . ." He also saw the need for supplemental programs in the liberal arts, such as English, foreign languages, history, economics, and music.

Simon could rankle the mildest mannered of men, and it appears that he indeed rankled the usually easy going Prados that evening. At least, Simon **thought** he had and on April 15 wrote Prados that "Hopefully by this time you have lost some of the irritation you displayed" when the resolution was introduced. While he had introduced the resolution, "surely by now, it has occurred to you that I am not the sole initiator of the proposal, that other responsible members of the Support Council approved of the proposal. You recall that the resolution was quickly seconded by **Ike Grizzell,** who . . . was not aware that it was to be offered. After you declared your disapproval, **Nelson Forrester** offered some cogent arguments why we should proceed with an effort to persuade the administration that the time had arrived for broadening the base and scope of the Institute." It had been suggested before to **Ed Boling** and **Joe Johnson** who had met it with "equanimity." It was not done to embarrass Prados, but Prados's reaction was a "surprise" and gave Simon the impression that Prados had "bowed up to protect turf." Simon was aware that the council and UTSI needed help from "the Hill" in fulfilling dreams. Success in efforts for fruition of the dreams of a vastly expanded Institute could "only redound to the benefit of the UT system" and Simon hoped that "bureaucracy does not succeed in stifling visions which already have been greatly beneficial to UT, this area, the state and nation."

Prados wrote back on May 3, saying that his only surprise was "that you thought I showed irritation." He usually was accused of being too diplomatic and indirect, he said. Perhaps he was "getting meaner in my old age." He said everyone at UT wanted to do the best for the Institute, and he had tried to give Harwell maximum autonomy. But he did not believe that the Space Institute "would be well served by a separate campus status." Under the UT system's organization, this would mean that the Institute would have to have a separate faculty and academic degree-granting authority. It would cut the UTSI program off from the accredited academic program of the Knoxville campus and require a separate evaluation by an accrediting agency. The change would "seriously weaken the academic support for programs offered at the Space Institute" and could place a burden of additional administration expenses for the Institute. **Prados** was, of course, part of the UT administration, and his appraisal of the proposed change was in line with others on "the Hill," including **Joe Johnson,** who was not present for the March meeting. Prados closed the letter by expressing gratitude for Simon's "strong support" and assuring him that "I did not bow up to protect 'turf.' I have more jobs than I can get to. I did not seek the administrative responsibility for UTSI."

Simon's impression that he had irritated Prados was somewhat supported by a brief letter dated March 25, 1983, that Prados wrote to Goethert in which he said, "I understand fully that you were not responsible for **Morris Simon's** unfortunate remarks at the Support Council meeting last week. I only hope that some of the Council members will open their pocketbooks as wide as their mouths!" His last comment missed the mark if it was directed toward Simon, who was one of the consistent contributors to UTSI and its projects. On November 14, 1983, Goethert wrote to Simon, thanking him for his "gracious gift" in the "greater than $1,000" category (actually, it was five thousand dollars) to the Honorary Professorship set up the previous year by the Support Council. Goethert closed warmly: "Dear Morris, you know how much I appreciate your personal efforts in supporting the Institute as shown again by your recent contribution. . . . In the early years prior to the establishment of the Institute, and afterwards as the Vice

Chairman of the UTSI Support Council, you initiated the frequent drives to update the Institute; **most recently with the push towards having the UTSI become a prime campus.** You have been certainly a spark in spearheading advancements. I sincerely hope that also in future years the Institute will not miss your so decisive leadership.''

The committee met a couple of times, but it was more than a year before the council received a report. Carter called a special meeting on December 8, 1983, with the report on the agenda. However, Simon's wife, **Lillian Tobe Simon,** was ill, and he missed the meeting as did Forrester and Swafford, so there was no report. Minutes of the meeting note that ''a large number (of those attending) urged that any recommendation to change UTSI's status be studied and approached very carefully so that UTSI's present excellent working relationships with UT at both the Knoxville and system level are not jeopardized.''

On August 21, **1984,** Simon introduced his resolution, ''respectfully'' requesting the UT president and board of trustees to ''begin taking the necessary steps to make UTSI a primary campus and that they appoint a vice president or chancellor as the leader of UTSI.'' The council, by voice vote, adopted the resolution. One member, **State Senator Douglas Henry,** invited **Dr. Joseph E. Johnson,** executive vice president, and **Dr. Ed Boling,** UT president, to react to the resolution. On September 11, 1984, Johnson wrote to Henry, listing ''our thoughts'' on the resolution. They were well pleased with the Institute as it was operating and saw no reason to make significant changes. The title of Dean was an ''honorable and appropriate'' one, and they saw no lack of respect, responsibility or authority in the title. The title of Vice President was used for administrators with statewide functions and responsibilities, but this was not the case at the Space Institute, and the title of Chancellor was used for chief administrators of the four primary campuses. It would not be appropriate to have a chancellor in charge of the Space Institute. Johnson listed five reasons why the administration thought the Institute should not be a primary, separate campus: Its mission was limited to graduate education and research related to specific academic areas, and it was not a broadbased institute; its graduate student enrollment was small; the academic programs and faculty were closely related to units at UTK, and accreditation of Institute programs was based on that relationship; the State of Tennessee ''certainly has enough, if not too many, primary campuses,'' and there was a fear that designation of the Institute as a primary campus would ''intensify some current efforts to extend this institution into **undergraduate education,** which is outside its mission and which might conflict with Motlow State, Middle Tennessee State, Tennessee Tech, and other institutions.

Finally, Johnson wrote, the Space Institute was ''alive, expanding, energetic, and sound. It has a separate budget for operating expenses; it receives separate consideration for capital outlay, and it has access to the statewide University administration through a Vice President.'' The changes proposed in the resolution appeared to be ''related to symbolism rather than substance.'' They never wanted to do anything to retard or harm the Institute, but ''we do not see that the proposed changes would be of value to the Space Institute.''

There the issue rested until March 12, **1985,** when the Support Council unanimously reapproved it. **Dr. Prados** said that as an ex-officio member, he could not vote. If he could, he said, his vote would be negative. Carter sent a copy of the resolution to President Boling, noting that the council understood it was a long-range project, but ''to accomplish it, we feel it is imperative to commence the thought process.'' On March 18, 1985, Boling replied that he understood ''to some extent'' the feelings of the council, but ''it is not easy to convert a higher education entity, such as the Space Institute, into a primary campus.'' UTSI had certain critical academic ties to Knoxville . . . it had very valuable students in important fields, but the enrollment level was ''small by design.'' Boling knew of no strong reason for making major changes—the title of dean was significant and meaningful in academic administration—and he, Prados and Johnson would be happy to discuss it further.

The ''thought process'' had commenced, but it would be a couple of years yet before the administration (Dr. Johnson in particular) would warm up to the proposal.

Hodges Construction Company of Ashland City had contracted in the spring of 1983 to replace framework and glass in the UTSI's lobby at a cost of $129,256. Several panes had developed cracks, caused in part by deterioration of the metal framework.

A former UTSI student, **Colonel Donald Peterson,** was ''walking on air'' in April as one of the astronauts who took a ''spacewalk'' from the shuttle Challenger—the first time in nine years that Americans had left the pressurized confines of their spaceship. This was the first trip into space for the native of Winona, Mississippi, who had studied at UTSI in 1969–70.

On April 21, **President Reagan** asked Congress to use twelve million of the remaining fifteen million dollar appropriation for the 1983 fiscal year to **shut**

down all MHD research. Three million dollars would be set aside, the president proposed, in case the project should be revived. It was an annual challenge that would come from the White House. The year before, a similar move was defeated on the floor of the House. Congressman **Jim Cooper** of Shelbyville said Regan was being "very short-sighted" and "penny-wise but pound-foolish." **Albert Gore Jr.** said such a shutdown would deal energy research a "critical setback." By a vote of 265 to 121, the House on May 26 rejected Reagan's proposal, and shortly afterwards, approved thirty million dollars for nationwide MHD research.

Cooper, who had taken office as a congressman in January, was complaining about the "outrageous" waste of the U.S. Synfuels Corporation that had been created during the Iranian hostage crisis in 1980 to search for alternatives to imported oil. Cooper, son of **Prentice Cooper,** a former Tennessee governor known for his tight-fisted control over public spending, said taxpayers would likely not see much return on the fifteen billion dollars that had been appropriated to Synfuels. The MHD facility at UTSI, he said, was doing its work for about the cost of just providing office space for Synfuels. Yet, the MHD project was "targeted for elimination by the administration before I and other members of Congress stepped in on its behalf," Cooper said.

A few months later, the Department of Energy approved a three-year extension on UTSI's MHD contract with a $21.5 million ceiling, contingent on funding by Congress. **Norm Johanson,** program manager, was expecting about six million dollars to be approved for the fiscal year starting October 1, 1983. A few days later, DOE approved a $6.9 million grant for the project.

The Institute's Soaring Club held a picnic and gave Dr. Goethert a plaque of appreciation. Members at that time included, among others, **Peter Solies, Peter Liver, Ed Amaulis, Pay Lynch,** and **Al Lowery.**

About thirty universities sent representatives to UTSI in May to discuss ways of converting university research into profitable private commercial ventures. Bob Young was chairman of the one-day seminar.

Eight foreign countries had participants in a flight testing course directed by **Ralph Kimberlin** in June. **Peter Solies** and **Lisa Hughes** were UTSI graduate assistants involved in the course.

Fan-Ming Yu and **K. Ramachandran,** engineering graduate students at UTSI, were among one hundred and forty students and eight professors inducted into Phi Kappa Pi National Honor Society that summer. **Dr. William Snyder,** a former UTSI professor and head of UT's Department of Engineering Science and Mechanics since 1970, was named dean of UT's College of Engineering effective July 17, 1983.

General James V. Hartinger, commander of the Air Force Space Command and commander-in-chief of the North American Aerospace Defense Command, visited UTSI in July for a briefing on the Institute's academic and research programs. With him were **Brigadier General Kenneth R. Johnson,** AEDC commander, and retired Air Force **Major General Frank T. McCoy** of Nashville.

After fourteen years at UTSI, **Dr. Fletcher Donaldson** retired as professor of computer science in the summer of 1983. With a doctorate in applied mathematics from the University of Texas, Donaldson had worked with Lockheed Missiles & Space Company, the Ramo-Woolridge Company, General Electric and General Dynamics, Convair division, before joining the Institute's faculty. In retirement, Donaldson would continue his fight to expand the use of the MUMPS computer language and operating system in hospitals and clinics.

In 1980, as a guest at the Fourth Annual Symposium on Computer Applications in Medical Care in Washington, D.C., **Donaldson** had persuaded the Veterans' Administration to accumulate work in MUMPS done by various veteran facilities into what became the VA Distributed Hospital Computer Program (DHCP). In 1988, DHCP was being used in one hundred and sixty-nine VA hospitals, one hundred and sixty Indian Health Service hospitals and clinics and one hundred and sixty Navy Occupational Health Clinics. However, Donaldson's sweetest victory would come in 1988 when MUMPS was selected by the U.S. Department of Defense (DOD) to be used in six hundred military hospitals and clinics throughout the world. This would come after Donaldson pushed for the use of MUMPS by DOD as a budget-saving device and for better coordination of health care facilities between the department and the Veterans' Administration. Donaldson, serving on the Office of Technical Assessment Committee of Congress to evaluate DHCP, would make his pitch at the sixteenth annual MUG (MUMPS User's Group) meeting in Atlanta in June, 1988. He also would become involved in implementing a DHCP system on a computer that could be used as a teaching device for VA trainees. **Donaldson** remembered that his dream "came true" in March, 1988, with an announcement that Digital Equipment Corporation (DEC) had been awarded a hundred million-dollar contract to continue and expand the Veterans' Administration's DHCP.

"This followed one week the announcement of DEC's four hundred million-dollar hardware and service contract award as subcontractor on DOD's contract to Science Applications International Corporation for their Composite Health Care System in the amount of one billion dollars over a ten-year period for DOD's seven hundred and fifty facilities **world-wide,"** Donaldson said. These would be MUMPS systems.

As part of his long campaign, **Donaldson** had presented numerous papers on the use of MUMPS, including one given in 1971 to the Second International Symposium of the World Organization of General Systems and Cybernetics at Blackburn, England. This invitation had come after an article extracted from Donaldson's first proposal (on MUMPS) to HEW in 1965 was published in a new journal, The International Journal of Bio-Medical Computing, edited by **Dr. J. Rose.** The next year, Donaldson was invited to present a paper at Oxford, and in 1973 he presented two papers to the Brazilian government at the First Brazilian Congress on Cybernetics and General Systems.

Unquestionably, Donaldson always felt that he had made a greater contribution than Dicks with his scene-stealing MHD program. The same idea was expressed in a letter to the History Committee in 1989 by **Dr. Bryan E. Burgess,** operations manager with BellSouth Services in Birmingham, who graduated from UTSI in March, 1976, with a Ph.D. in Engineering Science.

"Although I believe we made only small contributions as a result of working on some DOT and FAA contracts managed by **Ray Sleeper,"** he wrote, "I have seen changes in how the airports are designed to handle air and ground traffic. I have also seen changes in the automation of medical systems similar to those that Dr. Donaldson was telling all who listened." He also remembered that **Dr. Eugene Huebschman** had told him that "the micro processors would soon be in almost everything of our daily contact. **What I have not witnessed is social use of the research in MHD,"** he wrote, expressing disappointment that the bulk of UTSI's research money went for MHD "while transportation, electronics, and medical automation received so little attention and financial support, yet the latter three have produced so much more for society." Burgess, at the time Director of Career Education at Motlow, had started taking random classes at UTSI before being attracted by the new offering of a Ph.D. in Engineering Science. Donaldson, Sleeper, and Huebschman also had encouraged him to go for his doctorate.

Dr. Susan Wu went to Moscow in September for the eighth international conference on MHD electrical power generation as a guest of the U.S.S.R. Academy of Sciences. Wu was invited also to visit the academy's Institute of High Temperatures; this is the institute that had been "loaned" a two million dollar super-conducting magnet by the United States. Wu said Russia and the U.S. were negotiating for the return of the magnet. (The last joint experiment between the two countries was in 1979 just before Russia's invasion of Afghanistan, which had prompted **President Carter** to halt the exchange activities and place an embargo on the shipment of high-technology equipment to the Russians. This had drawn a response from **John Dicks** that close to twenty million dollars, invested in the cooperative program, had gone down the drain.)

In late August, Dr. Goethert suffered a stroke, three days before **Callie Taylor,** whom he had hired as his secretary, reported to work on August 31. In typical fashion, Goethert was soon back in the harness, but he confided that the illness had taken its toll on his endurance. While he was mending, his new secretary worked in the personnel office, but she would be there, just outside Goethert's office, on a cold morning a few years later when "Doc" truly started down the long hill.

UTSI welcomed seven new faculty members for the fall quarter, 1983. They were **Dr. S. I. Hariharan,** assistant professor of mathematics and computer science, **Dr. Boris Kupershmidt,** associate professor of mathematics, **Dr. Remi Engels,** associate professor of engineering science, **Dr. Moonis Ali,** professor of computer science, **Dr. George Garrison,** professor of mechanical and industrial engineering, **Dr. J. W. L. Lewis,** professor of physics, and **Dr. Bruce Bomar,** assistant professor of electrical engineering. Bomar's wife, **Kathy,** also came on board as a part-time instructor of computer science.

Engels, a native of Belgium, came to UTSI from Martin Marietta Aerospace in Denver, where he was a staff engineer on the MX missile program. He received his doctoral degree in engineering science and mechanics from Virginia Polytechnic Institute (VPI) in Blacksburg and his master's and bachelor's degrees from the University of Ghent, Belgium. **Hariharan** came to the Institute from NASA Langley Research Center where he was a staff scientist at the Institute for Computer Applications in Sciences and Engineering. He received his bachelor's degree from the University of Sri Lanka and a master's from the University of Stanford in England. He earned a second master's degree and his doctorate in mathematics at Carnegie-Mellon University. (His wife, **Jeanette,** was working toward a master's degree in electrical engineering at UTSI.)

Garrison joined the **UT** faculty in 1981, and he moved to the Institute from UT-Chattanooga. He had received three degrees, including his doctorate, in mechanical engineering from North Carolina State University and a master of Business Administration from Vanderbilt University. Garrison was one of the faculty members who helped in the early efforts to establish the Engineering Management program at

UTSI, and in the following years, he would teach EM courses. He also later developed a popular short course on Project Management. As a lead engineer, he had held several positions with ARO Inc., and later with Sverdrup Technology, Inc., at Arnold Center from 1966 until joining the UT faculty in 1981. This experience included his serving as Project Manager for the MHD Projects Section.

Kupershmidt, who had emigrated to the United States from Russia in 1978, received his master's degree in theoretical mechanics at Moscow University and taught and did research in Russia until emigrating. He earned a doctorate in mathematics in one year at the Massachusetts Institute of Technology and prior to joining UTSI taught at the University of Michigan. **Bruce Bomar,** a native of the Raus community near Tullahoma and Shelbyville, received his bachelor's in electrical engineering from Tennessee Tech, a master's in electrical engineering from UTSI, and a doctorate in EE from UT, Knoxville. He had taught at UTK and worked as a research engineer with Sverdrup Technology Inc. and with Calspan Field Services Inc. His wife, a Knoxville native, had earned both bachelor's (electrical engineering) and master's (computer science) degrees from UTK and had taught in the UTK computer science department.

Ali, a specialist in the development of robots and artificial intelligence, received his undergraduate, master's and doctoral degrees from Aligarh University in India. He was on the faculty of Old Dominion University in Norfolk, Virginia, prior to joining UTSI's faculty. He came to the United States in 1974 as a Fulbright, post-doctoral fellow at the University of Texas. He later joined the Texas faculty and conducted research. He spent three years teaching computer science and developing automatic translation systems at Mosul University in Iraq before joining Old Dominion in 1980.

Lewis would be in the Applied Physics Research Group—along with **Jim Few, Dr. Dennis Keefer,** and **Dr. Carroll Peters**—that would succeed in getting a state Center of Excellence established. He had been senior scientist and supervisor, Diagnostics Measurements Development Section, with Calspan, Inc., from January, 1980, through October, 1982. He obtained his undergraduate, master's and Ph.D. in physics from the University of Mississippi. Lewis had worked as a physicist at the U.S. Army Missile Command at Huntsville, and as senior scientist and supervisor, Flow Diagnostic Section, ARO, Inc., at AEDC. He spent one year as a post-doctoral fellow at Queen's University, Belfast, North Ireland.

UTSI and the Highland Rim Section of the American Society of Mechanical Engineers (ASME) hosted a regional student leadership conference in September that attracted about sixty students from universities in five southern states.

Ralph Kimberlin, UTSI's test pilot, was testing a "jetwing" aircraft that the Ball Corporation of Muncie, Indiana, had donated to UTSI. This Ball-Bartoe research jet—one of a kind—could take off within 1,250 feet and land in even less space. It had a liftoff speed of fifty-eight miles per hour and a stalling speed of seventy m.p.h. The jet engine was in the front of the fuselage, and the jet exhaust that drove the craft forward came out of "hoods" on top of the wings, which gave it much greater lift than other types of jets. **Dr. Frank E. Ashenbrenner,** a Ball vice president, visited the Space Institute.

At 4:30 P.M. October 11, **Dr. Hermann L. Jordan,** head of West Germany's counterpart of NASA, delivered the ninth annual Quick-Goethert lecture entitled "On New Research and Development Facilities for Aeronautics in Germany." Jordan was chairman of the board of the German Research Establishment for Aeronautics and Astronautics in Cologne, central organization for six major aerospace centers located throughout Germany. It was a position formerly held by **Dr. A. W. Quick.**

Jules Bernard, who for years had been in charge of public relations at UTSI while also managing the short course program, became a full-time manager of public relations in October, 1983, and **Sandy Shankle,** a principal secretary, became manager of the short course office.

By 1989, Sandy Shankle, left, was assisted by Betty Bright, standing, and Judy Rudder.

More than a hundred participants, including airline pilots, Federal Aviation Administration regulators, NASA and university researchers, airport operators and others, attended the seventh annual workshop on meteorological and environmental inputs to aviation systems at UTSI in late October. It was under the direction of **Dr. Walter Frost,** director of UTSI's atmospheric sciences division, and **Dennis W. Camp,** research engineer with the aerospace science division of NASA's Marshall Space Flight Center.

A milestone was reached in UTSI's MHD research late in 1983 with the announcement by **Dr. Susan Wu** that the team had completed one hundred hours of a "low-mass flow test run" in the year-old coal-fired facility. The man who had led the MHD research for years, **Dr. John B. Dicks,** was presented a plaque for his twenty years service as a physics professor as well as his leadership in the MHD work. Dr. Harwell, in a special ceremony, also gave service awards to more than one hundred other faculty and staff members. Dr. Goethert was cited for his nineteen years service and Dr. Young for twenty-five (including his time with the UT graduate program at AEDC preceding the founding of UTSI.) Three men and their wives each received fifteen-year awards. These were **Dr. Jimmy** and **Dr. Susan Wu, Dr. Arthur** and **Helen Mason,** and **Freeman** and **Ruth Binkley.** (All six actually had exceeded fifteen years at the Institute at this time.)

An eventful year lay ahead as the Space Institute prepared to celebrate its twentieth year. An early birthday present would be the funding of a state Center of Excellence that would open up entire new horizons.

Four Space Institute faculty members presented a three-day workshop at the National Chin Kung University in Tainan, Taiwan, Republic of China. They were greeted by two UTSI graduates, **Dr. C. H. Chen** and **Dr. Y. L. Chou. Dr. Kenneth Harwell,** dean, was accompanied by **Dr. Jimmy Wu** and **Dr. Mitsuru Kurosaka,** professors, and **Dr. Edward M. Kraft,** part-time associate professor and branch manager at AEDC's propulsion wind tunnel for Calspan Field Services Inc.

Dr. Walter Frost spoke before a U.S. House subcommittee about aviation weather research programs designed to minimize aircraft accidents. He told how a UTSI workshop had supplemented and validated a recent aviation weather system plan put forth by the Federal Aviation Agency. Frost was director of an annual, UTSI-hosted, workshop dealing with meteorological and environmental effects on aviation systems. That spring, Frost directed a workshop on aviation weather in Australia.

The Institute, in cooperation with the U.S. Naval Test Pilot School, had presented a course on helicopter performance flight testing at the Naval Air Test Center, Patuxent River, Maryland. Students from the United States, Canada, Italy and The Netherlands attended the course. **Ralph Kimberlin** from UTSI was one of the teachers. **Colonel Henry Sherborne** of UTSI assisted with the course.

On January 26, 1984, **John T. (Jack) Shea,** public affairs director for twenty-nine years at Arnold Center, suffered a fatal heart attack. The sixty-two-year-old Shea had been in the center of the fight to get UTSI established, working behind the scenes to support Dr. Goethert's struggle. He had moved from St. Louis to Tullahoma in 1953, joining ARO, Inc. as special assistant to the managing director and later was named director of public affairs. He established and was the first editor of **"High Mach,"** the base newspaper at Arnold Center. When the Air Force split its contract in 1981, Shea joined Sverdrup Technology Inc. as director of public affairs, and retired in May, 1982. He had served in World War II as a rifle platoon leader with the 115th Regiment of the Army's 29th Division under Major General **Norman D. (Dutch) Cota,** who led the assault on Omaha Beach at Normandy, France. After the war, Shea was chief of the Air Force press desk in Washington and then joined McDonnell Douglas Aircraft Corporation in St. Louis as vice president for public relations.

Robert J. (Bob) DuBray joined **Susan Wu's** organization on January 3, 1984, as manager of ECP's Project Support Services. This was a slightly new position, combining the Project Control Office, which had been managed for six years by **Bill Boss,** and ECP's business office. Several persons had succeeded **Howard Parker** after he resigned in 1978 as head of the business office. His successors included **Marshall L. Seymour, Al Camp,** and, briefly, **Ralph Cantrell,** who left in late 1983. Seymour took a master's degree in engineering administration (through special arrangements) at UTSI in June, 1979.

Albert J. Bart, with an M.B.A. in Industrial Management from Boston University, had joined the ECP team in 1980 as deputy program manager with responsibilities in budgeting, project control and special projects. **Clara Ferguson,** secretary, and **Charlotte Henley,** senior editorial assistant, were attached to Bart's office. In the years ahead, DuBray's business staff would include **Diane Chellstorp,** administrative assistant, **Lillie H. Stricklin,** administrator of properties, **Pamela B. Selman,** accountant, and **Dixie Bass Crawford,** accounting clerk. **Sherrie Perry,** principal secretary, **Madge Gibson and Mary**

Danek, administrative secretaries, also were part of ECP's Lower B Wing staff.

DuBray, whose last assignment before retiring as an Air Force major, was at Arnold Center, had been closely associated with ECP for three years. During this time as staff consultant for Decision Planning Corporation, DuBray provided on-site management and administrative support to the Department of Energy's project office at UTSI as the Coal Fired Flow Facility (CFFF) was completed and put into operation. In fact, DuBray shared an office at the Facility with DOE's **Jay O. Hunze,** who would soon be leaving the campus. After retiring from the Air Force, DuBray had stayed on at AEDC as a civilian for a year, and then for two years he was project director of the Community Education-Work Council at Motlow College.

In mid-February, 1984, the UT trustees assigned second highest priority to a proposal for a million-dollar Center for Laser Applications at the Space Institute. The trustees approved this and eight other proposals as candidates for ten million dollars in state funds to establish centers of excellence. **Governor Lamar Alexander** had included the concept of centers of excellence in his Better Schools Program, being considered by the legislature. **Ken Harwell** was confident of getting the center funded. The proposal, written by **Drs. Dennis Keefer, Carroll Peters, James W. L. Lewis** and **Mr. Jimmy D. Few,** called for more than four million dollars in state funds over five years, with the state contribution decreasing as the center grew more self supportive.

In April, the Tennessee Higher Education Commission (THEC) approved a $1.02 grant for UTSI's Center of Excellence—less than the $1.9 operating budget that had been projected, but **Keefer,** chairman of the council that would operate the center, was happy to get the grant. He said a chunk of the grant would go for new equipment.

Dr. Susan Wu received the Arnold Award in May in recognition of her "outstanding contributions" toward advancing the state of the art of the aerodynamic and astronautical sciences. Presenting the award was **Dr. John McLucas,** national chairman of the American Institute of Aeronautics and Astronautics (AIAA).

Dr. Michael Farmer a physics professor at UTSI since 1978, took a leave of absence in June to join the Science and Technology Corporation in Las Cruces, New Mexico. Farmer entered UTSI in 1967 as a student and received his master's degree in 1968 and a Ph.D. in 1973.

Dr. C. C. Chao, director of a new Institute of Aeronautics and Astronautics of the National Cheng Kung University in Taiwan, visited UTSI to discuss future cooperation between the two schools. He said the new institution in Taiwan was patterned after UTSI.

Colonel Henry Sherborne, who as a licensed professional engineer had helped in the construction of AEDC, retired on June 30 after ten years as purchasing agent/manager at UTSI, and **Wilma Kane** was promoted to the position of manager to succeed him. A graduate of Falls Business College in Nashville and a native of Tullahoma, **Kane** had joined **Paul West** in the Institute's business office in 1970. **Harold Little** of Bradyville—an aviation enthusiast and builder of airplanes—would continue in the office as assistant purchasing agent. He had joined UTSI's staff in 1975 as instrumentation supervisor of ECD's Electronics Shop.

Wilma Kane and Harold Little later were joined by Pam Ledford, standing.

Sherborne started on August 29, 1974, as director of flight operations. Shortly afterwards, he also became ECD's first purchasing agent, but part of his time was still devoted to flight operations. It was shortly before this that UTSI's first purchasing office was established with **Wilma Kane** in charge. Prior to this, UTSI's purchasing had been handled in Knoxville. **Lillie H. Stricklin,** who began working at UTSI in 1976, worked half-time for **Colonel Sherborne** and half-time for **Howard Parker,** who had been named ECD business manager in 1975. In August, 1977, **Henry Markant,** who had succeeded **Ken Tempelmeyer** as ECD program manager, made **Harold Little** assistant purchasing agent to Sherborne. (**George Charboneau** had been put in charge of instrumentation at the shop.) With the signing of contracts for the Coal Fired Flow Facility, Markant brought **Sherborne, Little, Stricklin** and **Ron**

Sanders together as a purchasing unit. Along with four secretaries, they worked together in a trailer in 1978 while **Wilma Kane,** as purchasing agent for the Institute, worked out of A-Wing. Eventually, the two purchasing units were combined into one for UTSI with **Sherborne** as manager.

In the summer of 1979, **Lillie Stricklin** began working for **Richard H. Smith** who was ECD's manager of planning and information. A native of Cincinnati, Ohio, Smith completed work on his master's degree in Engineering Administration at UTSI in 1974. He left on August 24, 1979, and was succeeded by **Al Camp. Mrs. Stricklin** later became administrator of properties for the energy conversion program.

Hank Hartsfield was chosen that summer to command a crew of six on the maiden flight of America's new space shuttle Discovery. The shuttle was launched August 30 from Kennedy Space Center and landed at Edwards Air Force Base on September 5. The Birmingham native received his master's degree at UTSI in 1971 and became a NASA astronaut in 1969. He had piloted the fourth and final test flight of the space shuttle Columbia, and was a member of the astronaut support crew of Apollo 16 and Sky Lab flights 2, 3 and 4. **Hartsfield** came back in October to help celebrate UTSI's birthday.

Frank Leslie Crosswy, Michele Macaraeg and A. K. Sinha completed work on their doctorates at UTSI during the spring quarter. Of the seven students receiving master's degrees, **Thomas Clay, Margaret Ann Crawford** (wife of UTSI professor **Dr. Lloyd Crawford),** and **Jack Welch** were the first to receive degrees in the new industrial engineering program. **James Dale Hull, Thomas Chris Layne, Jere Joseph Matty,** and **Montgomery (Monty) Smith** also earned their master's degrees. Monty later would earn his Ph.D. at UTSI and join the Institute's faculty.

The Support Council, in August, honored several of its members for their help in raising money for the "B. H. Goethert Professorship." **William R. Carter,** council chairman, said more than eighty thousand dollars toward the one hundred thousand-dollar goal had been raised. **Morris L. Simon, Dr. Robert L. Young, Mrs. Charlotte Parish, Dr. Ewing J. Threet, and Henry Boyd** were cited at the dinner meeting.

At 4:30 P.M. on Thursday, October 11, **Dr. Hans W. Liepmann,** Theodore von Karman Professor of Aeronautics and director of the Graduate Aeronautical Laboratories at the California Institute of Technology, delivered the tenth annual Quick-Goethert lecture. His topic was "Fluid Dynamics Research With and on Helium."

Tours of various laboratories preceded the lecture, which kicked off a four-day celebration of the Space Institute's twentieth anniversary. A series of technical lectures on research activities at the Institute were presented the next day, and on Saturday, a group of staff, students and faculty attended a brunch at the home of UT Chancellor Reese in Knoxville, watched the Tennessee-Florida football game, then returned home for a dinner at the Industry-Student Center and a ball in the lobby of the main building.

An open house and tour of laboratories was held on Sunday just before the anniversary ceremony. **Dr. Edward Boling,** UT president, welcomed guests, and **Dr. Goethert** presented a history of the Institute. **Colonel Philip J. Conran,** commander of Arnold Center, introduced **Hank Hartsfield,** the keynote speaker. A reception was held afterward for the astronaut. Others on the program were **Chancellor Reese, Dr. William Snyder,** new dean of UT's College of Engineering, Support Council **Chairman William R. Carter, General B. A. Schriever, State Senator Ernest Crouch, Morris L. Simon, Dr. Threet, R. M. Williams,** past president of ARO, Inc., **General Lee V. Gossick, Dr. Jack D. Whitfield,** president of Sverdrup Technology, **King D. Bird,** general manager of Calspan, and **J. G. Nuckels,** general manager of Pan Am World Services, Inc.

Hartsfield told his audience that the world was "on the threshold of an industrial revolution in space." He said the weightless environment of space was ideal for some types of scientific research. He predicted that a hormone isolated on shuttle flights would prove to be a major development.

Sadly, on the second day of the celebration, **Freeman Doyle Binkley,** who had been a part of the Institute's life since 1965, died after a short illness. He was sixty-three. Before joining the UTSI staff, he had been employed in the City of Nashville's engineering department and later worked for ARO at Arnold Center. He had served in the U.S. Army during World War II.

King Bird announced a tentative agreement between Calspan Field Services Inc. and UT to establish a joint aerospace research center at UTSI. UT officials and those from Calspan's parent company, Arvin Industries Inc., Columbus, Indiana, had agreed to proceed with the non-profit research center, Bird said. The Center for Aerospace Research (CAR) later was approved and in turn became the operator of the Center for Advanced Space Propulsion in the UTSI Research Park.

NASA and the Federal Aviation Administration (FAA) awarded UTSI a hundred-thousand-dollar contract that fall to develop a "flight expert system," designed to detect problems in an aircraft, warn the crew, and suggest remedies. **Dr. Moonis Ali,** UTSI professor of computer science, was principal investigator. Professor **Ralph Kimberlin** was to provide aviation data.

In November, **Dean Harwell** presented a plaque to **Dr. Arthur A. Mason** commemorating his twenty years at UTSI. Cited for completing fifteen years of service were **Dr. Kenneth R. Kimble, Mary M. Lo, and Patsy Solomon. Ann Wolfe, Keith Walker, Arnold Parks, Marjorie Joseph, Delia Burchfield, Dr. Jack Hansen, and Dr. Frank Collins** were honored as ten-year veterans. On board for five years had been **Dr. Charles T. N. (Ted) Paludan, Dr. Alfonso Pujol, Phyllis Adams, William H. Boss, Betty Bright, Judy Cooper, John Dace, Birdie Farris, Fred Galanga, Madge Gibson, Dr. Thomas Giel, Charlotte Henley, Betty Huskey, Marshall Hutto, Norman Johanson, Jackson Frazier, John Lineberry, Baw-Lin Liu, Odie Mann, Joseph Marler, Charles Martin, Edward Mirtes, Benjamin Morris, Roger Morris, James Roy, James Stephens, Maurice Taylor, Sandra Weinberg, and Cliff Wurst.**

Another Babcock & Wilcox Company old-timer, **Richard J. Dohrmann,** had joined the Institute's staff on April 1 as ECP technical coordinator. Dohrmann had been with Babcock & Wilcox since 1957 and brought a wealth of experience in technical supervision, project management, engineering and research and development in fossil energy conversion. In his new position, Dohrmann was responsible for identifying technological strengths and weaknesses of ECP and UTSI as well as developing and implementing plans for technology diversification. He received his bachelor's degree in mechanical engineering and his master's in engineering mechanics at the State University of Iowa. In 1990, he would be elected a Fellow by the American Society of Mechanical Engineers (ASME).

Late in 1984, the ASME named **Gary D. Smith** as a 1985 Congressional Fellow. Smith, manager of the facility laboratory for UTSI's Energy Conversion Research and Development Programs, would spend the next year in Washington, D.C., on the staff of **Senator Robert C. Byrd,** and on his return he would become Dr. Harwell's executive assistant. He had served as an officer of ASME's Highland Rim Section, and he also was a member of the American Institute of Aeronautics and Astronautics and of Sigma XI, national research society.

Susan Wu, back from a trip to China, was amazed at cultural changes since her visit there in 1979, when the situation had been ''very rigid,'' with a severe lack of merchandise and housing. In 1984, she had found the City of Beijing to be a ''crane city'' with high-rise apartment construction and road-widening flourishing and merchandise ''everywhere.'' She attributed the changes to the introduction of capitalism. She and her family moved to Taiwan in 1947, ten years before she came to the U.S. as a student.

Hopes were soaring in December when the UT trustees approved a proposal for a $1.4 million computer programming center specializing in engineering. This Center for Computational Mechanics was one of nine proposed centers of excellence at state universities that the trustees approved on December 7. **Dr. Harwell** was exuberant about the center's potential in helping engineers solve complex problems by drafting computer programs. ''With a computer, an engineer would be able to find structural flaws in a design before the structure is designed,'' the dean said.

The project had cleared its first big hurdle. The second lay just ahead, on January 24 of the new year.

✦

Chapter Twenty-Two

'LIKE A RED-HEADED STEP-BABY'

AN ICE STORM hit the mid-state area in early February, 1985, leaving more than fifteen hundred Coffee Countians without electrical power and holding some families in Franklin and Grundy counties prisoners for most of a week.

The Tennessee Higher Education Commission in January had burst the Space Institute's bubble, rejecting its proposal for the computer processing center. In recommending the center the previous month, the UT trustees had ranked the UTSI project seventh among the nine proposed centers. The proposal called for using $760,000 in state funds with $615,000 in matching money from UTSI during the 1985–86 fiscal year. **Dr. Kenneth Harwell** said the commission had run out of money. The dean said funding would be sought from other sources, but the financing never came.

At UTSI, **Joe Hane** and **Jo Ann Myers,** both of Estill Springs, and **David Aycock** of McMinnville were elected to the Employees Relations Committee. **Bob Kamm** was chairman, and **Jim Stephens** served as secretary. Aycock, at ECP, had been with UTSI five years and represented the technical services group. Hanes, with seven years experience with UTSI, represented skilled craft workers, and Myers, who had been at UTSI five years, represented the secretary and clerical group. She also was the Space Institute's representative on the university-wide employees relations board in Knoxville.

Tennessee state representatives voted 95 to 0 to abolish the University of Tennessee Energy Institute (TEI), which had the function of providing support and marketing alternative energy sources such as UTSI's MHD. **Representatives Lane Curlee** and **Ed Murray** said TEI's authority to issue five million dollars in bonds to match federal grants was hurting the state's bond rating. This institute had done nothing for the past three years, Murray said, adding that he was a member of the board, which had only met once during that time. Curlee quoted UT Vice President **Joe Johnson** as saying the energy institute was not necessary. Dean Harwell was hoping to keep the energy institute alive in the Senate "so it will be there when we need it." **Dr. Susan Wu** had hoped it would be kept for commercially marketing MHD at some future date. State **Senator Jerry Cooper** said he would study the matter carefully. In April, he helped get his fellow senators to take no action on abolishing the energy institute, which effectively nullified the previous action by the House. Cooper said this allowed proponents two years to prepare and present a case for continuance of the energy institute when the measure came up for another review.

Dr. William H. Heiser, a professor of mechanical engineering and aerospace engineering, who was on special assignment from UTSI as a distinguished visiting professor at the U.S. Air Force Academy, was chosen in February to become vice president and director in June of the Aerojet Propulsion Research Institute in Sacramento, California. Heiser had been affiliated with UTSI since 1970, becoming a full-time professor in 1983. **Dr. Susan Wu** was chosen to serve as an assistant editor of the Journal of Propulsion and Power, a new AIAA publication. **Dr. John B. Dicks** was presented the Ralph James Award for his "outstanding contribution to the petroleum division of the American Society of Mechanical Engineers" in ceremonies at Dallas, Texas. Specifically, Dicks was honored for his work as chairman of the division's synthetic fuels committee.

In March, 1985, the National Advisory **Group**—formed in 1965—voted to change its name to "UTSI National Advisory **Board.**" Members also voted to formalize the board's operations and unanimously elected **Dr. Jack Whitfield,** president of Sverdrup Technology Inc., as temporary chairman. He was to appoint committees to establish a charter, bylaws and policies. Dr. Harwell, acknowledging the

Advisory Group's contributions to UTSI's early growth, believed that "at this stage in our development, more significant contributions and inputs from industry and government agencies will enable us to meet the challenges of today's rapidly changing technology."

James Earl (Jim) Whiteford retired as senior energy research technician and supervisor in the MHD laboratory. He had joined UTSI's staff in October, 1972, and had been instrumental, as lead technician, in developing a joint study with the Soviet Union.

Faculty and staff donated two books in honor of **Freeman D. Binkley's** twenty years of service to UTSI. His widow, **Ruth,** and Librarian **Helen Mason** accepted the books from **Pat Pelton** and **Kathy Hice,** members with **Pam Turner** of a committee that had collected funds for the books: "Tennessee—A People and Their Land" by Lamar Alexander, and "Tennessee" by Edward Schell.

The Intel Corporation gave UTSI six System 310 advanced microcomputers, valued at more than $60,000. **Drs. Roger Crawford, Moonis Ali, Frank Collins, Remi Engels, Kenneth Kimble,** and **Mr. James Hornkohl** had plans for using the machines. **Dr. Al Pujol** had helped coordinate the transaction. (In April, **Dr. Emil Sarpa,** corporate manager for academic relations at Intel, visited UTSI.) Later in the summer Intel gave another twelve computers, which made a total of thirty-one computers that the firm had given to UTSI through the years.

The Support Council on March 12 adopted a resolution asking the University of Tennessee to begin the necessary steps to grant UTSI campus status and that either a vice president or chancellor be named. The proposal came from a committee headed by **Morris L. Simon** and including as members **Donald R. Eastman Jr., Dr. Ewing J. Threet, Ike Grizzell, William Crabtree, Franklin Yates** and **Bob Kamm,** secretary. The proposal received a cool reception in Knoxville. In a letter, **Dr. Joseph E. Johnson,** executive vice president of UT, wrote that the proposal appeared to be related to "symbolism rather than substance," adding that "we do not see that the proposed change would be of value to the Space Institute," and declaring that the "state has enough if not too many primary campuses." He also defended the title of dean, saying that it did not lack respect, responsibility or authority. But the issue was far from dead.

The annual struggle to keep MHD afloat surfaced in March when The Tullahoma News reported that MHD funds would be cut April 1—taking one hundred and forty jobs down the drain—unless Congress intervened. **Susan Wu** and **Ken Harwell** had been meeting with congressional delegates in Washington, trying to stave off the annual assault on the program. The battle see-sawed back and forth throughout the year, with Congress keeping the program afloat through the fiscal year, and in December, Congress approved a six-million-dollar appropriation for UTSI's MHD program for the 1985–86 fiscal year, and **President Ronald Reagan** signed it into law the following month.

UTSI and MEI Systems Inc. of Appleton, Wisconsin, had proposed to conduct a joint nine million dollar project for the Department of Energy for reducing sulfur dioxide emissions. If approved, **Dr. Atul Sheth,** principal investigator, said the MHD facility at UTSI would be modified and used in the project. By using this existing facility, he said, the work could be accomplished for nine million as opposed to twenty-two million dollars otherwise. This project was never approved. Sheth had joined the Space Institute in the fall of 1984 after having worked for Exxon in Houston since 1980. Like Dr. Young, he had received his Ph.D. from Northwestern University and for almost eight years worked for the Argonne National Lab near Chicago.

Governor Alexander and **Congressman Jim Cooper** announced in March, 1985, the appointment of a task force to study development and promotion of high-tech industries in southern Middle Tennessee. Members were **Cooper, Dean Harwell, Lee V. Gossick,** former AEDC commander and deputy general manager and vice president of Sverdrup's AEDC Group, **Dr. Jack Whitfield,** president of Sverdrup Technology Inc., **Dr. Alfred Ritter,** director of technology at Calspan Field Services Inc., AEDC Division, and **John Parish,** president of Lannom Manufacturing Company, Tullahoma, and former state commissioner of economic and community development. The next month, they attended a semi-annual meeting of the board of directors of the Tennessee Technology Foundation along with about one hundred and fifty state and area leaders at UTSI. Alexander stressed the need for "home-grown jobs" and cited the need to develop an awareness of the area's resources and to concentrate on improvement of education. Cooper said federal cutbacks should "intensify our efforts to pull ourselves up by our own bootstraps." Alexander and the legislature had succeeded in chartering the Tennessee Technology Foundation three years earlier as a non-profit, Knoxville-based, entity to develop "high-tech corridors" between Nashville and Tullahoma and in the Knoxville to Oak Ridge area. Up until 1985, most attention had focused in East Tennessee, so the activity in Middle Tennessee was well received. In June, **Thomas E. Bailey** of Fayetteville, a project manager with the Tennessee Valley Authority's office of natural resources and economic development, was appointed to the staff of the Southern Middle Tennessee High Technology Task Force. He was to administer Task Force problems related to new home-grown high-technology and entrepreneurial companies as well as programs geared toward the expansion of existing industry and attracting new industry. UTSI provided Bailey an office in Upper A Wing. Later, after Dr. Harwell moved into this space, Bailey would move to Upper B Wing, but this would be after he had left TVA and become an employee of the University. The Tennessee Technology Foundation and the Tennessee Department of Economic and Community Development also had pledged to support the task force with their respective professional staffs and other resources.

The Space Institute's efforts to get a $450,000 advanced technology lab ran into problems with THEC and with the state office of finances and administration, and in March, Governor Alexander recommended shelving the proposal. Dr. Harwell said this delay could curtail research at the Institute. Officials at the new laser center had initially invested heavily in equipment, and they were running out of space at the old lab building. Two months later, **Senator Jerry Cooper** and **Representatives Ed Murray** and **Lane Curlee** recommended budget amendments to provide money for the lab. In June, Curlee was confident that money would be in the 1985–86 budget for this first of six labs scheduled to be built by 1990, but it would be a couple of years before the folks at the crowded laser center would actually get relief.

Governor Alexander had recommended a $2.681 million budget for the Space Institute's next fiscal year, including $155,000 in five-percent salary raises, $47,000 to replace equipment, and a thousand dollars in new research funds. This was a six and a half percent increase over the 1984–85 fiscal year, but it contained no funds for new academic or research programs requested by UTSI. The Institute was asking for $3.096 million, to include $100,000 to fund space systems research involving the "star wars" defense program. Harwell wanted to develop the space systems program with two new faculty members and new equipment, which the dean said would help attract federal and industrial contracts and grants. It would include such projects as classified studies into the effectiveness of high-energy lasers in destroying targets, a commercial development of space projects, and studies on the motion of large structures in space. Cooper, Curlee and Murray said they would introduce budget amendments totaling $415,000 to fuel the space systems program, an engineering management program, and a computer applications project. Dr. Harwell also was looking toward forming a consortium of universities to submit a proposal to NASA for a center for commercial development of space, but it would be two years before this dream would materialize.

UTSI was seeking $130,550 for two full-time members for the Engineering Management program, plus additional video equipment. **Boyd Stubblefield,** UTSI photographer since 1982, and **Dr. Jerry Westbrook** had started the video-taping, nights, in the short course room, and **David Brandon** was employed on August 22, 1983, as a media technician, working with Stubblefield on the video program. A studio later was set up in Lower E Wing. Brandon transferred to UT Knoxville in April, 1986, and Stubblefield handled the chores until September 22, 1986, when **Scott Ulm** of Fayetteville was put in charge of the fast-growing video program. Ulm left in August, 1989, and **Christine Skelton** of Manchester succeeded him. Also in 1989, **Laura Payne** of Shelbyville joined the UTSI staff, dividing her time as assistant to Stubblefield and Skelton.

Three part-time faculty members were teaching the Engineering Management courses at this time, but the program was rapidly growing. Enrollment in the program had doubled from eighty-two in the winter of 1983 to one hundred and sixty-four by the next winter. Funding for the course initially had been obtained through student fees and cost recoveries such as contracts and grants. Another $184,400 was being sought to add faculty and equipment for computer research into such areas as artificial intelligence, computational fluid dynamics and computer graphics.

Jean Broadway, a former secretary for **General Alexander Haig,** began working on March 18, 1985, as an administrative assistant in the Center for Laser Applications. Three and a half years later, she would move to A Wing as administrative assistant to the Institute's associate dean for academic affairs, **Dr. Richard M. Roberds,** and serve with him when he became acting dean. Mrs. Broadway had worked for Haig in Belgium in 1975 and 1976 while he was supreme commander of allied forces in Europe.

Also in the spring of 1985, **Dr. Jack Hansen,** assistant professor of civil engineering, conducted a short course on data base project management, and UTSI entered exhibits in the annual session held by the Huntsville Association of Technical Societies (HATS) at the Von Braun Center. In May, **Dr. Harwell, Dr. Jimmy Wu,** director of gas dynamics research, and **Dr. John Prados,** UT's vice president of academic affairs, went to China to sign an agreement for the exchange of students and teachers with three universities of the People's Republic of China: Beijing Institute of Aeronautics and Astronautics, Nanjing Aeronautical Institute, and Northwestern Polytechnic University in Xi'an.

Dr. Frank Collins, who had joined UTSI's faculty in 1974, and **Dr. Marion Laster,** director of technology for the Air Force at AEDC, were named as Associate Fellows by the American Institute of Aeronautics and Astronautics (AIAA) that spring, and a UTSI graduate, **James C. Uselton** of Tullahoma, was elected as vice president of the Sverdrup Corporation at St. Louis and named corporate principal for advanced technology. He already was serving as vice president (since 1981) and general manager of the technology group with Sverdrup in Tullahoma. Uselton earned his master's degree in mechanical engineering from UTSI in 1966, and he, too, was an AIAA Associate Fellow.

Dr. Goethert's popular short course on aeropropulsion drew a record attendance in Phoenix in May with about seventy enrolled and another fifty-six applicants turned away. Goethert had enrolled fifty-two in the course in November, but fifty had been turned away, so he revamped it and offered it again in the spring. **Jules** and **Dorothy Bernard** were with the Goetherts in Phoenix.

Speaking to the Tullahoma Chamber of Commerce in May, 1985, **Congressman Jim Cooper** expressed concern that UTSI and Arnold Center were being taken for granted outside Coffee County. The Space Institute, he said, was one of the most picturesque sites to be found; it was a "magnificent showcase." But, Cooper said, the UT administration in Knoxville had not shown the proper attitude toward UTSI, and they often seemed to forget about the Institute. Cooper had "let the UT people know about it," he said.

UT Trustees in June approved **creation of the UT-Calspan Center for Aerospace Research (CAR).** The "not for profit" center was expected to conduct basic aerospace research in aviation safety, fluid mechanics and flight mechanics and was to be funded initially by Calspan, headquartered in Buffalo, N.Y., but it was expected to become self-supporting through research contracts and grants. **Dr. Arthur Mason,** associate dean, predicted it would grow into "an important research center in Middle Tennessee" and be "an economic benefit for the area." It was modeled upon a joint project in Buffalo between Calspan and the University of New York Research Foundation, and in due time, the new center would "hit a homerun" with NASA.

UTSI got a $97,000 NASA grant to continue research on turbine engines that had started under a $500,000 contract in 1982. **Dr. Carroll Peters** and **Dr. Roger Crawford** were principal investigators.

Dave Stephens reported in The Tullahoma News in July that the Space Institute had organized a consortium of universities and the military to submit a proposal to NASA for a multi-million-dollar center for the commercial development of space. The object was to stimulate high-technology research and to evaluate manufacturing in zero-gravity conditions of space. Motlow College's Industrial Business Institute was to provide locations for seminars. **Dr. Frank Collins,** acting director and chief organizer of the consortium, was expecting an answer by August. Educational partners in the consortium were UT Knoxville, Tennessee Tech, Tennessee State University, Memphis State University, Auburn University, and Alabama A&M University. This group's proposal was one of twenty-one submitted to NASA including a "for profit" proposal from a consortium organized by **Dr. Ted Paludan,** professor of geography and deputy director of the remote sensing division at UTSI. **Dr. Firouz Shahrokhi,** UTSI professor and director of the remote sensing division, would be director of this proposed "Center for Commercial Development of Earth and Ocean Observation From Outer Space" proposed by the Sun Belt and Coastal Consortium. This group proposed relatively little research, but would sell data about the earth such as mineral deposits. Neither of these proposals would be approved, but within a couple of years, **CAR** would be awarded a five-year NASA grant to establish the **Center for Advanced Space Propulsion (CASP)** at the UTSI Research Park. Because of CAR's contractual affiliation with the University of Tennessee and its proximity to the Space Institute, CASP would utilize talents, resources and facilities at UTSI.

Another honor came to **Dr. Susan Wu** that summer when she received the Achievement Award from the Society of Women Engineers—that group's highest honor—while she was a delegate to the national convention and student conference in Minneapolis, Minnesota. **Dr. Jack Whitfield** was elected executive vice president of Sverdrup Corporation, St. Louis, and would continue as president of Sverdrup Technology at Tullahoma.

Dr. Merritt A. Williamson of Nashville, the man who had worked with the College of Engineering at UT Knoxville to start an Engineering Management program at UTSI, died on July 19, 1985, at the age of sixty-nine. Williamson, also a professor of Industrial Engineering at the Space Institute, administered at UTSI the statewide program offered by the College of Engineering at UT-Knoxville. He had established the master's degree program at Vanderbilt University in 1966 and was its director for years before retiring from Vanderbilt in 1982. Dr. Williamson and **Dr. Jerry Westbrook** had pioneered in videotaping courses at UTSI for use elsewhere. Westbrook would assume the director's role in the program.

Dr. Remi C. Engels of Tullahoma, assistant professor of Engineering Science and Mechanics, received a $120,000 contract from the Air Force in July to conduct research into the dynamic response of a structure to applied force, the results of which could be applied to a large variety of space structures. Engels had been teaching courses in the field of structural dynamics and vibration.

Dr. Goethert, on July 8, 1985, gave a lecture on "Graduate Education in Propulsion" at the twenty-first Joint Propulsion Conference of the American Institute of Aeronautics and Astronautics, the American Society of Automotive Engineers, the American Society of Mechanical Engineers, and the American Society of Engineering Education in Monterey, California.

Callie Taylor of Decherd, senior secretary for the offices of **Colonel Ed Masters** and Dr. Goethert, passed the "certified professional secretary" examination on her first attempt. A few years later, seven other UTSI secretaries would join the CPS ranks. They were **Dixie Crawford, Vicki Hill, Paula Reed, Becky Stines, Carole Thomas, Madge Gibson, and Betsy Hastings.**

Schneider Services International named **James W. Garrett** of Pittsburgh as support services project general manager at AEDC in August and **John F. Stubbs** with the OAO Corporation in New Orleans as deputy general manager. Stubbs was no stranger to Tullahoma. He was a former base squadron commander at AEDC and after retiring from the Air Force, he had served on UTSI's staff as assistant for management after **Colonel Brantner** resigned.

The Institute bought a cryogenic high-vacuum space simulator chamber from the University of California at Berkeley, carrying a one million dollar "replacement cost." In August, 1985, it was believed that UTSI was the only university in the United States to own such a chamber. It measured thirty-five feet long and nine feet in diameter. It would eventually be put into first-class condition by personnel in the Center for Laser Applications.

A supplemental appropriation bill with some money for MHD research passed both houses on Congress in late July. It contained $8.35 million for various research by the Department of Energy (DOE) Fossil Fuel, but exactly how much would be allocated to UTSI's MHD program was not immediately known. However, Tennessee's congressmen and women (**Rep. Marilyn Lloyd** of Chattanooga was one of the Space Institute's strongest advocates) managed to keep the program alive.

UTSI got a two-year, $597,000 contract as a subcontractor with Pittsburgh Energy Technology Center on a DOE project dealing with sulfur dioxide and nitrogen oxide control. **Thomas E. Dowdy,** a chemical engineer in the Energy Conversion Program, was the principal engineer involved in developing and selling his concepts for the program. Object of the work was to confirm the feasibility of a system to control the emission of sulfur dioxide and nitrate oxide from a fossil fuel plant. Dowdy said the contract would require development, fabrication and erection of a new test facility at the Space Institute's ECP complex.

Students elected **Gordon Alan Lowrey** as president of the Student Government Association that summer. Lowrey, a native of Florida, was working on his doctorate under the guidance of **Drs. Robert Young** and **Roy Schulz.**

One of **Ken Harwell's** friends from their days at Cal Tech in the early 1960's—**Dr. Hans Gronig,** professor and director of the shock wave lab at the Technical University of Aachen, in the summer of 1985 lectured at UTSI. **John Rampy** (one of the first full-time students at UTSI), technical director for testing at AEDC, was promoted to the senior executive service, the highest rank that a Civil Service employee could attain. **Dr. B. H. Goethert,** at a meeting of National Aerospace Scientists in Bonn, West Germany in October, received the **Ludwig Prandtl** ring, established in honor of the renowned researcher and teacher at the University of Gottingen in West Germany. Goethert was cited for his contributions to aviation and aerospace.

A thirty-one month, $285,000 grant was awarded to UTSI as a subcontractor for Morgantown Energy Technology Center, which had a contract with the Department of Energy. **Dr. Thomas Giel** was program manager and **J. U. Son,** an employee with ECP, was principal investigator.

At 4:30 P.M., October 17, **Dr. Philipp Hartl,** director of the Institute for Navigation of the University of Stuttgart in West Germany, delivered the eleventh Quick-Goethert Lecture, which he titled: "Very Precise Navigation by Means of Satellites." At a banquet following the lecture, **Dr. Jimmy Wu** was selected as the first "B. H. Goethert Professor." Dr. Harwell presented the award, which had been created by the Support Council with pledges for endowing the professorship with more than a hundred thousand dollars in salary supplements. Dr. Wu, "very humbly" accepting the award, was warmly congratulated by Goethert.

A host of officials from Saturn Corporation, which was planning to build a five-million-dollar automobile plant at Spring Hill, visited UTSI on a fact-finding tour that fall, saying they wanted to establish a "working relationship" with UTSI.

Dr. Ed Boling, UT president, announced that UTSI's share of gifts raised by the UT Development Council during 1984–85 totaled $134,318. The UTSI Research Park, late in the year, got a new tenant: Tennessee Space Labs Inc., a firm specializing in testing technology for satellite-based, infrared sensors and other space system structures. **Dr. John G. Pipes** and **Willis Clark,** both employees at AEDC, were co-owners.

Dr. Remi Engels and **Dr. Raj Kaul** directed a short course in November dealing with the design and fabrication of aircraft and space structures, using composite materials. Twenty-six flight test engineers also attended a course on introduction to flight test instrumentation. **Dr. Neil Loeffler, Ralph Kimberlin,** associate professor of aviation systems (who was promoted to the rank of colonel in the Air Force Reserves in late 1985), and **Robert Jones,** instrumentation specialist, directed the course.

Dr. Engels, a native of Ghent, Belgium, and **Dr. Ahmad Vakili** became new American citizens in a naturalization ceremony held in Chattanooga in December, 1985. Engels had come to the United States as a graduate student at Virginia Polytechnic Institute, Blacksburg, Virginia, in 1974. He received his doctorate at VPI and was an assistant professor before joining Martin Marietta Aerospace Company at Denver in 1979 as a staff engineer. On Labor Day, 1983, he joined the Space Institute as an associate professor, Engineering Science and Mechanics. Engels' wife, **Jennifer Wendell,** later worked at the Institute as publications specialist. Vakili, a native of Iran, was a senior research engineer at UTSI, but he joined the faculty the next year as an associate professor of Aerospace Engineering and Mechanical Engineering. He received his doctorate at UTSI in 1978 and then joined the Gas Dynamics Research Division the next year, working with his former professor, **Dr. Jimmy Wu.** Vakili had been a student at the Technical University of Aachen in 1973 when he met Dr. Goethert, who convinced him that he should enroll at UTSI. When he completed his master's work at the Space Institute, Vakili was planning to enter Massachusetts Institute of Technology (MIT) where he had been offered a scholarship. However, Dr. Wu convinced him that he should stay at UTSI for his doctoral studies.

Dr. Seung-Chul Lee joined the UTSI faculty in December, 1985, as assistant professor of computer science. He had received his master's and doctoral degrees from the University of Florida after completing his undergraduate work at the Seoul National University in Korea.

As 1985 was coasting to its end, several UTSI employees were recognized for their years with the Institute, including four who had been with UTSI for twenty years: **Dr. Walter Frost, Drs. Jimmy and Susan Wu,** and **Dewey Vincent.** Finishing out their fifteenth year of service were **Wilma Kane, Joyce Moore, Dr. Trevor Moulden,** and **James Goodman.** Recognized for a decade of service were **Diane Chellstorp, Patricia Pelton, Harold Little, Luther Gilchrist, Ramon Cintron, James Knight, Lloyd Lester, Edwin Moulden, Sue Tate** and **Douglas Wurst.** Cited for completing five years were **Al Bart, Emmett Cox, Eddie Holt, Patrick Lynch, Elizabeth Meyers, William Millard, Jo Ann Myers, Dr. Harold Schmidt, Billy Jones, Stephen Chapman, Bruce Hill, Susan Murphy, Jayaraman Muthuswamy, Kenneth Coffelt, William Hardison, Ginger Housley, Paula Reed, Bettie Roberts, Pamela Selman, Betty Wilkinson, David Aycock, Billy George, William Lorance, Alton Webb,** and **Jo Ann Winkleman.**

While employment fluctuated at the Space Institute, especially with the ups and downs of the MHD program, UTSI apparently enjoyed a better-than-average stability with its staff. Starting in 1985 and continuing into 1987, UTSI spotlighted an "Employee of the Month," and most of those selected had been with the Institute for from five to ten years, and most would still be on hand for the Silver Anniversary in 1989. This list of honorees not only reflects longevity but also diversity of jobs.

For instance, **Frances Bauer,** who in 1972 had become the first woman hired as a **machinist** by the Allis-Chalmers company in Little Rock, had started working as an assistant in UTSI's printing office in 1980, and later was in charge of that office. Others and the year they joined the Institute included **Robert Darden,** energy research technician, 1980; **Ginger Housley,** grants and contracts clerk, 1980; **Judy Rudder,** senior secretary in the personnel office (later on Sandy Shankle's staff), 1983; **Bobby Terry,** energy research technicians supervisor in the facilities lab of ECP, 1980; **William (Eddie) Holt,** energy research technician, 1979; **Mrs. Patricia (Pat) Pelton,** switchboard operator, 1975; **David Henderson,** energy research technology supervisor, 1983; **Diane Chellstorp,** administrative services assistant, 1975; **Vicki McCullough (Hill),** principal secretary of the facility lab, 1978; **Mrs. Odie L. Mann,** chief cook, 1979; **Bobby Quick,** senior engineering technician in the machine shop, 1978; **Mrs. Lynne Curtis,** who started in the physical plant in 1984 and two years later joined the business office as data entry operator; **John C. (Charlie) Burton,** senior material control clerk for the physical plant and ECP, 1966; **Maurice Taylor,** lab photographer, ECP, 1979; **Kathy Hice,** senior secretary, 1977; **Jean Armstrong,** food department (Industry-Student Center), 1985, and **Mrs. Birdie Farris,** lab assistant, ECP, 1979. Only Taylor, Armstrong and Burton (who retired in 1988) had left the Institute by its twenty-fifth year. By that time, one hundred and thirty-two Institute employees accounted for an aggregate of more than thirteen hundred years at UTSI.

Birdie Farris represented one of the Institute's "success stories," having advanced from mopping floors in the MHD control room and laboratory for four years to becoming a technician in that lab. She joined UTSI's cleaning staff on December 10, 1979, and became a lab technician trainee in November, 1983. In a magazine story about her, **David Dunkleberger** wrote that she credited lab supervisor **Philip Sherrill, Reuben Daniels,** chemist, and **Kathy Hines (Parks)** and **Jo Ann Winkleman,** technologists, for inspiring her to learn new things and assume new responsibility. And when **Dr. Susan Wu** needed someone to wash the glassware in the lab, **Birdie Farris** seized the opportunity.

The Space Institute joined the nation in mourning the loss of seven crew members of the shuttle Challenger, which blew apart seventy-four seconds after lift-off on January 28, 1986. **Dr. Kenneth Harwell** noted UTSI's close ties with the space program, pointing out that many graduates of the Institute go on to work for NASA or for space-related industries.

The Franklin County Economic Development Corporation in January, sponsored a meeting at the Space Institute that focused on a closer relationship between development in Southern Middle Tennessee and Huntsville, Alabama. **Jim Morse,** editor of the Herald-Chronicle in Winchester, reported that area government leaders and officials of the Southern Middle Tennessee High-Technology Initiative (organized in 1985 by **Congressman Jim Cooper** and **Governor Lamar Alexander**) participated in the discussion. A primary purpose of the session was to discuss what procedures counties in Southern Middle Tennessee might follow to reap spin-off plants from such Huntsville projects as the Boeing Company's facility in Huntsville. **Robert W. Hager,** vice president in charge of Boeing's Huntsville work on a NASA space station, told the group that his firm was establishing a major permanent facility in Huntsville. One speaker, **Dr. Harwell,** stressed the need for a working relationship between Boeing and the Space Institute.

A former Boeing official—**Jules W. Bernard,** manager of public relations at the Space Institute—had retired at the end of 1985 after nearly fifteen years at UTSI during which time he had worked under three deans—Dr. Goethert, Dr. Weaver, and Dr. Harwell. He had been with Boeing in Huntsville before joining UTSI in 1971, having worked on the Apollo moon rocket program and the Lunar Rover vehicle. Now, he was returning to Boeing as an aircraft inspector. In Tullahoma, he had been active in the "Tullahoma Bunch," a group of pilots and their families, and the Experimental Aircraft Association Chapter.

Gary D. Smith, after a year in Washington, D.C., as an American Society of Mechanical Engineers Congressional Fellow, was back, this time as Dean Harwell's executive assistant. He had been facility laboratory manager of the ECP, and he would continue to be involved in the MHD work. However, he also assumed duties as media and governmental coordinator. In Washington, he had served on the staff of Senate Minority Leader **Robert C. Byrd** of Virginia. Since 1981, Smith had been an officer of the American Society of Mechanical Engineers' (ASME) Highland Rim section, and in 1986 was chairman-elect of the ASME national membership development committee. Four years later he would be installed as vice president of ASME's Southeastern Region XI.

James L. Roy of Rock Island resigned as senior mail clerk on February 28, 1986. He had assumed the postal duties—working under **Dewey Vincent**—on September 4, 1979. He was a retired postal worker when he joined the UTSI staff. **Charlie Burton** had been the first "postman" at UTSI, and he took turns "filling in" after Roy left. Of course, other employees of the physical plant, including **Charlotte Campbell** (senior administrative services assistant), **Robert Parson,** and **Dewey Vincent** himself also pulled postal duty from time to time. **Frances I. Bauer** of Estill Springs, who took over printing services in 1980 after **David Brown** had left the Institute, also was a backup to Charlie Burton in the post office. In August, 1986, **Carolyn Floyd** of Estill Springs was hired as full-time mail clerk. **Bea Hill,** also of Estill Springs, joined Vincent's staff as a custodian in 1983 and later would join **Frances Bauer** in the print shop and also as an alternate mail clerk. (Vincent was a strong believer in "cross-training.") At the time of UTSI's Silver Anniversary, the print shop and post office were in adjoining offices in Lower D Wing.

UTSI received a small grant from the National Science Foundation to study fluctuation in high interaction in a subsonic MHD generator. **John T. Lineberry and Dr. Mary Ann Scott** were the principal investigators.

Alan Lowrey, president of the Student Government Association, presented a portrait of **Freeman D. Binkley,** painted by **Robert Suttles,** an employee of UTSI. The portrait was hung on a wall in the lounge of the Industry-Student Center.

On February 1, 1986, **Dr. Richard M. Roberds** joined the Space Institute as associate dean for academic affairs and professor of engineering science. **Dr. Arthur A. Mason,** who had been appointed assistant dean in 1976 and associate dean and director of academic programs in 1980, was returning to full-time teaching and research as a professor of physics. Roberds, a retired Air Force colonel, came to UTSI from the University of Tennessee at Martin, where he had been professor and dean of the School of Engineering and Engineering Technology since September, 1984. From April, 1980, to August, 1984, he was associate professor and head of engineering technology at Clemson University. Roberds had earned his bachelor's degree in physics from Kansas University in 1956, a master's degree in nuclear physics from the University of Kansas in 1963, and a doctoral degree in nuclear engineering from the Air Force Institute of Technology in 1975. A command pilot with four thousand flying hours, Roberds flew one hundred and eighty-three combat missions over Vietnam and Cambodia and was awarded two Distinguished Flying Crosses and eight Air Medals. His last assignment before retiring in 1980 after twenty-four years with the Air Force was as chief of the reconnaissance and weapons delivery system at the Air Force Avionics Laboratory at Wright-Patterson Air Force Base, Ohio. Prior to this, he had served as chief of the experimental and applied physics branch at the Air Force Weapons Laboratory at Kirtland Air Force Base, New Mexico.

Mason said his "direct role in the growth and development of the academic programs of UTSI" had been rewarding, but he wanted to spend more time teaching. Exciting things were happening in the field of infrared physics, and he wanted to get back into research. His research previously had led to his publishing in numerous publications of the American Physical Society, the Optical Society of America, and the American Institute of Aeronautics and Astronautics. An experimental molecular spectroscopist, Mason's research had ranged throughout the electromagnetic spectrum from the far infrared to the vacuum ultraviolet. A charter member of the Institute's faculty, Mason had immediately established an active cooperative research effort with engineers and scientists of ARO, Inc. at Arnold Center. This effort had played a significant role in the development of several important techniques in the area of spectroscopic diagnostics of high-temperature non-equilibrium gas flows. Many Space Institute students participated directly in those investigations.

George (Zhi) Shi, a Chinese graduate student, was selected in the spring of 1986 as the second "Goethert Scholar." The scholarship award had been established in October, 1982, by the family of Robert L. Young. The Institute also hosted a five-month visit by **Dr. Muhamyankaka Damien Bambanza,** a Fulbright Scholar from the Republic of Rwanda in Central Africa. He was director of the Rwanda national airlines and taught fluid dynamics at the National University of Rwanda. At UTSI, he joined **Dr. Remi Engels** and **Dr. Robert Hackett** in research in the area of low-cycle fatigue factors of materials.

The Student Government Association and UTSI's Astronomy Club sponsored some viewing sessions for Halley's Comet, coordinated by **Patrick Murphy** and **Alan Lowrey.**

Two new Ingersoll-Rand centrifugal air compressors were installed at the Coal Fired Flow Facility (CFFF) which **Norm Johanson** said would pay for themselves that year because they greatly reduced the cost of testing. **Cliff Wurst,** program leader, said this was the first major facility modification in the CFFF.

Dr. Susan Wu was "thrilled" in April when **Senator Albert Gore Jr.** announced that the federal Government Accounting Office (GAO) ruled that the Reagan administration's annual attempts to eliminate funding for MHD research were illegal. The GAO, while specifically mentioning MHD research, ruled that the administration could not switch funds appropriated for one program to another. Because the MHD program was funded by a line-item appropriation, the funds could not be used "for other fossil energy research and development projects." A month later, **Senator Jim Sasser** announced that the Senate Appropriations Committee had overturned the Reagan administration's proposed deferral of $13.1 million in **nationwide** funding for MHD research for the remainder of the 1986 fiscal year. Without these funds, Sasser said UTSI's MHD work would have "dried up sometime this summer."

In the spring of 1986, **R. Gregg Schulte** joined the Space Institute as assistant dean and director of business and finances. He came from Northern Kentucky University where he had been an assistant professor of accounting and then director of personnel services. Dr. Harwell said Schulte's initial duties included responsibility for business, procurement and personnel activities, and that these would be expanded. In fact, Schulte was being groomed to assume responsibilities of **Bob Kamm,** who was planning to retire in 1988; however, Schulte left in 1987, and the following January 1, **Raymond P. Harter** filled a new position of director of administration and finance.

(Politics in Tullahoma heated up in May when **Mayor George Orr** resigned, blaming a lack of support by aldermen and The Tullahoma News. **Beth Bryant** was appointed by the board to fill the vacancy, becoming Tullahoma's first woman mayor. She later was defeated by **Doyle Richardson** in her bid for a full term.)

In June, newspapers reported that **Dr. Frank L. Wattendorf** had died in Washington, D.C., of cancer at the age of eighty. Wattendorf had been civilian chairman of the AEDC planning group in the late 1940's and also had been one of the first proponents of locating an institute near Arnold Center.

On the last Saturday in June—on General "Hap" Arnold's hundredth birthday—seventy-five hundred people toured Arnold Center during an "open house" commemorating its thirty-fifty year. A special "Camp Forrest Reunion" also brought old-timers back to Tullahoma that summer, and a "Camp Forrest Memorial" was dedicated on August 19, just outside Tullahoma.

The six-member Southern Middle Tennessee High Technology Task Force in June recommended that UTSI be upgraded to a "world class" institution by installing an upper level undergraduate program, improving highways, developing a regional airport and identifying all suitable tracts of land that might be used for industrial development. **John L. Parish,** president of Lannom Manufacturing Company, Tullahoma, was chairman of the group. **Governor Alexander** attended this meeting at UTSI on June 24, 1986. The task force recommended "the establishment of a select undergraduate 'feedstock' program for UTSI" with the "brightest and best students" being encouraged to study at the Institute. This proposal obviously reflected Goethert's own position. The group specified that this recommendation did not imply the establishment of another full undergraduate school of engineering in Tennessee. Nevertheless, the idea was received cooly in Knoxville with **Dr. Joe Johnson** advising Parish that UTSI could meet needs of the task force without changing its status. Because of UTSI's limited focus in the areas of secondary and vocational and technical education, Johnson suggested that the task force call on other area state schools, such as Motlow and Middle Tennessee State University. The Space Institute, he said, could help in many ways, such as offering expert consultation, but "We feel keenly that the Space Institute is appropriately organized and positioned within (UT) and that its academic ties to (UT-Knoxville) are of significant benefit." The Institute had to operate "within a mission that does not encroach on or duplicate efforts of the other institutions," the UT vice president said.

Congressman Jim Cooper allowed that UTSI had been treated "as a red-headed step-child," and he felt several things were needed to make sure those days were over. Mainly, he said that UTSI should be granted full campus status and its dean should be made a chancellor so he could promote the Space Institute on the level as the other UT campuses. And **Ned Ray McWherter,** swinging through Tullahoma with his campaign for governor, said he would ask the Tennessee Higher Education Commission and UT trustees for their recommendation on the proposal for full campus status for UTSI, and if feasible, he would support it. (It had been almost two years since the Support Council had adopted a resolution asking UT leaders to start taking steps to make UTSI a primary campus with either a vice president or a chancellor, but backers had known all along that it would be a long process.) McWherter said he also would recommend establishing a second Chair of Excellence for the Space Institute.

Doctors T. Dwayne and **Mary Helen McCay** signed on as faculty members with UTSI that summer and as members in its Center for Laser Applications. Both came from NASA's Marshall Space Flight Center in Huntsville where Mary Helen was a metallurgical engineer in the materials and processing laboratory, and Dwayne was chief of the propulsion division in the structures and propulsion laboratory. He had received his Ph.D. in aerospace engineering in 1974 from Auburn University and was principal investigator for advanced propulsion technology programs at Huntsville. Mary Helen received her Ph.D. in metallurgical engineering from the University of Florida in 1973. At Huntsville, she was a scientific investigator for several flight experiments and was a backup payload specialist for Spacelab 3 in 1985. Before long she again would be involved in a space mission.

The Space Institute hosted a week's seminar on "Airplanes of the Next Century" in August. Plans were under way for more than seven hundred and fifty UT alumni runners to carry a symbolic torch across the state in the fall to celebrate the UT National Alumni Association's one hundred and fiftieth anniversary. **Lee Roy Grizzle,** president of the Coffee County UT Alumni Association, invited alumni and friends to attend a "Torch Lighting Ceremony" and buffet at UTSI. Astronaut **Henry W. (Hank) Hartsfield Jr.,** a UTSI graduate, on October 15, wielded the torch at that ceremony in front of the administration building while **Ernest Crouch** looked on. The next day, at 4:30 P.M., Hartsfield delivered the twelfth Quick-Goethert Lecture on "The Space Shuttle Program." This lecture was the highlight of UTSI's "Homecoming '86" program.

"We must not give up our quest for space," the Alabama native and Auburn graduate said. "Man is not perfect, nor are his machines, but the benefits outweigh the risks, and there will be greater benefits than we can now imagine." He was happy that NASA scientists and engineers were taking what they learned from the Challenger tragedy to rebuild a space program. At a press conference after the torch-lighting ceremony the day before, Hartsfield revealed that after the Challenger blew up, he learned that problems similar to those that destroyed the Challenger had been detected on two of his own previous flights.

At the banquet following the lecture, three former graduates were honored as "Outstanding Alumni." These were **Dr. James Mitchell,** technical director and director of operations at AEDC, **Dr. Jack Whitfield,** president of Sverdrup Technology, and **Dr. Kinya Aoyama.** Dr. Harwell presented a plaque to Mitchell; **Marcheta Whitfield** accepted a plaque for her husband, and **Dr. Jimmy Wu** accepted the award on behalf of Aoyama.

On October 7, **Dr. Bernard (Barney) Marschner** died at Fort Collins, Colorado, at age sixty-five. Before retiring as an Air Force colonel, Marschner was project chief of the propulsion wind tunnel at Arnold Center. He was assigned to AEDC in 1950, first working at St. Louis while the center was under construction. In the summer of 1977, Dr. Marschner, on temporary duty from Colorado State University, served as technical manager for a summer program at **UTSI.** According to Bob Young, Marschner at one time might have become dean of the Institute had things gone a little differently.

Marschner saw combat in North Africa, Sicily, Sardinia and Corsica as a member of a ranger outfit and was on his way to Japan with an infantry division shortly after V-E Day when a talent scout from the National Advisory Committee for Aeronautics (forerunner of NASA) found him and had him assigned to the Flight Test section at Langley Field, Virginia. A year later, Marschner went to California Institute of Technology to start work on a master's degree in aeronautical engineering. After getting his master's at Cal Tech, he received his Ph.D. in aeronautical engineering in 1954. Marschner and **Bob Kamm** had become close friends while both were at Arnold Center. Marschner was one of the "Three B's," Kamm said, because of his close association with **General B. H. Schriever** and **Dr. B. H. Goethert.** Once, Kamm said, the trio of "Bennie, Bernie and Barney" were in New Mexico, and Marschner got tired of waiting for Goethert—who was notorious for being late—and flew off and left him.

Robert (Ricky) Morris, an engineer at UTSI, also died in October of that year after a brief illness.

Highland Rim Construction Company of Fayetteville, with a price of just under five hundred thousand dollars, was low bidder to construct an advanced technology laboratory. The state had appropriated only $450,000 for the project, but **Bob Kamm** said some changes might be made to bring Highland Rim's $498,900 bid in line. Some scaling down was proposed, lowering the cost to $380,000, but a few weeks later most of the alternatives had been restored after additional funds became available. This structure, eventually, would turn out to be the first stage or "shell" of the Center for Laser Applications (CLA).

The Tennessee Higher Education Commission (THEC) in November recommended an operating budget appropriation of almost three and a half million for the Space Institute's next (1987–88) fiscal year, an eleven percent increase. THEC also recommended $995,000 for CLA and another $89,000 to be used for minority programs. THEC members were projecting a seven-percent increase in student fees (going up $110 to $1,569 for a full-time load) to help finance the budget.

Late in 1986, **Dr. Kenneth Harwell** told of renewed efforts to raise funds for a million-dollar "Chair of Excellence" at the Institute—one of thirty-five allocated to the UT system after the program started in 1984. Harwell said at least $250,000 and maybe half a million dollars would be needed. This project would prove to be a larger undertaking than may have appeared at the time, but UTSI soon would get another "Chair."

In November, **Professor Hao Ran Bi,** Director of Nuclear Technology and Design Division, Power Machinery Engineering Department of Shanghai Jiao Tong University in China, visited the Institute and toured various facilities. He was accompanied by **Dr. Ed Keshock,** UT professor from Knoxville, where Bi was investigating cooperative research possibilities.

The Tennessee Technology Foundation Executive Committee met in Knoxville and appointed twenty-one members to a ten-county Southern Middle Tennessee organization to succeed the Middle Tennessee High Technology Corridor Task Force. The group was to implement recommendations developed by Congressman Cooper and Governor Alexander's task force in 1985. **John Parish** was to serve as temporary chairman, and **Tom Bailey** would serve as executive director. **Dean Harwell** was a member of the group.

Al Ritter was leaving Calspan to join a firm in Huntsville, but he would keep in close touch with UTSI as a future chairman of the National Advisory Board.

Dr. Susan Wu got another honor, receiving the Chinese Institute of Engineers—USA Distinguished Achievement Award at the Institute's 1986 annual convention in New York for her "outstanding contribution" to the development of MHD power generation technology. But the new year would bring a replay of money problems, so that, faced with twice as much testing and less money to spend for it, Dr. Wu expected to lay off eight of one hundred and fifteen MHD employees—two technicians and eight support people—by the end of January. Nationwide, MHD appropriations had been cut to twenty-six and a half million dollars, which was even less than the previous year's appropriation. The total pie had shrunk; a slight lay-off was inevitable.

✦

Chapter Twenty-Three

DECISIONS PROPEL INSTITUTE FORWARD

THE SPACE INSTITUTE'S twenty-third year (1987) would bring several significant developments including the creation of a vice presidency for UTSI by the UT Trustees—a major victory for the Support Council and a step closer to campus status for the Institute even though it would take more than two years to fill the new position. This also was the year that the Center for Advanced Space Propulsion was established, a million-dollar Chair of Excellence was created for UTSI, a Space Institute degree program was approved for Maryland, two old-timers retired, and several Institute professors were honored. The Space Institute saluted **Dr. B. H. Goethert** on his eightieth birthday, which, sadly, was to be his last. And on August 7, the Institute would lose the best friend it ever had "on the hill" with the death of **Dr. Andy Holt.**

Dr. Donald D. Tippett joined Dr. Westbrook's team on Jan. 1 as associate professor of engineering management. A 1969 graduate of the U.S. Naval Academy, Tippett received master's and doctoral degrees from Texas A&M University. For more than a decade, he served as navy carrier pilot, reaching the grade of lieutenant commander. Before joining UTSI, he had been a senior systems engineer with Union Carbide Corp. and had also served as program manager, advanced technology, with Newport News Shipbuilding.

In February, **Professor Mouji Liu** and **Zhiyong Lu** from the Beijing Institute of Aeronautics and Astronautics visited UTSI as part of an Academic Exchange Agreement between the two institutes that had been signed on May 8, 1985. Their host at the Space Institute was **Dr. Jimmy Wu.** In 1985, **Dr. Kenneth Harwell** and **Dr. John Prados** had visited Beijing to establish the agreement. (In October, **Dr. Yu Wei,** president of The Nanjing Institute of Technology in Nanjing, China—one of the leading universities in The People's Republic of China—also would visit UTSI.)

Dr. Susan Wu, professor and administrator of Energy Conversion R&D programs at UTSI, told a group of congressmen in March that the United States' lead in MHD technology was being threatened by many other countries.

"While the dark clouds of funding uncertainty continues to hang over the head of the U.S. MHD program," she said in a written statement, "the international scene is quite different." As examples, she cited substantial efforts on MHD in progress in China, Japan, the Soviet Union, Italy, Israel, France, Australia and Finland. She was concerned about the "rapid erosion of our technological leadership and international competitiveness."

U.S. Congresswoman **Marilyn Lloyd** of Chattanooga had invited Dr. Wu to appear before the House Committee on Science, Space and Technology's Subcommittee on Energy Research and Development, chaired by Lloyd. Dr. Wu had, the year before, led a team of researchers (**Brad Winkleman** and **John Lineberry**) from the Space Institute that participated in a series of MHD experiments in the People's Republic of China under a grant from the National Science Foundation. She also had attended the International Conference on MHD Electric Power Generation held in Japan.

In March, 1987, **Dr. Wu** was one of four women in science honored by Public Service of Colorado, a utility company. The company featured pictures of the women on "Great Minds in Science and Engineering" posters that were distributed to high schools throughout Colorado in honor of Women's History Week. In the text accompanying Dr. Wu's picture, it was noted that she had published more than seventy scientific papers on MHD and that her research of power generation and conversion techniques was "leading the way for the United States to produce electricity cleaner and more efficiently." In November, she was elected to the grade of Fellow by the American Society of Mechanical Engineers, who cited her "exceptional engineering achievements" and "notable contribution to the engineering profession." She had been nominated by the Highland Rim Section of ASME, and **Gary D. Smith** presented the award to Dr. Wu. He said only about one percent of ASME members was awarded the Fellow grade. Another UTSI professor, **Dr. John B. Dicks,** had earlier received the honor, and **Dr. Marcel K. Newman,** another of the founding faculty members at UTSI, who had retired August 31, 1971, was elected in May, 1987, to the grade of Fellow.

Dr. Goethert, who had brought Newman to UTSI in 1964, described him as a "man who did not strive for high-level, broad recognition" and as one who had "dedicated his professional life to develop advanced academic programs and, above all, excellent young students and engineers." Upon his retirement, Newman had been named **Professor Emeritus,** the Institute's first. He remained active in the engineering community, and in 1989 would be an enthusiastic participant in the Silver Anniversary Celebration. Newman had served as professor and head of the Mechanical Engineering Department at Notre Dame for eleven years before joining UTSI. During that time, he pioneered the nuclear engineering degree program at Notre Dame and established and equipped a nuclear engineering laboratory. He also had been instrumental in bringing the first computer to Notre Dame in the 1950's over objections of the Mathematics Department.

Officials and supporters of the Space Institute in March, 1987, launched a fund-raising campaign to endow a one million dollar Chair of Excellence in engineering management at UTSI. UT President **Edward J. Boling** appointed a committee to seek the required matching funds from private sources and named **John Parish** of Tullahoma, former state commissioner of economic and community development, as chairman. **Justin Wilson,** secretary of the Justin and Valere Potter Foundation, contributed the first $25,000 at an organizational meeting on March 9, contingent upon $100,000 in matching funds being raised locally. The Chairs of Excellence program, which originated with the Tennessee General Assembly in 1984, provided for a one million dollar endowment to support a distinguished professor or chairholder, with the institution providing matching funds, and at least half of those matching funds to come from private sources.

Other committee members were **Charlotte Parish,** UT trustee from Tullahoma; **Bill Hughes,** Boeing-Huntsville; **Dr. Jack Whitfield,** president of Sverdrup Technology in Tullahoma; **King Bird,** vice president and general manager of Calspan at Arnold Center; **Morris L. Simon** of Tullahoma, lifetime vice chairman of the Support Council; **William R. Carter,** Fayetteville, Support Council chairman; U.S. Representative **Jim Cooper** of Shelbyville; **Don Shadow,** Winchester, former UT trustee; **Robert M. Williams** of Tullahoma, retired president of ARO, Inc.; **Bob Wright,** president of Teledyne-Lewisburg; **John Greeter** of Monteagle, Support Council member; **Charlie Coffey,** Shelbyville, president of the Tennessee Valley Aerospace Region; retired Rear Admiral **Robert Owens,** program director of Grumman Data Systems at AEDC; and **Tom Bailey,** executive director of Tennessee Technology Foundation, Tullahoma.

While this project was still shy of its goal in 1989, a separate "Chair" was established in the fall of 1987 after the Signal Companies contributed five hundred thousand dollars. This was the **Edward J. and Carolyn P. Boling Chair of Excellence in Space Propulsion.** Dr. Boling, UT's president for seventeen years, had served on the Signal board of directors from 1977 until the merger that formed Allied-Signal Inc., and he had continued to serve on the Allied-Signal board.

Dean Harwell said the chair would help accelerate UTSI's effort to become the leading aerospace institute of technology by the year 2000. A national search was to be conducted for the chair-holder, he said, and later that year, he appointed a chair search committee with **Dr. T. Dwayne McCay** as chairman. Other members included **Dr. J. M. Wu, Dr. Roger Crawford, Dr. Roy J. Schultz, Dr. Mary Helen McCay, Dr. C. E. Peters,** and **Dr. George Garrison,** all UTSI faculty members, and **Dr. James Mitchell,** chief scientist at Arnold Center. But Dr. Boling had already announced that he would resign as UT president effective July 1, 1988, and former **Governor Lamar Alexander** would succeed him, and filling the chair at the Institute would be put on the back burner pending selection of a vice president for UTSI. Harwell said the person who filled the chair would be involved in work of the new NASA Center for Advanced Space Propulsion (CASP), which had been established at UTSI in July with a five-year (five-million-dollar) grant from NASA. Here is how that came about:

The University of Tennessee-Calspan Center for Aerospace Research (CAR), located in UTSI's Research Park, had proposed the propulsion center in the spring of 1987 in response to a solicitation from NASA for development of Centers for the Commercial Development of Space. In July, **Senators Jim**

Sasser and **Albert Gore Jr.** and **Congressman Jim Cooper** announced that NASA had designated the Center for Advanced Space Propulsion (CASP) as one of its centers for the commercial development of space. **Dr. Fred Speer** of Huntsville, a retired NASA executive with thirty years experience in NASA's space programs, was chosen to be the first director of CASP. **Dr. John William (Bill) Davis,** executive director of CAR, said that **Speer** was "extremely qualified and ideally suited to lead the efforts of the center toward commercial development of space propulsion products and services."

Working with **Davis** in leading the proposal development team were **Dr. Harwell** and **Dr. George Garrison,** a UTSI professor and proposal manager for CAR. Firm support for the center had been obtained from seven major industrial organizations, and the state of Tennessee had endorsed the proposal. CAR was chartered as a not-for-profit corporation, drawing upon resources of both Calspan and UT and fully supported by research grants and contracts. In 1987, **Dr. John W. Prados,** UT vice president for academic affairs and research, was chairman of CAR's board of directors and **Dr. H. Robert Leland,** president of Calspan Corporation, was vice chairman. Board members included **Dr. Ed Boling, James K. Baker,** chief executive officer of Arvin Industries, **Dr. Tom Collins,** UT Vice Provost for Research, **King Bird,** general manager of Calspan AEDC Division, **John L. Parish,** president of Lannom Manufacturing Company in Tullahoma, **Dr. David Patterson,** president of the Tennessee Technology Foundation, **Bruce Reese,** consultant, **Dr. Davis,** who also was a Calspan executive and director of the propulsion wind tunnel facility at Arnold Center, and **Dr. Kenneth Harwell,** associate executive director.

NASA was to provide the center with up to one million dollars annually for five years. Harwell said Marshall Space Flight Center at Huntsville would provide technical support for the center, which would perform cooperative research with NASA and private industry to develop commercial products and services related to advanced space propulsion systems.

Dr. Harwell was ecstatic, citing the "key role the center would play in completing the high technology corridor linking Nashville, the Arnold Engineering Development Center, UTSI, and Huntsville." The center, he said, meshed "perfectly" with goals of the Tennessee Technology Foundation and the Tennessee Valley Aerospace Region in providing the basis for development of new high-tech industry in Tennessee.

In May, the Maryland State Board for Higher Education had granted approval for the Space Institute to offer an off-campus program in Maryland leading to the Master of Science degree in Aviation Systems. **Dr. Richard M. Roberds,** associate dean for academic affairs, would serve as liaison with the Maryland board. **Professor Ralph Kimberlin** was directing the Aviation Systems Program at UTSI. Courses would be offered through facilities at the Patuxent River Naval Air Station near the nation's Capitol. Fourteen students—the maximum allowable per session—registered for the initial program beginning on June 15. These students consisted of civilian engineers attached to the Naval Air Test Center, who were seeking advanced training in flight test engineering.

Initially, the program consisted of an intensive twelve weeks of study taught on site at Patuxent by Kimberlin. He was assisted by **Lisa Hughes,** UTSI engineer, **Bob Jones,** assistant engineer from UTSI, **Dr. Trevor Moulden,** UTSI professor of aerospace engineering, **Dr. Harvey J. Wilkerson,** professor of aerospace engineering at UT Knoxville, and **William Kershner,** noted aviation author and consultant from Sewanee, Tennessee. Students would earn fifteen credit hours for successfully completing the session. Subsequent course work was offered through the videotape medium and via on-site instruction in selected courses. Students would be required to complete a minimum of thirty-six quarter hours of course work in addition to completing a thesis project to earn a master's in Aviation Systems from UTSI.

The Aviation Systems program involved studies of the aviation transportation system with emphasis on the technical, operational, and management aspects. Topics would include aircraft design, performance, stability and control, flight testing, and meteorological aspects of aircraft operations. Dr. Harwell, commending Kimberlin and his staff for the new off-campus degree program, pointed out that this was not the first off-campus degree program offered by the Institute. The Engineering Management program at that time was being offered to students in sixteen locations in a three-state area.

Colonel Edgar James Masters, who joined UTSI after a thirty-year career with the Air Force, retired as Manager of Student Personnel Services—the first person to fill this position at the Space Institute. At a party held in Masters' honor on May 18 before his retirement in June, Dr. Goethert praised Masters for his dedication and service to UTSI students for the past fifteen years. It was Goethert who had persuaded Masters, upon his retirement from the Air Force in 1971, to leave Hawaii to join the Institute's staff. He was assigned responsibility for student recruiting, academic assignments, and, as Goethert put it, "helping them with daily worries." He was, in effect, the first person in charge of admissions. Dr. Harwell said the welfare of the students was always Masters' "first concern."

Masters had played an important role in the establishment of the Institute. In the early 1960's, General Schriever assigned Masters as his special representative to the University Planning Board for the Space Institute as a liaison with the U.S. Air Force Systems Command. Masters served six years at the Systems Command Headquarters. On September 24, 1964, Masters—at that time officer in charge of the systems command's educational programs—attended UTSI's first academic convocation, and at his retirement party, Goethert presented him with a copy of a photograph showing Masters on stage with Drs. Holt, Goethert, and other dignitaries while General Schriever delivered the keynote address. **Mrs. Mary Groff,** representing the UTSI Student Government Association, and **Jiada Mo,** representing international students, thanked Masters for his assistance and contributions to the students. Later, Goethert would say that Masters was instrumental in getting the Air Force to donate the three hundred and sixty-five acres as a site for the Institute. Masters had grown up in Lowellville, Ohio, and received his bachelor's degree from Ohio University and his master's from the University of Florida. He began his military career in 1941 as a second lieutenant.

Colonel Masters had persevered in his post despite having cancer, and he died on September 23, 1987, at Harton Hospital in Tullahoma. He was sixty-nine. **Dr. Richard Roberds** observed that Colonel Masters "had a positive influence on the Institute far beyond the number of years he was with it." He had played an active role in "urging the creation of the Space Institute."

William H. (Bill) Boss, a senior ECP engineer and retired Air Force major, worked part time in the office during the last months before Colonel Masters Retired. In October, **Dr. Charles L. Lea** was named director of admissions and student services. Roberds said the position was being upgraded with additional responsibilities, and the next year the title was changed to assistant dean for admissions and student affairs. Lea had served for seven years as director of student services at Motlow College and from its inception in 1983, he had headed the Job Training Partnership Act (JTPA) program at Motlow.

In June, 1987, the woman who had nursed a UTSI library from a few volumes shelved at her house to a research facility with more than fifteen thousand volumes, thirty thousand research reports, and subscriptions to one hundred and sixty periodicals, retired. Dr. Harwell described **Helen Mason** as "an institution who will always be remembered as **THE** UTSI librarian."

A native of Maryville, Helen Burnette attended Knoxville Business College and UT in Knoxville. In 1940, she began her forty-seven-year career in the UT system as a secretary with the UT Engineering Experiment Station. Two years later she moved to the UT Physics Department, first as a secretary and later as office manager. This is where she met and married **Dr. Arthur A. Mason,** and shortly afterward she accompanied him to UTSI where he was to be one of the first faculty members.

At the retirement ceremony, **Gaye Goethert** (Winfried's wife), representing the technical library staff from Arnold Center, presented Mrs. Mason with a photograph of AEDC.

Effective July 1, **Mary Lo,** who had joined the UTSI staff in 1969 as assistant librarian, was named head librarian. She had her bachelor's degree in history from National Taiwan University and a master's in library science from Spalding College in Louisville, Kentucky, and had been assistant librarian at Ithaca College in Ithaca, New York, prior to moving to Tennessee in 1967. Her husband, **Dr. Chiang-Fang Lo,** at the time was on leave as staff scientist with Calspan Corporation at AEDC in order to serve as a research fellow at NASA Ames, Moffett Field, California, under sponsorship of the National Research Council. He later would become a full-time professor at the Space Institute.

Professional honors came to several members of the Space Institute family in 1987. The American Society of Mechanical Engineers (ASME) recognized **John Lineberry** of Manchester for excellence for presenting the "Best Paper" at a national technical symposium. Lineberry, supervisor of the ECP Gas Dynamics Section, **Dr. Susan Wu** of Tullahoma and **Brad Winkleman** of Moore County, a senior research engineer in the ECP Advanced Measurements Section, authored the paper: "MHD Generator Tests at the IEE Mark II Facility," which Lineberry presented at the twenty-fifth Symposium on Engineering Aspects of Magnetohydrodynamics held in Bethesda, Maryland. The paper presented results of a cooperative MHD research program between UTSI and the People's Republic of China, jointly sponsored by the U.S. National Science Foundation and the Chinese Academia Sinica. **Yang Changqi** and **Ju Zixiang** of the Chinese IEE were also listed as co-authors of the prize-winning paper.

The three UTSI researchers had traveled to mainland China in October, 1986, to participate for a couple of months in testing at the Institute of Electrical Engineering in Beijing with the Chinese high interaction MHD test facility, the Mark II. The testing resulted in the achievement of a major milestone in the Chinese national MHD program. In the spring of 1987, three experts from the Chinese MHD program spent three months at UTSI working on a similar project. They were **Gu Bozhong,** a chemistry professor at the University of Beijing, **He Hui-ying and Yang Peiyao,** also from the university. In the fall of 1987, Wu, Winkleman and Lineberry would go back to China for another month of testing.

Dr. George Garrison was honored by the Huntsville Association of Technical Societies (HATS) as 1987 Professional of the Year. He was nominated for the honor by UTSI in recognition of ''his broad range of significant contributions to UTSI over the past year,'' according to Dean Harwell. Garrison had joined UT's faculty in 1981 and was one of the original faculty members to help establish the Engineering Management program. He came to UTSI in 1983 from the faculty at UT-Chattanooga. Before entering the education field, he had held key management positions with Sverdrup Technology, Inc. in Tullahoma. In 1987, he was professor of mechanical and industrial engineering and manager of special projects at UTSI. He would soon become director of the Center for Advanced Space Propulsion with **Dr. Fred A. Speer** serving as associate director.

Dr. Roy J. Schultz, associate professor of mechanical and aerospace engineering, and **Dr. K. C. Reddy,** professor of mathematics, were elected to the grade of Associate Fellow in the American Institute of Aeronautics and Astronautics (AIAA). Schultz joined the Institute in 1979 as a staff research engineer and became an associate professor in 1983. He also was chairman of the AIAA Tennessee Section. Reddy had been on the Institute's faculty since 1966, and had taught a wide variety of courses in mathematics and computational fluid dynamics. Schulz and **Dr. Thomas V. Giel,** assistant professor and manager of the advanced measurements group, had developed a seminar program that the American Society of Mechanical Engineers in 1987 selected for national distribution. The program was entitled ''Surviving the Freshman Year in Engineering.''

Thirty-one Institute employees were recognized in December for providing a total of two hundred and fifty years of service to UT. Heading the group was **Dr. Robert L. Young** with thirty years. **Paul West** of Normandy, who joined the business office in 1967 as business assistant and in 1987 was budget officer, and **Charlie Childress** of Winchester, who had worked in the Physical Plant's maintenance section since 1967, were cited for completion of twenty years service. Recognized for completion of ten years of service were **Richard Attig, Kathy Hice, James Hornkohl,** and **Donald Huebschman,** all of Tullahoma; **Eunice Burton** of Hillsboro; **James Chapman** of Manchester; **Howard Isbell** of Estill Springs, and **Fred Schwartz** of Woodbury.

Paul West, a landmark in the Business Office, worked with Dee Dee Jones, right seated, and from left Ginger Housley, Jennifer Boyles and Lynne Curtis at the time of UTSI's Silver Anniversay.

Recognized for completion of five years of service were **Tom Dowdy, Jim Few, Donald Finger, Linda Hall, Terry Jobe, Dr. James W. L. Lewis, Linda McKelvey, Sherry Perry, Timothy Pitts,** and **Boyd Stubblefield,** all of Tullahoma; **Dr. Bruce Bomar** of Shelbyville; **John Casey, Dr. Roger Crawford,** and **Hubert Fry,** all of Estill Springs; **Carl Catalano** and **Ward Johnson** of Manchester; **Ben Counts** of Elora; **James Kinningham** of Cowan, and **Callie Taylor** of Decherd.

The Department of Energy's Pittsburgh Energy Technology Center awarded a $1.3 million contract to UTSI to develop a new coal combustion system to facilitate the use of coal in small industrial boilers. **Richard C. (Dick) Attig,** manager of the ECP Fuels and Environmental Department, was named program manager of this project, which would aim at making industries less dependent upon foreign oil by demonstrating that coal could be burned cleanly enough to be substituted in the boilers. **Dr. Schulz,** responsible for burner concept development, and **Charles (Chuck) Wagoner,** responsible for burner and systems evaluation, were to be principal investigators. Schulz had conceived the design for a new coal burner that used a novel concept of ''swirl''—a type of fluid spin or rotation—to stabilize and control the combustion process; he would be responsible for directing the development of the burner during the initial phase of the contract. (Eventually, a different design was used on this project.)

UTSI was commissioned, in September, 1987, to demonstrate high electrical power production in a compact MHD generator as a potential for providing the "burst mode power" requirements for the Strategic Defense Initiative (SDI) mission. **Dr. Harold Schmidt,** UTSI program manager for the project, said performance of the generators would require "substantial advances in order to render compact, portable burst mode power sources practical for SDI missions." The U.S. Department of Energy (DOE) would manage the two-year MHD program for SDI missions on behalf of the SDI office. **John Lineberry** would work with Schmidt on the project, with experiments conducted in DOE's Coal Fired Flow Facility at the Institute. Contract value totaled $328,000.

DOE also contracted with the Space Institute to investigate MHD "seed recovery and regeneration based on a formate process." **Dr. Atul Sheth** would be the principal investigator for the study that was to be done in two phases. The first phase was for a year, with a contract value of $333,276. The electrical power production in an MHD generator required seeding the combustion gas with potassium to create an electrically conductive plasma. In order for MHD to be accepted for commercial use, Sheth said, the spent seed material would have to be recovered, separated from the flyash, and recycled. **Dr. Joel Muehlhauser,** deputy program manager for ECP, hoped that the second phase—for twenty-nine months and with a contract value of $5,862,261—would be funded, and of course that the Institute would be involved in that phase. It was funded, but TWR, not UTSI, was chosen to conduct it even though results of the Institute's first-phase study exceeded the required achievements.

In November, the UTSI National Advisory Board elected **Dr. Alfred Ritter** with Booz-Allen & Hamilton in Huntsville to succeed **Dr. Jack D. Whitfield,** president of Sverdrup Technology, Inc. of Tullahoma, as chairman of the board. **James W. Garrett,** general manager of Schneider Services International at AEDC, was elected vice chairman, succeeding **Dr. Richard Hartman,** director of research at the U.S. Army Missile Command at Redstone Arsenal, Huntsville. **Gary D. Smith,** special assistant to Dr. Harwell, was elected as secretary, succeeding **Mrs. Ruth Binkley,** administrative assistant in the dean's office.

Five new members of the advisory board were welcomed to their first meeting. These were **Astronaut Hank Hartsfield,** who was deputy director of flight crew operations for NASA; **Dr. J. William Davis,** vice president and general manager of Calspan Corporation, AEDC Division; **Dr. William Kimzey,** vice president and general manager of Sverdrup Technology at AEDC; **Melvin Hartmann,** director of aeronautics at NASA Lewis Research Center, and **Dr. J. Mike Murphy** of Martin Marietta Aerospace, Denver, Colorado.

Whitfield commented on the Space Institute's "growing good image" and on indications of UT's upper administration's support of the "growing excellence of UTSI." He urged the Institute administration to keep its focus on strategic planning.

A major shift occurred during a meeting of the Support Council on the night of July 16, 1987, when **Dr. Joe Johnson,** vice president of development for UT, threw his support behind a proposal to change UTSI's position of dean to vice president with direct access to the UT president. **Morris L. Simon,** chairman of a Campus Status Committee, reported at the dinner meeting—with Johnson present—that Johnson had pledged his support during a meeting with the committee on May 29, and Johnson said Simon's report was "absolutely correct." July 16 was the first meeting of the Support Council since the committee had met. Simon reported that Johnson had told the committee that while campus status was very difficult—that it had to be approved by the Tennessee Higher Education Commission and the UT Board of Trustees—that he would recommend to authorities at UT and seek to have implemented the proposal for a vice president. Speaking with a Tullahoma News reporter the next day, **Johnson** said, "I think it (the proposal) certainly deserves consideration, and I wish to pursue it. I have agreed to discuss it with appropriate state officials and the UT board of trustees, and I hope to move it into the status of recommendation." The council had accepted the committee's recommendation, Johnson said, and "my response was that it is certainly a sound recommendation, and I wish to pursue it. It has a lot of merit." (Minutes of this meeting reflect that **Dr. John Prados** expressed support of Dr. Johnson's proposal to "enhance the status of UTSI.")

The Support Council, in August, 1984, had petitioned UT officials to start steps to grant UTSI campus status with either a vice president or a chancellor. Members "re-approved" this position in 1985, but the proposal engendered little enthusiasm from Knoxville, and the issue lay in limbo until the Support Council met on February 4, 1986. Simon brought it up again, saying that a majority of the council favored steps to upgrade the Institute's status, and they recognized it was a long-term process. He recommended that a committee be established to seek to bring campus status "to fruition" in the future. At that meeting, Dr. Johnson responded that while **Sena-**

tor Doug Henry had sought his and Dr. Boling's opinion on the resolution, the resolution was never formally submitted to the UT administration. He added that he and other UT officials were available to discuss the matter. The council agreed that a Campus Status Committee was needed, and Carter later named Simon chairman of the committee, **G. Nelson Forrester** of Tullahoma, vice chairman, and **State Senator Jerry Cooper** of McMinnville, **James C. Murray** of Tullahoma, and **I. J. (Ike) Grizzell** of Decherd as members.

Considerable one-on-one discussion took place between the chairman and **Joe Johnson** in 1987. Simon's wife of almost fifty years, **Lillian Tobe Simon,** had died in January in Indianapolis. The Space Institute and particularly the primary campus issue were high on Simon's priority list as he dealt with his personal loss. As president of Arnold's Furniture, he spent considerable time in an upstairs office at the store. One day **Dr. Johnson** visited him there and asked: "How important is this (primary campus) to you?" Simon assured him that it was of utmost importance. In the ensuing conversation, Johnson said that he and Boling would recommend to the trustees that a vice president be named for the Institute. There would be no chancellor, no primary campus, Johnson said, since either of those would require legislative action. They then drove to Fayetteville so Johnson could discuss his intentions with **William R. Carter,** Support Council chairman.

In addition to having direct access to the president, Simon said (at the July, 1987, meeting), "We also felt that having a vice president would immediately give UTSI more prestige in the national role it is taking in education." Johnson said vice presidents and chancellors were considered "co-equals" since both had direct access to the president.

Johnson's support of the proposal signaled the crumbling of opposition, and Simon and his fellow committee members knew that it now was just a matter of time.

Dr. Harwell was particularly interested in developing closer ties with NASA, and during the summer of 1987, UTSI had eleven research contracts with NASA, including several at Marshall Space Flight Center, and some UTSI faculty members worked at Marshall part-time that summer when a team from Marshall visited the Space Institute to brief officials on two new NASA initiatives for universities and to learn about the Institute's capabilities. **S. F. Morea,** director of NASA's Research and Technology Office at Marshall, discussed initiatives directed toward fostering innovative research by increasing involvement of engineering schools in NASA's research programs. NASA's space research and technology budget had declined since the mid-sixties, and the new initiatives were aimed at strengthening technologies in key areas.

Dr. Richard Roberds saw as a "major mission" for UTSI the production of high quality space engineering **graduates** to staff NASA agencies into the next century. Concern had grown that the base of technology available for space programs was inadequate to meet future needs, he said, and this was reflected in the critical demand for highly trained space engineers. UTSI wanted to help meet this demand.

Accompanying Morea were **Dr. George McDonough,** director of the Structures and Dynamics Laboratory, and **Robert J. Richmond,** head of propulsion technology programs in the Research and Technology Office at Marshall. **Dr. Dennis Keefer** showed UTSI's newest acquisition—a three kilowatt, carbon dioxide laser system to be used for research in laser materials processing and laser space propulsion. The laser, produced by Rofin-Sinar in Germany and costing more than $350,000, was one of the most advanced in the world and used radio frequency excitation to provide exceptional beam quality and pulse capability. Keefer explained that the system was integrated with a five-axis computer controlled work station that provided automated laser cutting and welding capabilities for a variety of metallic and nonmetallic materials.

Another NASA executive, **James B. Odom,** director of the Science and Engineering Directorate at Marshall, also was on campus that summer, speaking to **Dr. Donald Tippett's** engineering management class. Odom said one of the most serious problems facing the United States was the poor competitive position of U.S. industry in the world market place. One of the cures, he said would be renewed emphasis on research and development.

John Greeter, a member of the Support Council, arranged a visit by **Marvin Runyon,** president of Nissan Motor Manufacturing USA, for a briefing on ways that UTSI might benefit Nissan operations. UT **President Edward Boling** led a delegation from the UT system to participate in the briefing. Afterwards, Harwell said the Space Institute's advanced laser materials processing capability was of special interest to Nissan and other manufacturing companies in the area. Nissan had been pleased with a project conducted by Institute researchers for Nissan in support of a design of a waste incineration plant for Nissan's facility at Smyrna, Harwell said.

The Space Institute welcomed a visitor from a totally different field that fall when **Ben Vereen,** star of stage, screen and television, came calling. His interest in research going on at UTSI had been piqued during a visit to Nashville where he met **Dr. Al Pujol,** UTSI professor. Vereen promised that he would be back, and the next year he was, bearing gifts.

Dr. Susan Wu represented UTSI on an advisory board formed to counsel members of the committee searching for a UT president to succeed Dr. Boling the next year. The search committee, headed by **Bill Johnson,** consisted of seven UT trustees charged with screening candidates. This group visited UTSI in August, 1987. In late summer, **Dr. J. W. (Bill) Davis,** director of the propulsion wind tunnel at AEDC and executive director of the University of Tennessee-Calspan Center for Aerospace Research, was named vice president and general manager of Calspan Corporation, succeeding **King D. Bird,** who had been Calspan's general manager at AEDC since 1981. Bird was promoted to senior vice president and general manager of a newly formed Service Contracts Division.

By that fall, the UT trustees had assigned as top priority completion of the Center for Laser Applications' laboratory building. **Jim Few** said the shell of the structure had been "more or less" finished, but the completed facility was badly needed. Few noted that CLA, in four years, had gone from a few thousand dollars worth of research to more than one and a half million dollars in research. CLA's staff, including faculty and students, was up to thirty-four, and Few said the Applied Physics Group and Laser Application Center had equipment that was "as good as any university's in the country." It would be more than a year, however, before they would get to put that equipment into the new building.

On October 20, 1987, the Space Institute went all out to honor Dr. Goethert on his eightieth birthday—combining the celebration with the thirteenth annual Quick-Goethert Lecture. **Dr. William F. Kimzey,** vice president and general manager of Sverdrup Technology, Inc., AEDC Group, donated one thousand dollars to UTSI to support the birthday events. Kimzey—who had earned his doctorate under Goethert's tutelage—said the donation was in honor of Goethert and "in appreciation of the services UTSI, as an educational institution, brings to Sverdrup."

The celebration began on the evening before with a barbecue at the Institute's picnic area. The next morning, a "Festschrift" (a symposium featuring a number of technical papers that would be published in a bound volume) was held in two sessions with a birthday party in the lobby sandwiched in between. At mid-day, the UTSI faculty posed for a group photograph and, after lunch, a two-hour open house was held. At 4:30 P.M., **Dr. Hansjurgen Frhr. von Villiez,** director of the European Air Traffic Control Center in Maastricht, The Netherlands, delivered the Quick-Goethert Lecture entitled "The Process on Integration of European Air Traffic Services." Among guests from the Technical University of Aachen were **Dr. Horst Thomae,** director of the Institute of Aerospace Engineering at Aachen, who introduced Villiez; **Dr. Hans D. Ohlenbusch,** rektor; **Dr. Gerd Wassenberg,** chief public information officer; **Dr. Egon Krause,** who presented the first Quick-Goethert lecture in 1975, and **Dr. Rolf Staufenbiel,** director of the Institute of Aerospace Engineering.

The **Reverend A. Richard Smith,** pastor of Trinity Lutheran Church, offered an invocation, and **Dr. Edward J. Boling,** UT president, made opening remarks. **Dr. Joseph E. Johnson,** UT vice president for development, made a special presentation after the lecture, and **Dr. Jack Reese,** UT chancellor, gave the closing remarks. That evening, a dinner in Goethert's honor was held at the Industry-Student Center, attended by colleagues, governmental officials and industry leaders. Dr. Goethert was presented a citation from Tennessee **Governor Ned Ray McWherter** naming him an "Outstanding Tennessean." It was announced that the access road from the **Frank Wattendorf** Memorial Highway on the Arnold Center reservation into the UTSI campus was being renamed the **"B. H. Goethert Parkway."**

Goethert had received many honors during the seventy-five years since he had seen his first airplane. He had authored more than seventy technical books, papers and journal articles. He had served as a consultant to numerous organizations, both in industry and government, including the Air Force Wright Aeronautical Development Center, German Research Institute of Aeronautics, Sverdrup Technology Inc., National Academy of Sciences, General Dynamics, Martin Marietta Corporation, and the Federal Aviation Administration. In 1959, Goethert received the Air Force Scroll of Appreciation and was made an honorary citizen of Tennessee in 1965. The next year he received the Air Force Award for Meritorious Service, and in 1975 he was given honor membership in the German Association for Aeronautics and Astronautics. He was the first recipient of the American Institute of Aeronautics and Astronautic's Simulation and Ground Testing Award in 1976. In 1982, he was named "Akademischer Ehrenburger," or academic honorary citizen, of the Technical University of Aachen. This same year, the **B. H. Goethert Professorship** and the **B H. Goethert Graduate Study Scholarship Award** were established at the Space Institute. (**Dr. Jimmy Wu** was the first and only faculty member to hold the Goethert professorship during the

Institute's first twenty-five years; in 1990, **Dr. Dennis Keefer** would become the second.) In 1985, Goethert received the **Ludwig Prandtl Ring** in Bonn, West Germany—A German national award and one of the highest given in science and engineering. The AIAA also cited Goethert that year for outstanding contributions to air-breathing propulsion education during 1985.

Dr. Jimmy Wu, Goethert Professor.

Goethert was above five years old when he saw his first airplane—a primitive craft made of cloth and wood, powered by a tiny motor, wires dangling haphazardly. A few days after his seventy-fifth birthday in 1982, Goethert had described the experience to **Jeff Copeskey,** a writer with the Nashville Banner:

"It was a big sensation. I was just a boy—I guess I was about five—when I went with my father to see the first plane when it showed up in Hannover (northern Germany). I kept dreaming about being able to fly on the clouds—I was just fascinated by it, and I wanted to be up there, too. It was a tremendous boost to the imagination of young people . . . and I was one of them." He got his pilot's license when he was twenty-five.

Born in Hannover on October 20, 1907, Goethert received a bachelor's degree in mechanical engineering from the Technical University of Hannover in 1930, a master's degree in aeronautical engineering from Technical University of Danzig in 1934, and a Ph.D. in aeronautical engineering from the Technical University of Berlin in 1938. From 1934 to 1936 he had served as a research engineer for the German Research Institute for Aeronautics (DVL) in Berlin and then was Scientific Staff Engineer for Aeronautics there for three years. In 1939, he had been promoted to Department Chief, High Speed Aerodynamics at DVL—the position he held when the Russians came for him in 1945.

Contrary to his fears more than forty years before as Russian soldiers had marched him at gun point into the woods, Goethert had lived a full and eventful life. He and his wife were still residing in Manchester. Their children—including the three who had walked in sad bewilderment through a bombed-out and burning village one Christmas morning long ago—were living productive lives. The only daughter, Mrs. **Hella Lacy,** was a nurse in Linwood, N.J. **Winfried,** a senior research engineer with Calspan Corporation at Arnold Center, lived in Tullahoma while his twin, **Wolfhart,** retired from the Air Force, was living in Rome, New York, a senior computer scientist at the IIT Research Institute. The baby boy, **Reinhard,** was an architect in Boston. (Daughters-in-law were—truly—**Gaye, Joy,** and **Happy.**)

Winter was coming to the Space Institute, and the longer winter was not far away for the little doctor, but as he entered a new decade of his life, he had the satisfaction of knowing that his old friend, **Morris L. Simon,** had just carried the ball across the goal line for him one more time. The UT Board of Trustees in late September approved creation of a vice presidency for the Space Institute.

Ed Boling recommended that the Space Institute's chief administrator be a vice president. In his words, it was clear that the Institute had become a "more complex organization, and its external publics" had expanded significantly. It was time, the UT president said, to establish an administrative arrangement for UTSI that would "link it directly to the University wide administration, as was done with the Institute of Agriculture some seventeen years ago." The vice president would be a member of the president's staff, answering directly to him. The Institute should keep its mission focused on advanced graduate studies and research in the limited areas of science and engineering, Boling said, and thus be most effective in "supporting the federal aerospace agencies it was created to serve and in providing an effective catalyst for the high-technology development that the State of Tennessee desires."

Noting that the Institute's work in space propulsion had generated some "exciting developments,"

Boling called attention to the recent designation of a national center for advanced space propulsion at UTSI, and the establishment of a Chair of Excellence in space propulsion. UTSI also had become more active in encouraging and supporting economic development through high-technology initiatives such as the Tennessee Valley Aerospace Region recently formed through the Tennessee Technology Foundation.

For more than four years, the Support Council, and in particular, Simon, reading signals from Goethert, had agitated for a primary campus status headed by either a vice president or a chancellor. **John Prados** had said it would never happen. **Joe Johnson** had one compiled a list of all the reasons why it should **not** happen. Now, in the shank of the Institute's twenty-third year, it—at least part of it—indeed had happened. The Space Institute would get a vice president, but with a lame-duck president on the hill, it would be hard to say just when this would take place.

✦

Chapter Twenty-Four

IT HURTS TO SAY GOODBYE

FRIDAY, January 6, 1988, dawned bitter cold. **Sandy Shankle,** manager of short courses at the Space Institute and a friend of the Goethert family, called Dr. Goethert and suggested that he stay home. He insisted that he would go to work.

"You are a stubborn old man!" teased Sandy, one of the few people who could get away with such. She was driving to Manchester to take care of business at Beaver Press on the Square. She offered to stop by Doc's house and pick him up. He informed her that he could drive himself. Okay. All right. She proceeded about her business, and as she walked into the Beaver office, **Suzy Beall Robinson,** met her at the door with instructions: "Call Doc!"

His car wouldn't start, so Sandy was to pick him up after all. When she got to his house, she was astonished to see Goethert standing outside with no coat, no hat, while a young man attempted to jump-start Doc's VW, and Mrs. Goethert stood inside the house, fretting about his well-being. Sandy ordered Goethert inside. Mumbling, he reluctantly obeyed, got his overcoat and hat and went back outside "to help." Sandy offered to drive behind Goethert to the Institute, but he protested: "You don't have to follow me; I can do it myself."

Sandy Shankle had joined the Institute in June, 1980, as a senior secretary in Upper C Wing. After the building was expanded, she moved to Upper E as principal secretary. When **Jules Bernard,** who had come to UTSI in 1971 from the Boeing aerospace company, left the short course program to devote full time to public relations and moved into B-101, Mrs. Shankle was named short course manager. Initially, she shared the first office in B Wing (B-100) with **Betty Bright,** who had worked with Bernard since the late 1970's, and with **Becky White.** (Later, the office was moved to C-Wing and by 1988, **Judy Rudder** had joined **Sandy** and **Betty Bright** in the office.)

After retiring from full-time duty at the Institute, Goethert had continued teaching, but he also devoted considerable time to directing a widely-acclaimed short course in aeropropulsion. **Jules Bernard** had accompanied him in the past, but at the close of 1985, Jules retired. Sandy remembered Doc pacing in front of her desk in B-100. She thought: You will have to **ask** me. Finally he did: What was he going to do? Simple, Sandy answered. With Jules gone, he would have to find someone else to go with him to the short courses. Again, he hesitated and paced. Would **she** go? Yes. Would she be **allowed** to go? Allowed? Then she caught his meaning.

"Yes," she said. "Vern and I have discussed it. I can go; there will be no problem." ("Vern" was the **Reverend Vernon Shankle,** pastor of First Presbyterian Church, Tullahoma.)

On one of those trips to California, Goethert informed her: "You are bossy!" She disagreed, telling him: "I am independent; there is a difference." He didn't understand, so she explained that being a minister's wife was not easy, that he had to be gone a lot; therefore, she **had** to be independent. "Oh," Doc said, "I did not know this."

The truth was, in spite of her cheerful disposition and occasional drill-sergeant assertiveness, Goethert kept Sandy's stomach in a turmoil most of the time. She, like others who worked for Goethert, knew that he could be very difficult—that he could ruffle feathers and hurt feelings—but that he would quickly put harshness behind him at day's end and fifteen minutes after a heated disagreement he was just as apt to suggest dinner. **Sandy Shankle** knew that cold morning that further badgering was useless; Doc would drive himself to work.

To get to his office in Upper A Wing, Goethert had to walk through **Callie Taylor's** office. At this time, she was secretary for Goethert and for **Dr. Charles L. Lea,** who had come as director of admissions and student affairs in 1987, shortly after the death of **Colonel Ed Masters,** who had been the first manager of admissions. Callie had worked for the colonel since 1982 and also for Dr. Goethert. She noticed, as he passed her desk on his way into his office that morning, that "Doc" did not look well. He was inside his hot office for only a few minutes before he reappeared with his briefcase and told Callie that he didn't feel well and was going home. Callie, noting that he "looked blue," insisted that he have someone drive him, but Goethert shrugged and started down the hall. **Dr. Lea,** having heard the conversation, came out just as Doc, obviously having difficulty, reached for a chair. Lea steadied him. Goethert still insisted he was all right, but Lea protested: "You are not driving yourself!" Eventually, Doc said: "Call Sandy."

With Sandy at the wheel of the VW and **Judy Rudder** following in Sandy's car, Doc was concerned about two things. He kept asking Sandy: "Can you shift?" and he fretted that he would not be able to attend an aeropropulsion short course in Seattle. He had planned to go out and check on hotels and other details for the course that was slated for April or May. "I guess this means I can't go to Seattle," he said several times.

"I'll go for you," Sandy said. Goethert sighed as though a load had lifted, and he appeared calmer. All the way to Manchester, Sandy, while careful to avoid upsetting him, begged to take him to the hospital, but Doc said no. She saw him sneak a nitroglycerin tablet under his tongue.

Mrs. Goethert was away from home when they arrived. Doc went to bed, and Sandy was about to call an ambulance when Goethert's son, **Winfried,** walked in.

"He's bad, Wimp," Sandy said. He called an ambulance and then went to his father's bedroom. Outside, Sandy heard Doc fussing. Had Sandy called Wimp? No, **Bob Kamm** had. "Well he ought not have!"

Mrs. Goethert arrived just after the ambulance did. They took the doctor to the Coffee County Medical Center in Manchester, and then flew him to Nashville. It was his heart, but he would not give up easily.

At the Institute, **Dr. Fred A. Speer** of Huntsville, a retired NASA executive, had been named as the first director of the Center for Advanced Space Propulsion (CASP) effective January 1. Speer had retired from NASA in 1986 after thirty years service, and at the time of his retirement had been Associate Director for Science at the Marshall Space Flight Center (MSFC) in Huntsville. A native of Germany, Speer had become an American citizen in 1960—the year he joined the MSFC staff. Like Goethert, he was a graduate of the Technical University in Berlin, having received both master's and doctorate degrees in physics, and he had taught physics there for five years.

Raymond P. Harter, a certified public accountant in New York State, also had joined UTSI, assuming on January 4 the new position of Director of Administration and Finance. The position was created after **Greg Shulte** resigned. Shulte, who was an assistant dean for about a year, was slated to assume Kamm's responsibilities after his retirement in 1988, but instead, Shulte resigned. Harter had been dean of financial and business services at Corning Community College for ten years. **Dr. George Garrison** had headed the search committee that included **Dr. Roger Crawford, Mr. Jim Few, Mr. Norman Johanson and Dr. Al Pujol.**

Dr. Charles Lea was excited about enrollment for the winter quarter, which was nearing four hundred in mid-January. Final figures showed four hundred and seventy-six, a thirty-eight percent increase over the 1987 winter quarter. **Dr. Roy Schultz,** associate professor of aerospace and mechanical engineering, and **Dr. Tom Giel,** assistant professor and manager of the Advanced Measurements Group, were preparing another seminar aimed at cutting the high dropout rate among freshmen engineering students. The Highland Rim Section of the American Society of Mechanical Engineers co-sponsored the seminar on February 23, with a panel of engineers assisting.

In February, **Dr. Y. C. L. Susan Wu,** who with her husband, **Dr. Jimmy Wu,** had been a part of the UTSI faculty for twenty-three years, resigned as professor and administrator of the Energy Conversion Research and Development Programs. She had formed her own company, Engineering Research and Consulting, Inc., which was located in UTSI's research park. She had become administrator of the ECP in 1981, and was recognized as an international expert in MHD technology. She lectured at home and abroad on the MHD topic. In 1985, she became the first minority woman to receive the prestigious Achievement Award from the Society of Women Engineers. Two years later, she was elected to the grade of Fellow by the American Society of Mechanical Engineers. She also was honored in 1987 by Public Service of Colorado as one of four women selected as "great minds in science and engineering."

On March 1, 1988, Dr. Harwell announced the naming of **Dr. Joel W. Muehlhauser,** who had enrolled at UTSI in **1966** as a graduate student in physics, as acting administrator of ECP, effective March 16. He had been a key member of the staff of the

MHD project since **1974,** when he earned his Ph.D. in physics, and he had served as ECP's deputy program manager since **1986.** (It was as a student that Joel met **Sue Clifton** of Tullahoma, an ECP principal secretary with an office in B-101—across the hall at that time from a large classroom. She later became Mrs. Muehlhauser.)

Dr. Joel W. Muehlhauser

Dr. Wu, noting that Muehlhauser "has been with UTSI for nearly as many years as I have," was confident that he would provide the leadership necessary to maintain the ECP as a "vital part of our nation's energy research program." During the period of 1978–81 when the Department of Energy's Coal Fired Flow Facility (CFFF) was being constructed, Muehlhauser was manager of the Facility Laboratory. He was responsible for the fabrication and installation of the facility. Later, he was promoted to manager of the R&D laboratory where he was responsible for the project engineering functions on all ECP programs until his promotion in 1986. A graduate of Shelbyville's Central High School, Muehlhauser earned a bachelor's degree in physics from Auburn University and had served for four years in the U.S. Navy as main propulsion assistant and engineer officer.

Mainly because of MHD contracts, UTSI had ranked fifth among the nation's universities in DOE research contracts in the 1987 fiscal year. UT ranked fourth with thirteen DOE research contracts worth $8 million, but almost 88 percent of that total, including $6.6 million for the MHD power generation project, represented contracts at UTSI. UTSI's MHD contract in early 1988 exceeded $7.2 million, and the Institute also had six other research contracts with DOE that were expected to total $1.2 million in 1988. Proposals for other energy research projects promised to add another million dollars, according to UTSI spokesman **Gary Smith.** Muehlhauser noted that the Energy Department was the second largest contract research organization in the federal government, second only to the Defense Department. He credited success of the ECP research programs to the "dedication and diligence" of ECP employees and cited the "strong support provided by our Congressional delegation, especially Representatives **Jim Cooper** and **Marilyn Lloyd** and **Senators Jim Sasser** and **Albert Gore Jr.** (These officials had consistently fought to include MHD funds in the national budget for eight lean years.) **Dr. Kenneth Harwell** saw UTSI's high ranking as a "clear indication" that UTSI's strategic planning was working. Energy was one of the areas emphasized in the "UTSI 2000" plan, Harwell noted.

Fourteen winter graduates who had completed master of science degrees at UTSI were honored at a reception on March 14.

At this time, there was still hope that the state would land the Superconducting Super Collider (SSC) in Middle Tennessee, and **Dr. Steven Csorna,** Vanderbilt professor, was to speak on the subject at the spring meeting of UTSI's Sigma Xi Club, scientific research society. (Texas later won out in this venture.)

Meanwhile, Dr. Goethert was losing his private battle, but in early March he was at his home in Manchester. **Sandy Shankle** visited him, and Doc said to her: "I love you." Turning to Mrs. Goethert, Sandy said: "He's sick!"

"But he really **does** love you," Mrs. Goethert said.

"I know, but he would never **say** it. He's sick!"

Dr. Goethert laughed.

"I love you, too," Sandy assured him.

"Ours was a love-hate relationship," she was to say a year later. It had only been a few months before—in December—that she had been "mad enough to kill him." They had been at an aeropropulsion workshop at Wright-Patterson Air Force Base. Goethert always wore a microphone so he could interrupt the speakers, and when he wanted something, he would ask: "Where's my girl, Sandy?" On this occasion, a speaker had asked Sandy to turn view pages for him. Right in the middle of this, Doc's voice blared loud and clear: "Sandy! You did not give me the lecturer's notes."

"Shhh. Okay, I'll get them," she whispered. She flipped a page, noting that the slide appeared to be one that would require a bit of talk. She darted under the screen, rushed to the back, got the notes and returned, slipped them to Goethert, and turned the next page without missing a beat. The crowd applauded, much to Goethert's puzzlement. Later, he asked: "Why did they applaud? This is what you were supposed to do. Had you done it to start with, you would not have had to make a spectacle of yourself."

Mrs. Shankle recalled another time when Goethert publically chastised her for "contradicting" him. Back at the suite, preparing for a second short course session, Doc came in and continued his criticism with his wife as the only witness. Sandy stopped him, explaining that she had not contradicted him. Somewhat astounded, Goethert observed: "You're **mad**." His Short Course manager assured him that he was correct. "Okay," Doc said, and went to his room. A few minutes later he returned and, cheerfully, said, "Let's go to dinner." They did.

It was often said by Goethert's associates that he did not realize how his curt manner of dealing with others came across to them. They spoke, too, of the difference in his method and manner when he was engaged in "business" and when it was "time to go to dinner."

(Some months after his death, Sandy Shankle read a memorandum in which Dr. Goethert had thanked her for things she had done, and then also criticized her on some point. "As I read it," she said, "I suddenly realized that my stomach was in turmoil. I thought: He's **still** doing it!" Later, she told his son, Wimp, about it. He smiled and said: "I know what you mean!")

Dr. Jimmy Wu remembered seeing Goethert intimidate a secretary one morning soon after Wu had joined the staff. According to Wu, Goethert kept asking: "How am I going to work? How am I going to work?" while the trembling secretary tried to figure out what was wrong. It turned out that Goethert was asking how could he work—**without his pencils being sharpened.** Wu also liked to tell about the time—soon after Wu's arrival—that the director wanted Wu's help with an urgent problem. As Wu remembered it, Goethert wasn't sure how to assign grades on the ABC system.

A complex man, possessed by a strong drive to achieve, Dr. Goethert maintained his interest in his work even after the heart attack had sapped his strength. Each time his short course manager visited, he quizzed her about sign-ups for the Seattle short course, fretting and fuming that more had not signed up—"Why isn't someone doing something?" Eventually, Sandy began exaggerating the numbers so he wouldn't get upset. She told him that she and **Roy Schulz** would take care of it.

At 11:30 A.M. on March 29, 1988, Dr. Goethert died at Coffee Medical Center in Manchester. He was eighty years old, survived by his wife, Hertha, four children, and ten grandchildren.

Bob Young, who had worked closely with him in founding UTSI and later as the first associate dean, said that Dr. Goethert had "played a key role in the economic development of Middle Tennessee, the enhancement of the nation's defense capability and the advancement of graduate level and continuing aerospace education." **Ken Harwell** said Goethert's "distinguished career as a researcher, educator, administrator, and leader has earned him a place in history with the great minds of aeronautics and astronautics." He was, Harwell said, "truly a citizen of the world who loved his work and was dedicated to the education of scientists and engineers," and his "imprint on UTSI and on the technology of flight will live on forever."

Memorial services were held at 2 P.M. On Thursday, March 31, at the Trinity Lutheran Church in Tullahoma, and on the following Monday in the UTSI auditorium. The family suggested that contributions might be made to the B. H. Goethert Fund at Central High School in Manchester to further the education of gifted students. It was the kind of thing that would surely have pleased "Doc."

Sriniva Anday, who later would become president of the Student Government Association, had begun graduate studies at the Space Institute in January, but his entrance was hardly typical. A native of Hyderabad, India, Anday arrived at the Greyhound bus station in Tullahoma on a cold and rainy night, having traveled from Oklahoma. He asked a woman how far it was to the Space Institute, and she answered, "About ten minutes," A taxi driver said he would take Anday to UTSI for thirty dollars. Still under the impression that the Institute was "in" Tullahoma, Anday reasoned: "This is a small town. Why should I pay thirty dollars when I can surely walk?"

After taking a wrong road, Anday eventually headed out the AEDC access road that leads to the Institute. He took a blanket from a small handbag and covered his head against the rain as he trudged along the dark highway. Hearing noises in the underbrush and catching an occasional glint of eyes from the roadside forest, Anday "didn't know what kind of wild animals might attack me. I began to wonder if there really was a space institute." Signs informing him that a military installation was in the vicinity raised fears that he might run afoul of authorities. Then a state trooper passed, skidded to a stop, and backed up to ask the miserable student where he was going. Convinced, the trooper gave Anday a ride into the almost abandoned Institute.

Entering the lobby, Anday "only saw one person. She was a Chinese student, and I did not speak Chinese." Belatedly, he figured out that the stranger in Tullahoma who had told him UTSI was ten minutes away "thought I had a car."

UTSI hosted a conference in April for undergraduate students and faculty from across the southeast in conjunction with National Science and Technology Week. It was co-sponsored by student chapters of the American Society of Mechanical Engineers, the Institute of Electrical and Electronic Engineers, and the American Institute of Aeronautics and Astronautics.

Dr. Harwell told the Support Council that fundraising to match a second "chair of excellence" at the Institute was gaining momentum under the leadership of committee chairman **John Parish** of Tullahoma, who also had made a major donation. Existing chairs of excellence in the UT system had been matched, but Governor McWherter's proposed budget contained money for eight new chairs for UT. **Dr. John Prados,** UT vice president for academic affairs and research, said the Space Institute had a "high priority" within the UT system for a second chair if the legislature funded the eight new ones that had been proposed.

Support Council members adopted resolutions eulogizing Dr. Goethert and recognizing and commending Mr. Kamm, who had retired after twenty years of service. Dr. Roberds, associate dean for academic affairs, and Dr. Lea reported that minority recruiting at UTSI had been very successful. The Institute had a minority enrollment of 9.3 percent of the full-time student body, Roberds said, compared with a national average of 1.4 percent.

Dr. Frank Collins, Dr. Atul Sheth, Mr. Bradley Winkleman and Mr. William Boss represented UTSI as judges for the thirty-ninth International Science and Engineering Fair in Knoxville in May, 1988.

The Department of Energy awarded the Space Institute $9.2 million to continue developing MHD technology, bringing to a total contract value of MHD funding at the Institute $67.9 million since 1977. Actually, other energy conversion research contracts for the fiscal year 1988 pushed the total figure for that year beyond $10 million.

Sun Microsystems of Mountain View, California, donated computer equipment to UTSI with the potential to speed the development of three-dimensional pictures of the body for medical diagnosis. **Dr. Al Pujol,** associate professor of electrical engineering and director of UTSI's microprocessor laboratory, said the special computer graphics work station would be used as a teaching device and a research tool. He said the concept might eventually be applied in the engineering field to aid design engineers in visualizing three-dimensional concepts. Actor **Ben Vereen,** who had become interested in research being done at UTSI after he met Pujol at a party in Nashville the previous year, helped the Institute obtain the equipment. Vereen visited the Institute in June to see the new equipment demonstrated. He also joined **Dr. Arliss Roaden,** executive director of the Tennessee Higher Education Commission, in commending the Institute on its success in recruiting black graduate students. UTSI had increased its number of full-time black students from two in 1985–86 to seven. While this number was small, Roaden said, it was "very large" in relation to the number of blacks in highly technical, scientific fields. A month before, a black student, **Dennis Campbell** of Memphis, a master's degree candidate in the engineering and science curriculum at UTSI, was elected president of the Student Government Association.

The UTSI Sigma Xi Club presented awards for excellence in science and math education to three area teachers: **Linda Giltner** and **Ann Epley,** teachers at Tullahoma's Middle School, and **Juliet Sisk** from Coffee County Junior High in Manchester.

Gary D. Smith of Tullahoma—whose 22-year career at the Space Institute ranged from graduate student to special assistant to the dean—was awarded the 1988 meritorious service citation in the spring of 1988 by the American Society of Mechanical Engineers (ASME). Smith was one of the **founders** of the Highland Rim section of ASME and had held several important positions at the local, regional and national levels. For six years, he had been a member of the regional operating board that governs ASME activities in Tennessee, Mississippi, Alabama, Georgia and Florida. **Bobby L. Green,** ASME regional vice president, who presented the award, said that Smith's most significant contribution to the society may have been in 1985 when he was an ASME Congressional Fellow and spent the year in Washington, D.C., as a science advisor to **Senate Majority Leader Robert C. Byrd** of Virginia. Smith had earned his master's degree in mechanical engineering at UTSI in 1971. After **Jules Bernard** retired at the end of 1985, Smith became principal news media spokesman and liaison with Tennessee's congressional delegation. He had served on the Coffee County Commission since 1978.

Dr. Harwell hosted a reception on June 1 for 18 graduates; including one recipient of a doctorate (**Sudheer Nath Nayani** of Hyderabad, India), and 17 who had earned master's degrees.

On July 1, 1988, former two-time **Governor Lamar Alexander** became president of the University of Tennessee. After being succeeded as governor by **Ned McWherter,** Alexander and his family had taken a six-month sabbatical to Australia. Now he was back in the public eye. As far as UTSI and its new position of vice president were concerned, Alexander said he would have to study the mission of the Institute before taking any action to fill the new position.

Dr. Charles R. (Rick) Chappell, chief scientist and associate director for science at NASA's Marshall Space Flight Center, toured various UTSI labs during a visit in July. Dean Harwell told him that the Space Institute had eleven research contracts with NASA in addition to housing the NASA-sponsored Center for Advanced Space Propulsion.

Dr. Charles Lea resigned as assistant dean for admissions and student affairs in order to accept his appointment as dean of the School of Arts and Sciences at Chattanooga State Community College effective August 1. The move put Lea together again with **Dr. Harry Wagner,** president of the Chattanooga College, who had been president of Motlow College during most of Lea's tenure there. In 1989, Lea became dean of the Chattanooga college.

Dr. Dennis Keefer of Manchester, who was chairman of the Center for Laser Applications (CLA) at the time, received the General H. H. Arnold Award from the Tennessee Section of the American Institute of Aeronautics and Astronautics (AIAA) in 1988. **Dr. Roy J. Schultz,** immediate past president of the section, said Keefer was recognized for "sustained and significant contributions to the advancement of laser-powered space propulsion technology." Schulz said Keefer had contributed to the comprehensive understanding of the "interaction among the thermal, fluid mechanical, and optical mechanisms that control laser sustained plasmas in convective flow." Keefer, a professor of engineering science, joined the UTSI faculty in 1978 and helped found CLA. A Floridian, he held three degrees from the University of Florida.

Late in July, the Tennessee Higher Education Commission (THEC), meeting in Memphis, designated CLA as an "accomplished center for excellence." This designation meant that the center—which **Keefer, Jim Few, Carroll Peters** and **James Lewis** had gotten established in July, 1984—had reached a level of accomplishment to merit more stable funding and less detailed annual review, Harwell said. A center had to have been in operation for at least three full years and receive a top evaluation in order to be eligible for "accomplished center" status. External consultants ranked CLA at the "highest level of accomplishment," the dean said. THEC's action meant the center was given the green light to continue for the next five years (1993), when it would undergo another review.

Jim Few, CLA program manager, cited several specific accomplishments, including the following: Cumulative sponsored research income of $4.3 million, which exceeded the benchmark goal by one hundred percent; cumulative refereed publications that exceeded the goal by thirty-three percent; cumulative presentations that exceeded the target by one hundred and fifty percent, and a doubling of faculty and staff to twenty-four. The sponsored research income per faculty member was one of the highest, if not the highest, in the U.S., Few said, as compared with physics departments in the major U.S. universities. The new CLA building would be ready for occupancy by the following February.

Dr. Harwell announced that UTSI would offer a new program in Space Engineering in the fall, 1988, semester, with a curriculum leading to a master's or a doctoral degree in aerospace engineering. Early course offerings included one in rocket propulsion by **Dr. Roy J. Schulz,** one in propulsion testing by **Dr. Carroll Peters,** and "Special Topics in Aerospace Engineering: Space Engineering," by **Dr. Firouz Shahrokhi,** who would be in charge of the new studies.

Shahrokhi and Young, chairman of the Mechanical and Aerospace Engineering Department, had initiated the project, and in June of the following year (1989), the UT Trustees approved it as a concentration in Space Engineering within the Mechanical and Aerospace Engineering Department. Several new courses were proposed. (In October, 1989, **Shahrokhi** and **Dr. Edwin Gleason** visited the U.S. Air Force Academy in Colorado in response to the Academy's interest in the new program. They worked out a plan for the Academy to select a few of its top aerospace graduates and enroll them in the program at UTSI starting in the fall of 1990.) In late 1989, Shahrokhi would say that he had received more calls about this program than for any other.

In September, 1988, Schulz set up a one-week short course—"Fundamentals of Solid Rocket Motors"—conducted by **Dr. Robert Kruse** from the Huntsville Division of Morton Thiokol, Inc.

The Institute got a two-year, $530,000 contract in August to help the Westinghouse Corporation develop a combustor—more efficient and less polluting—than systems normally used in coal-fired power plants. **Norman R. Johanson** of Manchester, principal investigator and ECP program manager, said UTSI would conduct performance testing of a topping combustor that Westinghouse had built. The combuster—labeled a ''multi-annular swirl burner''—was to be a component in the pressurized fluidized bed combustor (PFBC) system.

Charles L. (Chuck) Wagoner of Tullahoma, a senior research engineer at UTSI, was chosen to serve a four-year term on the technical advisory committee for the Coal Quality Development Center, a division of the Electric Power Research Institute, in Homer City, Pennsylvania. Wagoner, who joined UTSI in 1987 after thirty-two years with the Babcock and Wilcox Company in Alliance, Ohio, was principal investigator on a Space Institute project to develop an advanced combustion system for industrial boilers.

Sixteen summer of 1988 graduates—all completing work for master of science degrees—were hosted at a reception on August 12.

Successful tests of Doppler radar at Denver's Stapleton airport in the summer of 1988 were called by **Dr. Walter Frost** a ''major step'' toward helping airline pilots avoid windshear microbursts, which had killed an estimated five hundred American air travelers over a twenty-year period. Frost had been one of the first windshear investigators to recommend the use of Doppler radar to warn pilots of windshear. It was Frost who, twelve years earlier, developed one of the first computer programs that could simulate windshear conditions, and those computer models were used to train pilots in cockpit simulators. Frost joined the Space Institute on March 1, 1965, after receiving his doctorate from the University of Washington. FWG Associates Inc., founded by Frost, was located in the UTSI Research Park.

Dr. Bruce A. Whitehead, associate professor at Drexel University in Philadelphia, joined the Space Institute faculty in the fall of 1988 as associate professor of computer science. Discussing his research interest in neural networks, Whitehead said they apply to computers ''the same principals that the human brain uses to function.'' Just like the human brain, he said, neural networks have the ability to learn on their own. The Estill Springs resident was a National Merit Scholarship recipient and received his doctoral degree in computer science from the University of Michigan in 1977. He had been a member of the technical staff of AT&T Bell Laboratories before going to Drexel University.

On September 2, 1988, a UTSI research team completed a record one hundred and eleven-hour **uninterrupted test run** of the coal-fired magnetohydrodynamics power plant. **Dr. Joel Muehlhauser** said the team operated the facility for a total of two hundred and fifty hours over a two-week period, including the span of continuous operation. This made a total of five hundred and fifty-seven hours of coal-fired MHD testing during the 1988 fiscal year, and a grand total of one thousand, twenty-three hours since the program began. Preliminary inspection of the facility and the MHD test train components revealed no major problems resulting from the record test, **Marvin Sanders,** facilities operations manager, said. **Norman Johanson,** ECP program manager, said the testing at the Space Institute was proving the feasibility of the concept over long periods of operation while providing critical design data required by equipment manufacturers for production of commercial scale power plant equipment.

UTSI also had been selected to provide technical support to an Italian consortium developing the MHD technology for commercial electric power plant applications. Johanson said the consortium—Ansaldo, Snamprogetti, and Franco Tosi—planned to eventually launch commercial MHD projects throughout the world and had contracted with the Institute for specialized technical support in helping them design an MHD retrofit system. The Italians planned a pilot plant that would be practically identical to the one UTSI had built for the Department of Energy.

The American Institute of Aeronautics and Astronautics in the fall of 1988 published a book edited by **Dean Harwell** and **Dr. Firouz Shahrokhi** entitled ''Commercial Opportunities in Space.'' The book emphasized how scientists and engineers from developing nations might capitalize on commercial opportunities in space, Harwell said, and reflected recent studies on the subject by colleges and universities.

With the Space Institute's silver anniversary approaching, Dr. Harwell appointed a ten-member committee to develop and catalog historical documents of the Institute and to compile a written history. He put **Dr. Arthur Mason** in charge of the committee, which included the following members: **Gary Smith,** special assistant to Harwell; **Dr. Robert L. Young,** professor of aerospace and mechanical engineering; **Robert W. Kamm,** assistant dean emeritus; **Helen Mason,** first librarian at UTSI; **Virginia Richardson,** first UTSI registrar; **Ruth Binkley,** administrative assistant, and **Weldon Payne,** writer. **David M. Hiebert,** U.S. Air Force historian at AEDC, and **Dr. Milton Klein,** UT Knoxville University historian, were named as ex-officio members.

Harwell noted that UTSI had conducted more than $114 million worth of sponsored research in support of national priorities, including the U.S. manned space flight program and advanced energy conversion technologies since it was established.

Following the example of **UT President Lamar Alexander,** who, after taking office on July 1, had spent two days with a UT Knoxville student on that campus and one day with a professor at UT Chattanooga, **J. Steven Ennis** of Tullahoma, a member of the UT Board of Trustees, spent October 6, 1988, with **Ronald Litchford,** A Winchester native and UTSI graduate student seeking a master's degree in mechanical engineering. Ennis first accompanied Litchford for a lecture by **Dr. S. M. Jeng** in heat and mass transfer modeling. He then went with Litchford to the Center for Laster Applications, and then attended **Dr. C. E. Peters'** class in propulsion ground testing and a class in rocket propulsion with **Dr. Roy J. Schultz.**

UTSI co-sponsored an Aerospace Technical Conference in October, designed to match needs of federal contractors in the Huntsville area with qualified manufacturing capability in Tennessee. Other sponsors were the Tennessee Valley Aerospace Region (TVAR), the State of Tennessee Department of Economic and Community Development, and the Tennessee Valley Authority.

Fourteen U.S. Navy personnel, including one lieutenant commander and thirteen civilian flight test engineers, graduated from the second twelve-week UTSI flight test engineering program taught on site at the U.S. Naval Air Test Center (NATC) at the Patuxent River Naval Air Station. (The first class graduated in the summer of 1987). **Captain Richard Sidney,** director of the NATC rotary wing directorate, in remarks at the graduation ceremony, praised the program, saying it should not be considered as an alternative to test pilot school, but rather "as a prerequisite . . . for NATC engineers."

Dr. Harwell said UTSI expected that UTSI's first two graduates from NATC would receive master's degrees in aviation systems at the end of the fall semester. Three flight test engineers at McDonnell Douglas in Long Beach, California, also were enrolled in the Institute's aviation systems off-campus program, he said.

Fred Watts, UTSI program manager and associate professor in aviation systems, said that by tailoring the program to the needs of flight test engineers, UTSI was able to graduate twelve to fourteen engineers in about a third of the time and at about the same cost as the test pilot school required to graduate one engineer. Subject matter covered in the course included propulsion, high and low-speed aerodynamics review, performance, performance flight testing, stability and control, stability and control flight testing, flight test instrumentation, handling qualities, and high angle of attack testing. Also participating in the program were **Dr. Peter Soiles** and **Betsy Smith** from UTSI, and **Dr. Harvey Joe Wilkerson** from UT Knoxville. Three UTSI aircraft, including the newly acquired Saberliner T–39, were used to provide actual flight instruction in the program.

At 4:30 P.M. on Thursday, October 20, 1988, **Dr. William H. Pickering,** president, Pickering Research Corporation, Pasadena, California, delivered the fourteenth annual **Quick-Goethert Lecture** at UTSI on "Who in the World Will Dominate Space?" Pickering, professor emeritus of electrical engineering at California Institute of Technology, had served as director of Cal Tech's Jet Propulsion Laboratory from 1954 to 1976. He said the United States must quickly establish popular new space program goals or the Soviet Union would become the leader in space exploration. The Soviet space program seemed to have stronger popular and governmental support, he said, and was "characterized by steady evolution, building on successful components," while the U.S. Apollo mission was set up as a dead-end program.

"We must either accept a leadership role and invite others to be on our team," Pickering said, "or we will stand aside and watch the Soviets leading mankind off this small planet into the universe." Joint ventures with the Soviets and other nations to explore Mars and to build a lunar base should be high on the U.S. agenda, he said. The speaker had no doubt that Americans would support commercial and military space efforts, but he was not sure about support for research and exploration. He said man's imagination and curiosity led him out of the Stone Age and that space research and exploration was the next challenge. Pickering was to deliver the same lecture at the Technical University of Aachen in Germany the next May. (And the following October, at the fifteenth Quick-Goethert Lecture, listeners would hear a different plea for cooperation in space ventures from the head of the European Space Agency, **Dr. Reimar Luest.**)

Also in October, 1988, Dr. Harwell named **Thomas E. Bailey** of Fayetteville as manager of technology assistance programs at UTSI. Bailey was to be the "focal point" for transferring the advanced technology developments resulting from UTSI's research to Tennessee companies. He would continue serving as executive director of the Tennessee Valley Aerospace Region (TVAR), with offices in the Upper B wing of the Institute. A civil engineer, Bailey had worked for years with the Tennessee Valley Authority before heading up TVAR.

With **Dr. Firouz Shahrokhi** as program director, the Space Institute joined in a unique satellite tracking program designed to accurately monitor the movements of migrating bald eagles by using a solar-powered radio transmitter. Small enough to be worn by the birds, the transmitters were so powerful that their signals could reach a satellite. In December, two

American bald eagles were captured by the Tennessee Wildlife Resource Agency (TWRA), fitted with the transmitters, and released at Reelfoot Lake while being monitored by the UTSI staff. The program attracted a lot of attention, partially because of its possibilities in other areas, such as tracking down stolen vehicles. Other sponsors of the "eagle project" were TWRA, the U.S. Fish and Wildlife Service, Tennessee Technological University, and Cumberland Wildlife Foundation. **Mike Luke** was UTSI project technical director.

Before the year ended, **Dr. Kenneth E. Harwell,** who had become the third dean in UTSI's history in 1982, announced that he was leaving the Space Institute after thirteen years. **Dr. Louis Padulo,** president of the University of Alabama at Huntsville, appointed Harwell as vice president for research and associate provost at UAH, effective January 1, 1989. Harwell, an Alabama native, was no stranger to Huntsville, and as he noted, many of his former students were living and working in Huntsville. Years before, Harwell had spent time in Huntsville as an Auburn faculty member with NASA research contracts and as a special assistant to the director of the Army's Research, Development and Engineering Laboratory. He said one thing that made the UAH offer attractive was its "space initiative" developed under **Dr. William Lucas,** former Marshall Space Flight Center director, to support the nation's space efforts into the next century.

Harwell was leaving behind an impressive record that included an eighty-eight percent growth in enrollment, a thirty-two percent increase in full-time faculty, and a seventy-seven percent increase in sponsored research revenue.

On December 13, UT **President Lamar Alexander** said he would appoint an interim dean for the Institute in early January, and that this person would serve until a vice president was chosen for the Institute. Alexander said that he and **Lewis Branscomb,** former chief scientist for the International Business Machines, would visit UTSI in January to help the president "get a clearer picture of what the next steps for the Institute ought to be." Shortly after that, he would accelerate the search for a vice president, Alexander promised.

Another change affecting UTSI came in mid-December when **Dr. Edward M. Kraft** was named an executive director of the University of Tennessee-Calspan Center for Aerospace Research (CAR), succeeding **Dr. J. William Davis,** Calspan vice president and general manager of Calspan's AEDC Operations. The announcement came from **Dr. John W. Prados,** UT vice president and CAR board chairman. Davis was to continue on the board but had asked that his CAR duties be reduced so that he could devote his efforts to increased responsibilities at AEDC. Kraft (who had received both master's and Ph.D. degrees from UT in aerospace engineering) was serving as manager of the technology and analysis branch for Calspan/AEDC Operations and also was an adjunct professor of aerospace at UTSI. He brought with him a reputation as an international expert in aerodynamics and fluid mechanics. CAR was the parent firm of the Center for Advanced Space Propulsion (CASP), and **Dr. Fred A. Speer** of Huntsville was serving at this time as CASP's first director. In February of the next year, **Dr. George Garrison** of Tullahoma, a UTSI professor, became CASP director and Dr. Speer stayed on as associate director.

President Alexander chose **Dr. Richard M. Roberds,** who had been associate dean for academic affairs since 1986, as acting dean of the Space Institute, putting the former Air Force colonel and combat pilot in the cockpit as UTSI zoomed into its twenty-fifth year.

Dr. Richard M. Roberds

✦

Chapter Twenty-Five

1989: A Year of Celebration

DR. RICHARD M. ROBERDS, who had logged more than four thousand hours of flying time during his twenty-four years with the Air Force, took over the controls at the Space Institute on the first day of **1989,** the year of the **Silver Anniversary Celebration.** It also was the year that a long-awaited decision would be made as to UTSI's vice president. Roberds had flown one hundred and eighty-three combat missions over Vietnam and Cambodia, winning two Distinguished Flying Crosses and eight Air Medals, before retiring as a colonel in 1980. He had joined UTSI's staff in 1986 as associate dean for academic affairs, and he became acting UTSI dean at the start of a busy and eventful year.

Bob Kamm was brought out of retirement in early 1989 to head the Silver Anniversary Committee, which put together a string of events that would stretch from February into November and bring in guests ranging from an astronaut to two defecting Romanian violinists, and from an expert on the National Aero-Space Plane to the congressman son of the late Governor **Frank G. Clement.**

But plenty was happening apart from the celebration. It was the "year of the astronauts" with six of them visiting the campus, and it was a good year for aviation with construction starting on UTSI's Flight Research Center at the Tullahoma Airport and a contract to teach flight testing to FAA pilots and engineers. The MHD team succeeded in removing ninety-eight percent of sulfur dioxide from burning coal, a bald eagle named **Phoenix** helped a UTSI graduate student with her thesis, funds were approved for a new building that would house UT-Calspan's Center for Aerospace Research (CAR), which operated the Center for Advanced Space Propulsion, and computer experts from all over the world came for a second conference on artificial intelligence. The Center for Laser Applications finally got into its new, 11,000-square-foot building in early 1989. The building had been put up in two stages at a total cost of around nine hundred thousand dollars. It was built near the laboratory area—between the main campus and the Coal Fired Flow Facility. Before, CLA facilities and personnel had been housed in an older, tin building.

It also was the year when the world watched in amazement as Chinese students demonstrated for freedom on Tiananmen Square and then the sadness and shock of the Beijing massacre was keenly felt on the Space Institute campus as Chinese students agonized over the fate of friends and relatives. And **Norm Johanson** worried that the turmoil in China had interrupted the Space Institute's timetable for helping China solve its industrial energy shortage. However, in early December, four top officials came from China with a seventy thousand dollar check and entered into a second agreement with UTSI to study retrofitting two coal-fired power plants near Beijing.

In April, **Dr. Edwin M. Gleason,** a former Air Force colonel, assumed duties as assistant dean for admissions and student affairs, filling a vacancy left in 1988 when **Dr. Charles L. Lea** resigned to join the staff of Chattanooga State Technical Community College. Gleason received both his master's degree and an Ed.D. in Educational Psychology and Guidance from UT and graduated from Baylor University with a bachelor's in psychology. He came to UTSI with more than eighteen years experience in higher education faculty and administrative positions. He had served the previous five years as Dean, Civilian Institution Programs, at the Air Force Institute of Technology (AFIT) at Wright-Patterson Air Force Base, Dayton, Ohio. Gleason arrived in time to join in a reception for the twenty-three spring graduates on May 11. Two of them—**Jiada Mo** of China (aeronautical engineering) and **Karen Seiser** of Huntsville (engineering science)—were getting Ph.D.'s.

Dr. Edwin Gleason and Callie Taylor, left, later were joined in the office by Melinda Branch.

Gary Smith had joined **Dr. Kenneth Harwell** at the University of Alabama at Huntsville in March as director of governmental relations. **Weldon Payne** of Manchester, a former newspaperman and Motlow College teacher, assumed some of Smith's public relations duties at UTSI on a part-time basis for the anniversary year and would become full-time coordinator of media and governmental relations on the following January 1.

UTSI acquired a mini super-computer that **Dr. Kenneth Kimble,** director of the computer center, said saved tremendous time. A former UTSI student, **James R. White Jr.,** helped UTSI get the equipment—a demonstration model—from Alliant in Huntsville at about half the normal price. **Linda Hall,** coordinator of the computer center, and **Linda Williams,** principal secretary for three divisions, were among two hundred and fifty professionals from around the world attending the tenth annual meeting of the TEX Users Group in California in 1989. Both women taught classes at UTSI in the highly sophisticated computer typesetting system.

In May, **Dr. Joel Muehlhauser,** who received his Ph.D. in physics at UTSI and in 1988 became acting administrator of the Energy Conversion Research and Development Programs (ECP) at the Institute, was chosen by UTSI as "Professional of the Year." He was recognized by the Huntsville Association of Technical Societies (HATS) at its annual meeting. **Dr. Ted Paludan,** chairman of the UTSI nominating committee, said Muehlhauser was nominated because of his "excellent performance" at the helm of ECP. (The following January, Muehlhauser was named to the permanent position of administrator, and in late 1990 would become UTSI's first Research Dean.)

Ruth Binkley, who had been with the Institute since its beginning and had filled in at the switchboard before the lobby was finished, retired on May 31. **Al Ritter,** chairman of the National Advisory Board, gave her a dozen roses in honor of her varied contributions during the Institute's first twenty-five years.

Paludan, professor of geography at UTSI, was the Institute's alternate representative to HATS, and he was one of five from UTSI who presented papers at the session in May. Others were **Dr. Ching F. Lo,** professor of aeronautical engineering; **Dr. George Garrison,** director of the Center for Advanced Space Propulsion and a professor; **Dawn Utley,** a former UTSI student and administrative assistant in the Industrial Engineering Department, and **Thomas E.**

Al Ritter has roses for Ruth Binkley on her retirement after 25 years.

Bailey, director of Tennessee Valley Aerospace Region.

The U.S. Department of Energy renewed UTSI's contract for MHD research and development on May 1, extending the agreement through July 31, 1990. The accompanying $11,411,903 appropriation represented the second largest contract award received by UTSI since construction of the MHD facility and brought to nearly seventy million dollars the amount spent on the program at the Institute since 1977. "Our goal," Muehlhauser said, "is to be able to add MHD technology to existing coal-fired plants in the mid-1990's."

The Tennessee Section of the American Institute of Aeronautics and Astronautics honored **Dr. Firouz Shahrokhi** with the "Arnold Award" in ceremonies at AEDC on May 30, 1989. Among Shahrokhi's accomplishments cited by **Johnny M. Rampy,** awards chairman, was his research to demonstrate the feasibility of using remote sensing and telemetry techniques for determining the migration activities of eagles and other wildlife. (He and **Carey Roberts,** Manchester graduate student, created a stir with the "Eagle" project, which involved equipping bald eagles with tiny, solar-powered transmitters in order to keep close track on the whereabouts of the birds. Phoenix, a young eagle that had once been shot, surprised the researchers by settling down on a Missouri farm shortly after being released with one of the transmitters, but Shahrokhi and Roberts were pleased—Roberts had been "dead on" Phoenix's trail and when he established residence in a barn lot, he made it easy to confirm that the transmitter indeed was working.) Shahrokhi also had been program director for a four-day symposium held in Nashville in March on "Space Commercialization: Roles of Developing Countries," which was attended by more than five hundred space scientists, corporate executives and government officials from thirty-two nations.

In early June, **Dr. Moonis Ali,** as general chairman, brought more than two hundred computer experts from about twenty countries to UTSI for a four-day conference on artificial intelligence. One of the speakers, **Dr. Y. T. Chien** of the National Science Foundation, said that to speculate about robots "making our coffee" was equivalent to the 19th Century scientists saying everyone should have an electric motor. Other keynote speakers included **Dr. James Bezdek** of Boeing Electronics, **Dr. Jaime Carbonell,** Carnegie Mellon University, **Dr. B. Chandrasekaran,** Ohio State University, and **Dr. C. R. Weisbin,** Oak Ridge National Lab.

Also in June, the UT Board of Trustees approved a new concentration in Space Engineering within the Mechanical and Aerospace Engineering Department at UTSI. The program, following a curriculum leading to a master's or doctoral degree in Aerospace Engineering, was initiated by Dr. Young and Dr. Shahrokhi. Roberds said the new emphasis was in keeping with a major mission of the Institute to provide high quality space engineering graduates to staff the aerospace industry into the next century. Shahrokhi said he had received more calls for the program than for any other, and the Air Force Academy was planning to send some of its top aerospace graduates to the program in the fall of 1990.

Srinivas Anday from Hyderabad, India—the young man who had walked for hours trying to find UTSI on a cold January night in 1988—was elected president of the Space Institute Student Government Association, succeeding **Dennis Campbell. Carey R. Roberts** of Manchester was elected to succeed **Herbert Thomas** as vice president, and **Christian M. Norton** of Murfreesboro followed **Selena Espy** as vice president for finance.

Anday was the twenty-first SISGA president since SISGA was established in 1968 with **Hrishikesh Saha** as the first president. Other presidents had been **Roger Crawford,** 1969; **Luther Wilhelm,** 1970; **Michael Varner,** 1971; **Frank Ianuzzi,** 1972; **Walter Harper,** 1973; **Raymond J. Ricco, Jr.,** 1974; **Donald Ey,** 1975; **Tran My,** and **Don Welch,** 1976; **John Harris,** 1977; **Bill Holt,** 1978; **Robert W. Clemons,** 1979; **K. Suryanarayanan,** 1980; **Mike Guthrie,** 1981 and 1982; **Peter Liver,** 1982, 1983, and 1984; **G. Alan Lowrey,** 1985; **Shannon Byrd,** 1986; **Kim Nelson,** 1987, and **Dennis Campbell, 1988.**

Dr. Ernst Messerschmid, a German astronaut and director of the Institute for Space Systems at the University of Stuttgart, on July 19, 1989, lectured at UTSI, recalling his experiences as a payload specialist aboard the last successful flight of the space shuttle Challenger. That flight, launched on October 7, 1985, was commanded by UTSI graduate **Hank Hartsfield.** (U.S. Air Force **Colonel Guion S. Bluford Jr.,** a mission specialist on that flight, spoke at UTSI in February, 1989.)

Messerschmid said passengers on the Challenger "got younger" by 26 microseconds a day during that seven-day mission. The calculation was based on comparing atomic-clock measurements of elapsed time in orbit with atomic clocks on earth. The astronaut said the difference was caused by the higher speed of the shuttle compared to the earth's rotation and by the lower pull of gravity on the shuttle compared to that on earth.

Messerschmid and his wife, **Gudrun,** flew from NASA's Johnson Space Flight Center in Houston for the lecture at the urging of the astronaut's personal friend and fellow aviator, **Thomas Gogle** from

Stuttgart, who was at UTSI on a one-year German scholarship. Gogle had studied under Messerschmid at the University of Stuttgart. Another of Gogle's former teachers, **Dr. Herbert O. Schrade,** professor and vice director of the Institute of Raumfahrt Systems, at the University of Stuttgart, conducted a seminar at the Space Institute a few days earlier on electric propulsion research.

Four other astronauts spent the last week in July at the Space Institute, being checked out on an experiment that was scheduled to fly aboard a space lab mission. A husband-wife team from Decherd, **Drs. Mary Helen** and **T. Dwayne McCay,** were hoping to determine whether superalloys could be more effectively made in a reduced gravity environment. Two astronauts—**Dr. Roberta L. Bondar** and **Dr. Kenneth Money**—were from Canada. A third, **Dr. Ulf D. Merbold,** was from the European Space Agency, and **Dr. Roger K. Crouch,** from NASA headquarters in Washington, D.C., was the only Tennessean; his mother, **Mrs. Maxine Crouch,** drove over from Jamestown and had lunch one day with her son, who joked: "She wanted to meet a **real** astronaut." Bondar and Merbold eventually were chosen as payload specialists with Crouch and Money as alternates. (The mission would be delayed for at least two years. Navy Captain **Manley Lanier (Sonny) Carter** and **Dr. Norman E. Thagard** later were chosen as mission specialists. However, Carter was killed in a plane crash that also claimed the life of **John Tower,** and Marine Lieutenant Colonel David Hilmers was chosen to take Carter's place.)

Registrar **Bettie Roberts** presented three candidates for Ph.D.'s and sixteen for master's degrees at a reception held August 10 to honor the UTSI students, who would receive degrees the next day in ceremonies at Knoxville.

The Institute also acquired in the summer of 1989 an Electron Microprobe Analyzer that the McCays said would be useful at a later stage in their search for a stronger metal. The sophisticated equipment, which was bought from the University of Texas for less than a hundred thousand dollars, was immediately put to use in welding studies at the Center for Laser Applications. Normally valued at about $700,000, the EMA was capable of analyzing with a high degree of accuracy the composition of metals.

✧ ✧ ✧

On August 2, 1989, **Dr. John W. Prados,** chair of the vice president search advisory committee, notified UTSI faculty and staff that his committee, meeting on July 27, had reviewed resumes of fifty-two applicants for the new position, and hoped to identify five finalists when it met again on August 18. Prados was hopeful that **Lamar Alexander** could recommend an individual to the UT Trustees at their September 29 meeting. On August 24, Prados identified the following finalists: **Dr. Ernest J. Cross Jr.,** dean of engineering and technology, Old Dominion University, Norfolk, Virginia; **Dr. Wesley L. Harris,** dean of engineering, University of Connecticut at Storrs; **Dr. George A. Hazelrigg Jr.,** deputy director of electrical and communications systems, National Science Foundation, Washington, D.C.; **Dr. Brian L. Hunt,** technology manager of the advanced tactical fighter program, Northrop Corporation, Hawthorne, California, and **Dr. Y.C.L. Susan Wu,** resident, Engineering Research and Consulting Inc., Tullahoma.

In September, Dr. Roberds announced the schedule for candidates to visit the Institute. By this time, the list had dwindled to four because Dr. Hunt had asked that his name be withdrawn from consideration. The other four visited in September, and each spoke during an open meeting in the auditorium.

Five UTSI faculty members who served on Prados' committee were **Dr. Robert L. Young, Dr. K. C. Reddy, Dr. Al Pujol, Dr. Mary Helen McCay,** and **Dr. James W. L. Lewis. Dennis L. Campbell** and **Kim D. Nelson** represented Space Institute students, **Eddie Washington** of ECP and **Joseph P. Hane,** physical plant, were UTSI staff representatives. External members included **J. Steven Ennis,** Tullahoma trustee, **Amon Carter Evans** of Columbia, Middle Tennessee trustee, **Mrs. Charlotte Parish,** Tullahoma, former trustee, **William R. Carter,** Fayetteville, chairman of the Support Council, **Dr. Jack Whitfield,** former chairman of the National Advisory Board, **Dr. J. William Davis,** vice president and general manager, Calspan Corporation, AEDC Division, and **Tom Bailey,** director of Tennessee Valley Aerospace Region. Members from the Knoxville campus were **Dr. William T. Snyder,** dean, College of Engineering, and **Dr. William M. Bugg,** professor and head of the UT Physics Department, Knoxville.

While speculation ran high at the Space Institute, The Tennessean on Saturday, October 7, reported that appointment of a vice president would be considered at a meeting of the UT trustees' executive committee on the following Tuesday morning. **Beauchamp Brogan,** UT general counsel and secretary of the board of trustees, was quoted as saying the committee would hold a telephone conference-call meeting to consider Alexander's recommendation.

UTSI's campus buzzed with questions on Monday and Tuesday, but the answers were not to come until Wednesday morning, October 11, when newspapers reported that President Alexander was recommending **Dr. Wesley L. Harris,** dean of engineering at the University of Connecticut and a former professor at the Massachusetts Institute of Technology.

Dr. Wesley L. Harris

UT Knoxville had put out a news release late Tuesday, embargoed for 9 P.M. (EDT). The executive committee would, the story said, meet by telephone conference call the next morning.

At 8 A.M. on October 11, **Dr. Harris** and his wife, **Sandra,** along with **Lamar Alexander** attended a breakfast/news conference in Suite 4-B of the Hyatt in Nashville. About ten news men and women were present. Alexander said Harris would officially assume full-time duties on July 1, 1990. (After the breakfast, a UT spokesman said that the executive committee had approved Alexander's choice.)

Dr. Harris said he would concentrate on identifying UTSI's "preeminent" areas of expertise. As examples, he cited the Center of Excellence for Laser Applications, the Energy Conversion Program's influence on environmental problems, and Computational Fluid Dynamics.

Alexander had described Harris' coming to UTSI as "an important step forward for the University of Tennessee." Harris was, he said, a "first-rate scientist and academic leader" who had earned the "highest respect for his work at the University of Virginia, Princeton, and MIT."

A native of Richmond, Virginia, the 47-year-old Harris had earned the bachelor's degree in aerospace engineering in 1964 from the University of Virginia, and the Ph.D. from Princeton University in 1968. He had headed his own planning and analytical firm and served as director of Massachusetts Institute of Technology's helicopter rotar acoustics group from 1980 to 1985. He also had been professor of aeronautics and astronautics at MIT from 1981 to 1985, at which time he was named dean of engineering and professor of mechanical engineering at the University of Connecticut. He was a member of the Accreditation Board for Engineering and Technology for aerospace engineering and a member of the academic affairs committee of the American Institute of Aeronautics and Astronautics.

After reporters had engaged in political banter with Alexander, one writer inquired as to Harris' "politics." Harris responded that he felt that one's politics should be shown by working hard to see that the taxpayers got a good return on their money, by providing elected officials with sound advice, and by voting. This drew laughter and at least one comment that Harris sounded "like a great politician!"

At any rate, he had made history, becoming the first vice president the Space Institute had ever had. And, as some news reports noted, he would be one of the highest ranking Blacks in the UT system. **Ann Wolfe,** who had been with UTSI for fifteen years, would become the vice president's administrative assistant, assisted by **Linda Willis Crosslin.**

Graduation ceremonies were held at Patuxent River U.S. Naval Base in Lexington Park, Maryland, at 6 P.M. August 9, 1989, for sixteen test flight engineers—graduates of a UTSI aviation engineering program that started in 1987. **Fred Watts,** course director and director of Flight Operations at UTSI, said the twelve-week course was distinguished by its intensity and educational offerings and that UTSI was the only university in the country offering this training. The 1989 class—eleven men and five women—brought to forty-four the total number of graduates since the program began.

When NASA's Voyager 2 rendezvoused with Planet Neptune in August before heading into the oblivion of outer space, UTSI mounted a dish satellite atop the administration building so that the event could be broadcast through the medium of a communications satellite, and the public was invited to come watch. **Dr. K.C. Reddy** coordinated the project and served as host for the visitors.

In the fall, architects gave Whitaker Construction Company of Tullahoma the go-ahead on a $219,000 contract to construct an addition, including new hangar facilities, for UTSI at the Tullahoma Regional Airport. Measuring about ten thousand square feet, the addition included an office, briefing room, mechanical and instrument repair areas, and the contract included enclosing an existing aircraft shelter. A March, 1990, completion date was set. Watts said the structure was needed to house the Institute's seven airplanes—including a twin-jet Sabreliner that would not fit under the roof of the open-sided shelter—and two sail planes. Earlier in the year, UT had leased enough additional space at the airport to double the size of UTSI's operations there, making way for the expansion.

This was the first of a three-phase $2.5 million construction program approved about a year before. Also being planned as the Institute moved into its twenty-sixth year was construction of a sixty-five-hundred square-foot propulsion "hot flow" laboratory building for combustion research, expected to cost about $500,000, which officials hoped would be completed by June, 1990. Construction of an advanced technological research laboratory and an office building complex also was planned, with a target completion date of August, 1990.

In October, the Federal Aviation Administration agreed for UTSI to teach flight testing to FAA pilots and engineers under what Institute officials hoped would be a three-year pact. Initially, FAA appropriated $184,718 for UTSI to develop and conduct a prototype course. The first six-week sessions was scheduled for the following January. Professor **Ralph Kimberlin,** Aviation Systems program chairman, said four courses were planned for 1990, and he hoped to teach four courses a year for two additional years, which would make the contract total $1.6 million. Later in the year, the Air Force contracted with UTSI to teach flight-testing to twelve officers from Wright-Patterson AFB in 1990, with options for continuing the courses two additional years. Summer enrollment hit two hundred and ninety—an increase of forty over the 1988 summer session—with one hundred and five full-timers. Nineteen students got their degrees at the end of the summer—three doctorates and fifteen master's—to bring the total number of graduate degrees earned at the Institute since it opened to eight hundred and sixty-four (719 master's and 145 Ph.D.'s).

In December, twenty-five more master's degrees and three doctorates were added for a total of eight hundred and eighty-nine degrees—148 Ph.D.'s and 741 master's degrees—earned in the twenty-five years since the Institute was founded on September 24, 1964. Enrollment in the fall of 1989 totaled five hundred and ten, including eighty full-time, two hundred and eighty part-time, on-campus students, and one hundred and fifty off-campus part-time students.

Dr. Max L. Hailey joined **Dr. Jerry Westbrook's** Engineering Management Program at the start of the Fall Semester as associate professor, coming from the School of Engineering at UT Chattanooga. The Jackson native had received his bachelor's degree in electrical engineering at UT and his Ph.D. from Texas Tech University. Before Chattanooga, he had been a senior consultant with the ORU Group in New York, and full-time consultant to LaJet Energy Co., Abilene, Texas, working in San Diego. (Hailey's wife, the former **Sarah Catignani,** had once been an administrative services assistant at UTSI.)

Another new faculty member—a native of Chattanooga—also assumed his duties as assistant professor of Electrical and Computer Engineering in the fall, but **Dr. Monty Smith** was already well known at the Institute. **Dick Roberds** said Smith was "perhaps the brightest student ever to graduate from the Space Institute." A graduate of the stellar McCallie School for Boys in Chattanooga, Smith had earned a bachelor's degree in math and physics from Rhodes College in Memphis and another bachelor's in Electrical Engineering from UT Knoxville before enrolling at UTSI in the fall of 1982. After earning his master's degree in EE in 1984, Smith joined the laser center as a staff engineer while pursuing his Ph.D. in Electrical Engineering. He got it in December, 1988, and stayed on as an engineer until being named to the faculty. Along the way, he also worked for Southern Research Institute in Birmingham and for the Tennessee Department of Public Health in Nashville.

Richard C. (Dick) Attig and **Charles (Chuck) Wagoner** got good news in the fall of 1989 when the Department of Energy gave the Institute permission to proceed with the second part of a $1.3 million research project that might reduce the dependence by U.S. industries on imported fuel. Attig, manager of ECP's Fuel Environment Department, was project manager for the DOE contract, and Wagoner was principal investigator. The purpose of the program was to show that industrial firetube boilers could burn dry, ultra-fine, ultra-clean coal in an efficient and environmentally acceptable way instead of having to rely on oil, propane or natural gas. Attig said the federal program was developing applications for residential and commercial boilers, too. The $1,312,000 contract with DOE's Pittsburgh Energy Technology Center began on June 11, 1987, with $761,000 allocated for the first phase. The second phase, expected to last about two years, was allocated $551,000. A major objective of the second phase was to prove the commercial feasibility of the process. Also heavily involved in the project were **Dr. Roy Schulz,** UTSI associate professor, **John P. Foote** (who had just received his Ph.D. at UTSI), **Bill Millard,** and **Jack Frazier.**

This (1989) also was the year that the UT-Calspan Center for Aerospace Research (CAR) got funding to erect a building near the causeway a short distance from the unit it had rented from Carter Ltd. Inc. The Appalachian Regional Council put up $525,000 and with **Governor McWherter's** urging, the State of Tennessee allocated $320,000. Bids for construction of the 10,500 square-foot building were taken in November, and **George Garrison,** director of the Center for Advanced Space Propulsion (CASP), hoped that the building would be completed by the next fall. (The building was dedicated in December of the next year, and eventually Garrison was named director of CAR). It would be owned by the Tennessee Technology Foundation, a non-profit state organization based in Knoxville, and would house CASP as well as the Center for Science & Engineering, also operated by CAR. **Tom Bailey** also would move his TVAR office into the new building.

Garrison was pleased with response to CASP's first technical symposium in October, which drew about seventy participants, including several industrialists from the West Coast. Speakers from the Institute included **Dr. Frank Collins, Dr. S. M. Jeng, Dr. Roger Crawford, Dr. Moonis Ali, Dr. Ching Lo, Dr. Basil Antar, Dr. Dennis Keefer, Dr. T. Dwayne McCay, Dr. Bruce Whitehead, Harry Ferber, Uday Gupta** and **Herbert Thomas.** Presentations also were given by **Dr. Edward G. Keshock,** UTK professor, and **Dr. John Brophy** from EPL Inc., Lancaster California.

On December 14, 1989, a reception was held for the final graduates of the first twenty-five years. Three had completed work for doctorates, making a total of one hundred and forty-eight Ph.D.'s; twenty-five were to get master's degrees, pushing that total to seven hundred and forty-four for a grand total of eight hundred and ninety-two.

Just after Christmas, **James P. Rhudy,** program manager with ECP since 1980, died in Manchester. Rhudy, who had earned his master's in Mechanical Engineering from UT, joined UTSI in July, 1970, as counselor/engineer in the Remote Sensing Division.

✧ ✧ ✧

The Space Institute spiced its birthday celebration with a variety of activities ranging from concerts to road races. A two-day Science Exposition, the 15th annual Quick-Goethert Lecture, and a Homecoming Ball and Banquet helped put the icing on the cake.

Serving with **Bob Kamm** on the Silver Anniversary Committee were **Bettie Roberts, Bill Boss, Judy Rudder, D. John Caruthers, Dr. Arthur Mason, Dr. K. C. Reddy, Dr. Walter Frost, Dr. Robert L. Young, Herbert Thomas, Patrick Sherley, retired Air Force General Lee V. Gossick, Srinivas Anday and Weldon Payne.**

A lecture by **Colonel Guion S. Bluford Jr.,** NASA astronaut, at 1:30 P.M. on February 22 got the celebration under way. Bluford's lecture on the space shuttle program served a dual purpose: It commemorated Black History Month at UTSI while also initiating the Silver anniversary Lecture Series and inaugurating the celebration. Bluford, an astronaut since 1979, had flown on two space shuttle flights, including the last successful flight of the shuttle Challenger (STS 61-A) prior to the explosion on January 26, 1986. His first flight was as mission specialist on STS-8, launched on August 30, 1983. This was the third mission for the Challenger and the first night launch and night landing in the space shuttle program.

Colonel Hank Hartsfield, who graduated from UTSI in 1971 with a master's degree, was commander of the shuttle on Bluford's **second** flight, in 1985—the first dedicated mission under the direction of the German Aerospace Research Establishment, and the first U.S. mission in which payload control was transferred to a foreign country (Germany).

On the weekend of May 5–7, the UTSI Tennis Club sponsored a Silver Anniversary tennis tournament, with **Dr. T. Dwayne McCay,** president of the club, heading up arrangements. Through the efforts of **Dr. Caruthers,** two concerts were held in the Space Institute auditorium during the summer. The Tennessee Bassoon Quartet from Knoxville performed on June 25. **Dr. Keith McClelland,** a UT professor and member of the Knoxville Symphony, and three of his former students, formed the quartet in 1985. Performing with McClelland were **James Lotz, James Lassen,** and **Michael Benjamin.**

The Air Force and operating contractors at AEDC—Calspan Corporation, Schneider Services International and Sverdrup Technology Inc.—hosted a banquet in honor of the Space Institute on the night of July 14. Among those attending were Dr. Goethert's widow, Hertha, and their son, Winfried. **Dr. Ewing J. Threet** and **Ernest Crouch,** two of the three men who worked closely with **Dr. Goethert** in early planning for the Institute, also were present; the third member of the trio, **Morris L. Simon,** was out of town.

Capping the affair was a speech by retired Air Force **General Bernard A. Schriever,** who reminded the audience that the Institute was really about forty-five years old if one traced the concept back to the days of **Hap Arnold** and **Dr. Theodore von Karman.**

"We need visionary people such as Arnold and von Karman," Schriever said, expressing frustration that Americans "are not looking far enough ahead." He stressed his belief that national security is dependent upon space superiority.

Schriever, who as commander of the Air Force Systems Command had been so helpful in the early efforts to found the Space Institute and continued as a strong supporter after it opened, had remained active after his retirement in 1966. He had held several government advisory assignments, including being chairman, by executive order, of the President's Advisory Commission on Management Improvement, a member of the National Commission on Space, the President's Foreign Intelligence Advisory Board, the Strategic Defense Initiative Technical Advisory Committee, and chairman of the SDI Institute.

Colonel Stephen P. Condon—who shortly afterward would relinquish his post as AEDC commander to **Colonel Richard H. Roellig** and be promoted to the rank of brigadier general—was master of ceremonies for the banquet. The vice commander, **Colonel Warren L. Riles,** gave the invocation. Other speakers included UT **President Lamar Alexander, Dr. Roberds, Dr. William F. Kimzey,** Sverdrup's vice president and general manager, **Joseph R. Tully,** general manager of Schneider Services, and **Larry Trimmer,** representing Calspan's general manager, **Dr. J. William Davis.** Trimmer said that forty Calspan employees had earned graduate degrees at UTSI during Calspan's nine years at Arnold Center.

Schriever and Condon teamed up with Roberds and **Dr. Don Daniel,** chief scientist at AEDC, early the next morning (July 15) to compete in the **UTSI Silver Anniversary Golf Tournament,** held at the AEDC Golf Course. Organized by **Harry Kowal,** a UTSI graduate student and officer in the Canadian armed forces, the Four-Ball Scramble tournament attracted seventy-two players.

More than a hundred people attended a **"UTSI DAY"** luncheon at noon on Saturday, July 29, at the Andy Holt Industry-Student Center. It was hosted by the same area Chambers of Commerce that had lent their support to initial efforts to found the Space Institute: Fayetteville/Lincoln County, Manchester, Lynchburg/Moore County, McMinnville/Warren County, Shelbyville-Bedford County, Tullahoma, and Winchester-Franklin County. House Speaker **Ed Murray** of Winchester was moderator. **Dr. Edward Boling,** president-emeritus of the University of Tennessee and keynote speaker, said chambers were "key elements twenty-five years ago" in making the Institute a reality. Congressman **Jim Cooper** of Shelbyville predicted a "long and bright history" for UTSI, saying, "This small, beautiful campus is sure to help lead us forward."

Floyd Heath, president of the Winchester-Franklin County Chamber, read a resolution from all the chambers, saluting UTSI, and he presented a plaque, honoring the Institute, to Dr. Roberds. **Dr. Ewing J. Threet** spoke briefly, paying tribute to the perseverance of Dr. Goethert. Other speakers included Roberds, **Dr. John Crothers,** director of high-tech development for the State Department of Economic and Community Development, who represented **Governor Ned McWherter,** and **Srinvias Anday,** president of the Student Government Association. State Senator **Jerry Cooper** surprised **Bob Kamm** with a resolution praising Kamm for his contributions through the years.

As part of the celebration, two lectures were held in August—one by **Dr. Robert F. (Bart) Barthelemy** and one by **Congressman Bob Clement.** Barthelemy, manager of the National Aero-Space Plane (NASP) Program at Wright-Patterson, came on August 4 to address the topic: "NASP—The Sky Is No Longer the Limit." **Dr. John B. Dicks** introduced Barthelemy. The two had gotten acquainted years before when Barthelemy, as a second lieutenant, had been contract manager for the first MHD contract that UTSI ever obtained.

Clement spoke on August 30, recalling the role his father, the late **Governor Frank G. Clement,** had played in establishing UTSI. "If my dad could be here today," he said, "he would say to each of you 'I'm very, very proud.' I would like to say for him, I'm very, very proud."

The congressman noted that his father's death in a one-car accident in Nashville occurred on November 4, 1969—exactly seventeen years after he was first elected as governor of Tennessee on November 4, 1952. Clement commented on UTSI's "large role" in bringing about technological changes an advancements. He quoted remarks made by the late governor in 1966 when he spoke at the ceremony at which the Institute's administration building—named for the governor—was dedicated. The governor had said the Space Institute "stands astride the path of the future, uniquely qualified in a very special way to be a vital contributor to man's search for new knowledge. . . ."

William H. (Bill) Boss, retired Air Force major and a senior engineer in the Energy Conversion Program, as chairman, packed everything possible into the Science Exposition, which was held on a Friday and Saturday (September 15–16). Response was overwhelming to his invitations for high school math and science students across the state to visit the Institute on Friday for a variety of tours; consequently, Boss had to draw the line at six hundred and try to reschedule visits for another two hundred. With the help of many volunteers, students toured wind tunnels, Center for Laser Applications, and MHD facilities in addition to getting first-hand looks at holograms, remote sensing techniques, and computer graphics.

Threats of rain slowly subsided on Saturday and by nightfall, several hundred visitors had watched sky divers float from the heavens, attended shows in the auditorium, toured various facilities, observed demonstrations, and stood in line to buy hot dogs from Tullahoma Boy Scout Troop 142. Late in the afternoon, a hot air balloon was filled and tethered in the circle in front of the Institute, and about one hundred persons rode it as darkness closed in.

An electrical show by **Robert Krampf** of Memphis, a Birds of Prey and Bald Eagle show by **John Stokes** of Tennessee Wildlife Resources Agency, and a NASA show by **William O. Robertson,** aerospace education specialist, were well attended in the auditorium. Boss's demonstrations of superconductivity also were popular.

An entry by **Anne Simmons,** Franklin County High School junior, won first place among Science Fair entries, and **Richard Knowles,** eight-grader from North Franklin Junior High, placed first in the junior high category with his entry. Anne also won the poster contest for high-schoolers, and **Christopher Call** from Tullahoma was grand-prize poster winner. Winners from elementary schools included **Zane Owen,** and **Aaron Bibb.** Winners of the paper airplane contest included **Matthew Kologinczak, Crystal Corky, Russell Goodman, Will Robertson, Srinivas Anday** and **Don Herslam.**

Highland Yacht Club sailors **Speed Baranco, Tom Brown, Jeff Hague, Marvin Sellers** and **Cliff Wurst** rode their sailboards in close to shore while **David Brown's** and **Dale Holasek's** sailboats decorated the horizon during the day.

A flyby of ultralight, antique, stagger wing, experimental and jet aircraft and a radio-controlled aircraft show by the Bama Flyers drew a large crowd of spectators. And the last sky diver to alight on the soccer field—**Chris Martin** of McMinnville—delivered a small package, labeled "Sky Divers Special Delivery From Out of the Blue," to Bob Kamm for forwarding to Dr. Roberds. The package contained a silver cup, inscribed with a message from the Silver Anniversary Committee, thanking Roberds for his support.

"This is also to express our appreciation for all that you have done for the Space Institute," Kamm told Roberds.

The next major event was on September 23, when Motlow College invited UTSI to join in a birthday celebration. Rain forced the celebration into the gymnasium where an enthusiastic crowd heard various musical groups and watched as **Dr. Frank Glass,** president of the twenty-year-old college, and Dr. Roberds cut a birthday cake and exchanged congratulatory resolutions. Performances were given by the **Hamilton Family** (led by UTSI's **Walter Hamilton**), an Air Force band from Maxwell Air Force Base, Alabama, the Motlow Jazz Ensemble, "Us Two and Him," Charlie Martin and his Tennessee Cloggers, and the **UTSI Singers,** directed by **Darryll Rasnake** and accompanied by pianist **Diane Chellstorp.** Singers included **Bettie Roberts, Sandy Shankle, Lillie Stricklin, Pam Selman, Xiaoling Fan, Joe Dalton, Dr. Kenneth Kimble, Dr. Lloyd Crawford, William Baucum, Thomas Dowdy, Dr. Robert L. Young, and Dr. Richard M. Roberds.**

Franklin and Coffee County UT Alumni chapters brought UT's new chancellor, **Dr. John Quinn,** to UTSI on the night of October 5 as guest speaker. This was the first visit to the Space Institute by Quinn, who had assumed duties in July as the fourth chancellor in the University's history. He came to Tennessee from Brown University in Providence, Rhode Island, where he had served as dean of faculty and a professor of physics. He said a desire to be part of the University's "quest for excellence" had influenced his decision to move to Tennessee.

Under the leadership of **Harry Kowal** and **Jim Stephens** (UTSI personnel director), two Silver Anniversary road races drew 120 participants on October 14. **Guy Giles** of Franklin and **Sherrie Bowers** of Nashville took the big trophies in the 10K run. Guy covered the 6.2-mile course on the Institute campus in 33:56 and Sherrie was top female finisher with a time of 38:51. **Walter Bruce** of Tullahoma was second among men with a time of 34:09 while **Pallie Jones** of Eagleville finished second among female runners at 39:27. This event drew forty-one male and twenty-one female contestants. **Terry Pickett** of the Lincoln County community of Kelso led the men in a 3K fun walk/run with a time of 12:20, and **Rachel Willis** of Tullahoma finished first for the women at 13:11. Fifty-eight persons participated in this 1.9-mile event.

Many dressed for the occasion on October 31, but Halloween at the Space Institute was all "treat" as the public came for two special events.

At 3 P.M., under an autumn sun, three men stepped from the past to pay tribute to their friend, **Dr. B. H. Goethert,** with whom they had "fought in the trenches" to bring the Space Institute into being. Several members of Goethert's family, including his wife, **Hertha,** two sons—**Winfried** and **Wolfhart**—and a grandson, **Kurt,** were present for the ceremony on the patio overlooking Woods Reservoir, and the invocation was given by Goethert's pastor, the **Rev. A. Richard Smith.** The occasion was dedication of the **Goethert Memorial Garden.**

Dr. Richard M. Roberds, acting dean, presided, standing next to a sun dial that had been donated by retired Air Force **Colonel** and **Mrs. Jean Jack.** Roberds pointed to the "thousands of dollars worth" of trees and plants that had been donated by three members of the Support Council—nurserymen **Henry Boyd** of McMinnville and **Hubert Nicholson** and **Don Shadow** of Franklin County. **Dr. and Mrs. Arthur Mason** had given a compass rose for the garden, and **Kenneth Perry** and **John Warren** of Manchester had donated their time to build a stone holder for a plaque at the base of a "Goethert Oak."

It was Goethert's "dynamic and enthusiastic personality" that brought the Space Institute into being, **Ernest Crouch** of McMinnville said, praising him as a "great scientist and teacher." Crouch, who had first gotten involved in Goethert's dream for an Institute while serving as a state representative, was in the Senate when final legislative action was taken to establish the Institute. He said his involvement on behalf of the Institute was a "highlight" of his thirty years in the legislature. Crouch was also the first of two Support Council chairmen.

Morris L. Simon, first vice chairman of the Council, and who, as publisher of The Tullahoma News, had played a major supporting role in getting UTSI established, described Goethert as a "complex man, consumed by his passion to achieve his goals for this Institute." Goethert was, he said, indeed a "tough taskmaster," but also a "kind-hearted man with a sense of humor, who found it difficult to tolerate incompetence." Simon attributed to the late **Jack Shea** a story that he said shed light on Goethert's personality. According to the story, Goethert, piloting his plane toward the airport at Huntsville, received instructions from the tower to delay plans to land. Goethert, not at all interested, switched to his native tongue, giving the impression that he could not understand the instructions, and bullied his way in for a landing.

Dr. Ewing J. Threet of Manchester, first Support Council treasurer, remembered Goethert as "a good friend, a good neighbor, and a good (dental) patient—he always paid his bills!" Three words—excitement, affection, and service—described Goethert, said Threet, who was state senator when he first was attracted to Goethert's Institute project.

Winfried (Wimp) Goethert and his mother expressed thanks for the tribute. The son said the garden was a "very appropriate way to honor" his father because Dr. Goethert appreciated having a place to reflect "how we fit into the scheme of things."

William R. Carter of Fayetteville, who had succeeded Crouch as Support Council chairman, closed the ceremony by wishing that Goethert "were here." Some, commenting on how the sun had broken through just moments before the ceremony began, confessed to having doubts that "Doc" was totally absent.

At 4:30 P.M., Professor **Dr. Reimar Luest,** director general of the European Space Agency (ESA), Paris, gave the fifteenth Quick-Goethert Lecture in the UTSI auditorium. He emphasized that a primary goal of the agency was cooperation in space research, technology and applications for "'exclusively peaceful purposes." Luest said the thirteen member states also wanted autonomy. He said ESA could match the United States and the Soviet Union in all areas of space activity except manned flights. The speaker acknowledged how the United States and NASA had helped Europe become a space power by opening institutes to foreign scientists and offering flight opportunities and launch services.

Luest noted that both **Professor August Wilhelm Quick** and **Professor Bernhard Hermann Goethert** left Germany in 1945 since they could no longer work in their fields in Germany. Quick went to France, became heavily involved in the construction of the Caravell airliner, and returned to Germany in 1954 to accept a chair at the Technical University of Aachen. Luest said many other scientists, particularly those in the fields of aerodynamics and nuclear physics, returned to Germany after Allies lifted restrictions on certain research areas. He said scientific cooperation increased greatly between the United States and West Germany during this time and many young German scientists (including Luest) were invited to U.S. universities.

At the beginning of the 1960's, Luest said, Goethert also had to decide whether to return to Germany where a chair at the Aachen university and directorship of a big institute had been offered to him. Goethert declined the offers from his native land for "various reasons," Luest said, but he accepted an honorary professorship and maintained strong links with Dr. Quick, with whom Goethert had once studied at the University of Danzig.

Dr. Arthur Mason, in closing remarks at the lecture, reminded the audience that the "beautiful technology" depicted in slides that accompanied the lecture represented a "**human** endeavor by many people from many nations." Scientists and engineers were shaping the "way we live and the way our children will live," Mason said, concluded that "It is **in the work** that we leave behind something for future generations."

Dr. Kenneth Harwell, former UTSI Dean, introduced Luest. Others on the program included Professor **Dr. Rolf Staufenbiel,** co-director of the Institute for Aerospace Engineering in Aachen, **Dr. Joe Johnson,** executive vice president of UT, **Dr. Roberds** and **Pastor Smith.** Professor **Dr. Horst Thomae,** co-director of the Aachen institute, became ill shortly before the lecture and was unable to attend. An informal dinner was held afterwards.

Late in the evening of November 4, 1989, friends and family of **Freeman D. Binkley** gathered on the sloping picnic grounds at UTSI while leaves were still falling and an occasional mallard skidded to a whispered landing on the nearby lake. **Wilma Kane,** UTSI purchasing manager, had attached a plaque, bearing the likeness of the late manager of student affairs and a written tribute, on a post of an unfinished Freeman D. Binkley Memorial Pavilion. As an ailing **Dewey Vincent,** who had started out with Wilma on the pavilion committee, watched from his parked truck at the top of the hill, the ceremony began with a prayer by Pastor Smith. Then, with Dr. Roberds presiding, persons who had known Binkley well, spoke of his love and concern for students. **Dr. Peter Liver,** with REMTECH in Huntsville, who was president of the Space Institute Student Government Association (SISGA) from 1982 till 1984, said Binkley went far beyond his duties as manager of student affairs and was a friend to students. **Gordon Alan Lowrey,** with Rockwell International in Huntsville, agreed that Binkley was ''one of us.'' Lowrey was SISGA president in 1955; however, he had accompanied Binkley to the picnic area, ''scouting'' for a site for a pavilion, just before Binkley entered the hospital for the final time before his death on October 12, 1984. Binkley had often discussed with students the need for two covered pavilions in the area. Students had played a major part, Roberds said, in starting a drive to build a pavilion, which Roberds said was made possible through the generosity of UTSI students, staff, faculty, alumni and friends. **John Greeter** of Monteagle, a member of the Support Council, had made a sizeable contribution of materials. **Harvie P. Jones** of the Huntsville architectural firm of Jones and Herrin, donated his services in preparing plans for the pavilion.

Lowrey said Binkley not only was a strong ''figure of authority,'' but also had been a kind of substitute father who always had time to listen to students. It was Binkley's conviction, Lowrey said, that students were the **purpose** for a university's existence.

Binkley's widow, **Ruth,** of Tullahoma, their son, **Mark** of Cary, North Carolina, and daughter, **Roxanne Binkley Garrett** of Carencro, Louisiana, each expressed appreciation for the memorial. Roxanne said if her father were there, he would be ''so moved that he would be speechless.'' But she could imagine him, with hands in his pockets, ''jingling his change,'' looking around and saying: ''**There's** a good place to put the golf cart.'' He would, she said, with ''a lot of 'attaboys,' say 'Let's keep going' '' with special concern for the continued excellence in teaching ''the students he loved so much.''

Dr. Robert L. Young, noted that Binkley, as a young man growing up in Nashville, had been friends with **Dr. Charles H. Weaver,** who was present at the ceremony and would speak at the banquet that night.

Freeman Binkley joined the staff of UTSI on December 19, 1964, as its first business manager, and he had worked in various capacities, including administering the Institute's insurance program. For the seven years before his death he had been manager of student affairs, and it was as a friend to students that most remembered him on that November evening as night's shadows crept in.

✦

Astronaut Hank Hartsfield lights the torch at UTSI in ceremonies in 1986. On the right is Ernest Crouch.

From left, Gaye, Kurt (front), Winfried and Mrs. Hertha Goethert stand with Bob and Shirley Kamm who donated this bench in Dr. Goethert's memory, and Dr. Richard M. Roberds.

With Dr. Roberds at the pavilion dedication are Ruth, Mark and Roxanne Binkley.

Bob Clement is surrounded by friends after speaking during the Silver Anniversary. From left are William R. Carter, Morris L. Simon, Richard Roberds, Bob Young, Clement, W.R. Davidson, Lee V. Gossick, G. Nelson Forrester and Bob Kamm.

Dr. Charlie Weaver, on his first return to UTSI on November 4, 1989, chats with Bob Kamm, left, and Morris L. Simon and Dr. Walter Frost.

Retired General B. H. Schriever found plenty of friends when he returned for UTSI's 25th birthday. From left are Lee V. Gossick, Dr. George Garrison, Dr. William F. Kimzey, Dr. Richard Roberds, Dr. Robert L. Young, Schriever, Dr. K. C. Reddy, Bob Kamm and Dr. Edwin M. Gleason.

The UTSI Singers, led by Darryll Rasnake, seated by pianist Diane Chellstorp, begin rehearsals in preparation for the annual Christmas dinner.

Chapter Twenty-Six

AND THE BAND PLAYED ON

LATER, after Nature's ink had blotted the blue powder of the Cumberlands, and scraps of sky patched the otherwise hidden water, punch flowed red from a fountain in the lobby, and bits of blue light splattered endlessly against the glass panes as couples danced to the music of Bill Sleeter's band. Outside, crisp paper sacks glowed with candlelight, lining the walk across the campus to the Industry-Student Center, and tiny white lights freckled the bare limbs of little trees. Employees of the physical plant, inspired by **Robert Parson** and **Charlotte Campbell** and working with **Dee Dee Frost,** had the place looking like a fantasyland. They lighted the candles while the **Silver Anniversary Alumni Banquet** was going on in the cafeteria so that the lighting would have maximum impact when the people strolled from the banquet to the Homecoming Ball. **Boyd Stubblefield,** UTSI's photographer for the past seven years, set up shop in the corridor of Lower A Wing, snapping couples' pictures as though they were at a high school prom.

This was a different face of the Space Institute from the stereotypical image of a staid citadel of equations and technological computations even as the physical surroundings of the Institute had always belied the dispassionate concept. Dr. Goethert himself, taking inventory of the assets during the Fifth Anniversary and Founder's Day Celebration on October 5, 1969, had said: "**. . . And above all, there still is the unspoiled beauty of the campus' natural landscape, of the lake, of the mountains in the background, and the wild life around us. . . .**"

And what of the other, the **mission** of the Space Institute? How had it fared since that September 24 in 1964 when it officially began? There had been evidence presented at the Banquet that very evening—the first Saturday night in November of the Institute's twenty-fifty year—as five persons were honored as "1989 Distinguished Alumni" of the Space Institute. They had come back from San Diego, Washington, D.C., Los Angeles, Tullahoma, and The Netherlands, graduates of UTSI who had gone into the world to make a difference. **Charlie Weaver,** in his first return to the Institute since 1981, had confirmed his reputation as an after-dinner speaker, keeping the banquet crowd laughing as he recounted "low lights" in the life of the Space Institute, yet there was between the lines of his humor enough truth to bring a nod and a knowing look from the old-timers—as when he said that the history of the Institute had been a "series of end runs and center rushes."

Ken Harwell, who had followed Weaver as dean, was at the head table, and **Morris L. Simon**—one of Goethert's "Three Musketeers" from the pioneering days—was there, too, chiding the speaker whom he was to introduce about the fact that he had been given "seven single-typed pages" of Weaver's accomplishments. And as **Walt Frost,** master of ceremonies, pointed out, a goodly number of faculty members who were on the staff when or shortly after Frost came in 1965, were present at the banquet. These included some—like **Art Mason, Bob Young, Jimmy Wu** and **Firouz Shahrokhi**—who were still teaching. **Marcel Newman** was among those who had retired, but he had taken an active part in most of the Anniversary events, and he was there in black tie this night. **Bill Snyder, Arsev Eraslan** and **Susan Wu** also were singled out by Dr. Frost as early faculty associates.

Miss Virginia Richardson, who was with the Institute as its first registrar when it opened, and **Dewey Vincent,** the first and only physical plant manager in the Institute's first twenty-five years, were there, too. It would be the last time Vincent would see many of his old friends for at 3:15 A.M. on December 11, 1989, he died at Harton Regional Hospital in Tullahoma. He was sixty-one. (Soon afterward, **Robert Parson** was named supervisor of the physical plant, and the next year the physical plant building would be named for Vincent.)

Dr. John I. Shipp, one of three physics students to earn the first Ph.D. degrees from the Space Institute in March, 1967, offered a toast for UTSI's continued prosperity. And **Bob Kamm** was there to give cups to former SISGA presidents recognized at the banquet. These included the first (1968)—**Dr. Hrishikesh Saha**—professor of computer science at Alabama A&M University, Huntsville—and the second (1969), **Dr. Roger Crawford,** of Estill Springs, associate professor of Aerospace and Mechanical Engineering at his alma mater. Also recognized were **Mike Guthrie** (1981 and 1982), from Teledyne Brown Company, Huntsville, **Dr. Peter Liver** (1983 and 1984), with REMTECH, Huntsville, and **G. Alan Lowrey** (1985), with Rockwell International, Huntsville, who had participated in the Binkley dedication but had been unable to stay for the banquet.

Weaver joked about how he would run to Simon with a request, Simon would go to the governor, the governor would call Knoxville, and so on. An example, he said, was when he told Simon he "needed a budget," and UTSI eventually got its own line-item budget. From this point forward, Weaver said, the Institute had "prospered." He also quipped that it had been necessary to differentiate between what **Dr. Andy Holt** said and what he meant. For instance, Holt might say something like: "I thought they were going to build a little block building down here (at UTSI), and just look what they've got!" What he meant, Weaver said, was: "You dirty dogs! You went behind my back to the legislature!"

In one serious line, Weaver said that in retrospect, he realized what a "tremendous contribution" had been made to the Institute by **Dr. John B. Dicks.** Dicks, the first full-time faculty member hired specifically for the Space Institute, and one of those who was still on the job as professor of physics, was not there to hear the praise.

Dick Roberds presented the five "1989 Distinguished Alumni," and sponsoring professors congratulated them. Cited were **Dr. Ola Brevig,** chief engineer, General Dynamics, San Diego, who had received his doctorate in 1970 under **Dr. Shahrokhi; Dr. James M. Lents,** executive officer, South Coast Air Quality Management District, Los Angeles, who received his doctorate in physics in 1970 under **Dr. Mason** and was getting a lot of attention for his West Coast war on pollution; **Dr. John S. Theon,** chief, Atmospheric Dynamics and Radiation Branch, Earth Science and Applications, NASA headquarters, Washington, D.C., who received his Ph.D. in engineering science in 1985 under **Dr. Frost; Dr. Jan Vreeburg,** director of Aerospace Research, National Aerospace Laboratory, The Netherlands, who earned his doctorate in aerospace engineering in 1969 under **Dr. Snyder,** and **Dr. W. F. Kimzey** of Estill Springs, vice president and general manager, AEDC Group, Sverdrup Corporation, Tullahoma. Kimzey had received his Ph.D. in aerospace engineering in 1977 under Dr. Goethert, having also earned his master's degree at UTSI. Dr. Theon recalled how Dr. Frost and others had convinced him that age fifty was not too old to start working toward his doctorate. A few days later, in a letter to **Dr. Roberds, Theon** recalled that his first contact with UTSI a decade before was when he attended one of Dr. Frost's workshops at "one of the most beautiful sites I have ever seen." He soon discovered, however, that the Institute's "real resource is its people." Based upon his experience as a federal research program manager, interacting with many university scientists and engineers, Theon wrote: "I can confidently say that **Professor Walter Frost** is among the brightest, most energetic, and most hardworking people I have encountered." He also praised **Dr. Robert Turner** as one who "personifies the best in the teaching profession." (Turner was an adjunct professor from Huntsville closely associated with Frost.)

These five honorees joined a list of ten others who had been singled out as "Distinguished Alumni" in earlier years. Cited in 1984 were **Dr. Dieter Jacob, Colonel Henry W. (Hank) Hartsfield Jr., Dr. Philip T. Harsha, Dr. Kenneth Tempelmeyer,** and **General Dr. Chwan-Haw Chen,** in 1986, **Dr. Kinya Aoyama, Dr. Jim Mitchell,** and **Dr. Jack Whitfield,** and in 1987, **Dr. Samuel R. Pate** and **Dr. Juergen Bitte** (a former UTSI faculty member, later a high official in Germany's space program).

As 1989 ended, **Dr. Pate**—a former Thorsby, Alabama, boy who had joined Sverdrup Corporation in 1960 as an engineer at AEDC and one of the full-time students when the Space Institute opened—had been promoted to president of Sverdrup Technology Inc. Pate had received his master's degree in mechanical engineering at UTSI in the winter of 1965 and his doctorate in aerospace engineering from UT in 1977. He was succeeding **Dr. Jack Whitfield,** who was to continue as Sverdrup Technology's chief executive officer. **Dr. Harsha,** who had received his Ph.D. under **Dr. Snyder** in 1970, was deputy director of the National Aero-Space Plane's National Program Office at Seal Beach, California. (Another graduate who was "making a difference" in a rather unusual area was **Earl C. Ruby,** with the world's largest—and perhaps smallest—almond shell fueled cogeneration plant at Sacramento. Blue Diamond Growers, an agriculture cooperative, was composed of five thousand grower members. **Dr. Young** had been Ruby's thermodynamics instructor and thesis advisor as he earned his master's degree in 1969.)

Eight hundred and eighty-nine degrees had been earned at UTSI in twenty-five years. The Space Institute had been entrusted with more than one hundred twenty-five million dollars worth of research contracts—including close to twelve million dollars in the 1989–90 fiscal year. Lasers were being used as tools to search for practical solutions to real problems in the world. MHD was gaining in stature as a technology that could make a difference on behalf of a cleaner environment. The Center for Advanced Space Propulsion (CASP) was bridging the gap between classroom and industry, with special emphasis at that time on electric propulsion, flight experiments such as the space shuttle and station, and small businesses and soon would be in charge of an ambitious effort with private industry to develop a space launch system (COMET). The Center for Laser Applications (CLA) was branching into new areas of research with various industries becoming involved in laser processing, and a strong relationship was developing between CLA and CASP. Tennessee Valley Aerospace Region was developing a conduit to transport advanced technological learning from theory into "high tech" jobs, and the Space Institute was sitting in the middle of a proposed corridor linking Middle Tennessee with Huntsville.

The Space Institute had forty-six full-time faculty members. ECP had a staff of one hundred and forty-four, including six part-time employees and fourteen graduate research assistants. **Administrator Joel Muehlhauser** (soon to become research dean) informed the National Advisory Board at a meeting on November 30, 1989, that at least twenty-four doctorates and sixty master's degrees had been earned by full-time graduate research assistants supported by ECP. The Center for Laser Applications had forty employees, and its professors were planning several new courses for the next year. CASP counted nine major industries and five universities as partners, along with AEDC and three NASA centers: Marshall Space Flight Center, Lewis Research Center, and Ames Research Center. The Engineering Management program was growing fast, helping private firms with "capstone" projects and reaching students all around the country with its video program—engineers teaching management to engineers. The Institute's annual payroll totaled almost ten million dollars with the bulk of this being spread among thirty-three cities and towns in Tennessee and only slightly more than one hundred thousand dollars going outside the state. Personnel records that fall showed a total employment—including part-time employees and graduate research assistants—of about three hundred and eighty persons. This included two hundred and fourteen full-time staff members.

Ernest Crouch looked back to the beginning and said that UTSI had proven its worth, not only in aerospace academics and research, but also in regional development. He was convinced that "UTSI was and is an important factor in attracting industry to our area, the Carrier plant being one, and that UTSI and AEDC also were factors in getting one of the first community colleges. UTSI definitely has contributed to a better quality of life for the people in our area."

As in the beginning, the Institute was dedicated to twin missions: Graduate education and research, and not surprisingly, with research paying an estimated seventy percent of the freight, there was some tension between the two. However, this was tempered by the heavy concentration of research-oriented faculty. The new vice president—**Dr. Wesley L. Harris**—was due on board by the next summer, and he had expressed a desire to make the most of the Institute's long-standing relationship with Arnold Center—a relationship that had held such promise and one that had been vital in the creation of the Institute. Dr. Goethert had staked a lot on this relationship, emphasizing the potential of the joint use of the facilities at Arnold Center. How had that actually turned out?

Bob Young, who had guided UT's graduate program for seven of its eight years at Arnold Center and whose life had been interwoven with the life of the Space Institute for every one of its twenty-five years, summed it up: **"Without the support of AEDC and its contractors, there would be no UTSI."**

Young recognized that the "envisioned use of AEDC facilities by UTSI personnel did not materialize to the extent expected." Overall changes in Air Force policy such as service funding and export of data regulations served to prevent the expected large joint research efforts. AEDC's prime mission was testing. The Air Force's first mission was to keep the peace and fight wars. Of course, research played a role in helping accomplish those missions.

Even so, through arrangements with AEDC "many of our full-time students have had a most meaningful and beneficial experience working under our overall guidance in AEDC facilities with AEDC personnel," Young noted. He had been amazed at how well AEDC had treated the Institute after it was firmly established in its new building. Employees continued to receive time off to attend classes—even daytime classes—at the UTSI campus. In its twenty-fifth year, the Institute still had many AEDC employees coming to the campus as part-time students.

"The AEDC students and the presence of AEDC provide a most stimulating environment for UTSI faculty," Young said. "The students are engaged in real-world problems, which makes technical class material most meaningful to the class."

After his retirement in June, 1990, Young would conclude: "For me personally, the interaction with AEDC was most rewarding and stimulating. It was a great pleasure to watch several of our AEDC employee-students advance to important positions at AEDC. The cooperation of AEDC in our Short Course efforts, both in terms of faculty and attendees, was outstanding and much appreciated. Several of the AEDC people served as adjunct faculty at UTSI and thus much enriched our academic program."

Despite difficulties in working with a government facility, Young felt that the Space Institute's experience probably represented "the most successful venture into this type of operation within the United States. This certainly is to the credit of administrators of the Institute, the University, and the Air Force." Perhaps some things could have been better, but "here over a fairly long period, we have worked in reasonable harmony with a governmental test center. The AEDC-UTSI relationship represents one of the most successful governmental-university relationships in the country, and I have been privileged to participate in it."

One indication of the "harmonious relationship" indeed was found in the Short Course program, which began in 1967. More than eleven thousand persons had attended those courses by the end of the Institute's fiscal year on June 30, 1989, and two thousand, five hundred and thirty-three of those came from AEDC. And Arnold Center had provided seven hundred and eighty-one of the four thousand, four hundred and sixty short-course lecturers.

As for the Space Institute's relations with the University, these had been "variable." Young felt that enthusiasm for the concept of an institute had come only from the very top levels of the University, largely for political reasons, and that they had not expected it to ever amount to much. ("In my opinion, the University was virtually forced to back the Institute because of the tremendous public opinion.") But the Institute's standing with the University had improved through the years and its value to UT was better appreciated. When **John Dicks** landed a forty-million-dollar MHD contract, "that impressed them," Young said. Knoxville also had been impressed by the "political clout" of the Institute, due to the interest shown by local legislators. **Charlie Weaver** also was an asset—known and trusted at Knoxville. UTSI benefitted in later years by reporting directly to **Dr. John Prados,** who was vice president for academic affairs and held in high esteem in Knoxville. And, Young said, the UT administration trusted **Ken Harwell** far more than they ever trusted Dr. Goethert. Goethert was like a stick of dynamite, respected by all those around him, but he also made them a little nervous—one never knew, exactly, the status of the fuse. (Charlie Weaver, at the Silver Anniversary banquet on that first Saturday in November, had "joked" that in Knoxville, when Goethert was expected, "we would hold meetings" to decide how to deal with him and his litany of complaints about all the things the Institute needed.) But it was Goethert, after all, who had made it possible for **Bob Young** to say: "It has been a real thrill to watch the Institute and to be involved in something which started from very little but has become quite significant. . . ."

Now the dancers were gliding in the blue shadows. The tapers were burning low on the Silver Anniversary Celebration. **Bill Pickering's** comments about UTSI, expressed in a latter to **Dick Roberds,** seemed appropriate: ". . . you have had a unique opportunity to motivate students and develop programs allied to space problems in the real world. **The record shows you have done well."**

Pickering, who had come from his research corporation in Pasadena, California, to give the 1988 Quick-Goethert Lecture, had sent his congratulations on the Institute's celebration. Since the lecture, he had become more optimistic about the United States' role in space. He liked what **President George Bush** was saying about it; he liked the public response to the Neptune flyby; he was encouraged by the successful launchings of the Magellan and Galileo spacecraft and of the Space Shuttle. The Challenger disaster was history. And the recent national disasters of hurricane Hugo and the San Francisco earthquake each showed the value of satellite communications and satellite observations.

"We can be encouraged to believe that out of those occurrences a strengthened earth observation program will be supported," Pickering wrote. The future for space research was bright. And, he concluded, "UTSI will have an even better report when you celebrate your 50th anniversary."

The music was starting again. **Morris Simon,** who had undergone open heart surgery a few months before, took a lively turn around the tiled floor with a former UTSI student: **Chris Barret,** who received her master's in Aerospace Engineering in 1987 and soon would come back for her Ph.D. **Walt** and **Dee Dee Frost** floated gracefully somewhere in the center, and **Art** and **Helen Mason** looked like the honeymooners they had been when first they had walked beside Woods Reservoir.

✦

APPENDIX A

GRADUATES OF UT-AEDC PROGRAM
JUNE, 1959—JUNE, 1964

KINSLOW, MAX	M.S.	Mechanical Engineering	6/59
LEWIS, CLARK H.	M.S.	Mechanical Engineering	6/59
PINDZOLA, MICHAEL	M.S.	Mechanical Engineering	12/59
TEMPELMEYER, KENNETH E.	M.S.	Mechanical Engineering	12/59
WHITFIELD, JACK DUANE	M.S.	Mechanical Engineering	12/60
HUMPHREY, RICHARD LEE	M.S.	Mechanical Engineering	8/61
McGREGOR, WHEELER K. JR.	M.S.	Engineering Science	8/61
DIETZ, ROBERT OTTO H. JR.	M.S.	Mechanical Engineering	3/62
ARNEY, GEORGE DONALD JR.	M.S.	Electrical Engineering	6/62
DUKE, JAMES LEE	M.S.	Electrical Engineering	6/62
WALKER, REX RAY III	M.S.	Electrical Engineering	8/62
SMITH, ROBERT EVERETT JR.	M.S.	Mechanical Engineering	12/62
McKEE, MARVIN L.	M.S.	Electrical Engineering	3/63
DAVID, LEROY JOSEPH	M.S.	Mechanical Engineering	6/63
GALE, JOHN GODDARD	M.S.	Engineering Science	6/63
HERRON, RICHARD DEAN	M.S.	Mechanical Engineering	6/63
ANSPACH, EARL E.	M.S.	Electrical Engineering	8/63
BREWER, JAMES HERBERT	M.S.	Electrical Engineering	8/63
COMER, WILLIAM HUGH	M.S.	Electrical Engineering	8/63
HENDRIX, ROY ERNEST	M.S.	Electrical Engineering	8/63
DESKINS, HAROLD EUGENE	M.S.	Mechanical Engineering	6/64
YOUNG, RAYMOND PRESTON	M.S.	Mechanical Engineering	6/64

APPENDIX B

UTSI GRADUATES
FALL 1964—FALL 1989

GALIGHER, LAWRENCE LEE	M.S.	Mechanical Engineering	12/64
KINGERY, MARSHALL KENT	M.S.	Electrical Engineering	12/64
ROBERTSON, JACK EDWARD	M.S.	Mechanical Engineering	12/64
BREWER, LeROY EARL JR.	M.S.	Physics	3/65
CURRY, BILL PERRY	M.S.	Physics	3/65
PATE, SAMUEL RALPH	M.S.	Mechanical Engineering	3/65
SUMMERS, WILLARD E.	M.S.	Mechanical Engineering	3/65
CLEMENT, FRANK L. JR.	M.S.	Electrical Engineering	6/65
LEE, OLIVER B. JR.	M.S.	Electrical Engineering	6/65
SHANKLIN, RICHARD V. III	M.S.	Mechanical Engineering	6/65
STRIKE, WILLIAM T. JR.	M.S.	Mechanical Engineering	6/65
WILSON, DONALD RAY	M.S.	Aerospace Engineering	6/65
BALL, HENRY WEBSTER	M.S.	Mechanical Engineering	8/65
LACY, JOHN JAMES, JR.	M.S.	Mechanical Engineering	8/65
MATZ, ROY JOSEPH	M.S.	Mechanical Engineering	8/65
THORMAEHLEN, KARL F.	M.S.	Mechanical Engineering	8/65
WHITED, JAMES BRAKE	M.S.	Electrical Engineering	8/65
GOINS, ELMER ELDRIDGE JR.	M.S.	Aerospace Engineering	12/65
JACOCKS, JAMES LAVERNE	M.S.	Mechanical Engineering	12/65
MADDOX, WENDELL VINCENT	M.S.	Electrical Engineering	12/65
MOODY, THOMAS LINDSEY	M.S.	Physics	12/65
RAMPY, JOHNNY MERRILL	M.S.	Aerospace Engineering	12/65
YOUNG, JIMMIE DON	M.S.	Aerospace Engineering	12/65
HALE, MAURICE GRIMES	M.S.	Engineering Science	3/66
SANDFORD, THADDEUS HOWES	M.S.	Mechanical Engineering	3/66
WHITE, JAMES J. III	M.S.	Mechanical Engineering	3/66
McKINNEY, CHARLES W.	M.S.	Physics	6/66
SHAFFER, WALTER L.	M.S.	Mechanical Engineering	6/66
WILSON, ROBERT EARL	M.S.	Aerospace Engineering	6/66
IANNUZZI, FRANCESCO A.	M.S.	Aerospace Engineering	8/66
NEAL, CAROLINE BENZ	M.S.	Mathematics	8/66
TRIMMER, LARRY LEE	M.S.	Aerospace Engineering	8/66
USELTON, JAMES C.	M.S.	Mechanical Engineering	8/66
DAVIS, MONTIE G.	M.S.	Physics	12/66
JACKSON, ROBERT H.	M.S.	Mechanical Engineering	12/66
KIMZEY, WILLIAM F.	M.S.	Aerospace Engineering	12/66
DOSS, DELANO A.	M.S.	Electrical Engineering	3/67
EDENFIELD, E. E. JR.	M.S.	Mechanical Engineering	3/67
FROEDGE, DON T.	M.S.	Physics	3/67
JONES, RAYMOND L.	M.S.	Aerospace Engineering	3/67
POWERS, WALTER L.	Ph.D.	Physics	3/67
SHIPP, JOHN T.	Ph.D.	Physics	3/67
TROLINGER, JAMES D.	Ph.D.	Physics	3/67
ANDERSON, KENNETH L.	M.S.	Mechanical Engineering	6/67
CASSANOVA, ROBERT A.	M.S.	Aerospace Engineering	6/67
CHASE, CHARLES J.	M.S.	Aerospace Engineering	6/67
CLAPP, NED E. JR.	M.S.	Electrical Engineering	6/67
CROSSWY, FRANK L.	M.S.	Electrical Engineering	6/67
LANDCASTER, JAMES R.	M.S.	Electrical Engineering	6/67
LE CROY, ROY C.	M.S.	Aerospace Engineering	6/67
MATTHEWS, RICHARD K.	M.S.	Aerospace Engineering	6/67
POWELL, THOMAS C.	M.S.	Engineering Science	6/67
SHIRLEY, BURLEY H.	M.S.	Mechanical Engineering	6/67
SPROUSE, JOSEPH A.	M.S.	Physics	6/67
BUCHANAN, TONY D.	M.S.	Aerospace Engineering	8/67
BURGESS, ERNEST G. III	M.S.	Mathematics	8/67
FITCH, CLARK RAMER	M.S.	Mechanical Engineering	8/67
HODAPP, ALBERT E. JR.	M.S.	Aerospace Engineering	8/67
MYERS, ALAN W.	M.S.	Aerospace Engineering	8/67
WEST, ROLAND H. A.	M.S.	Aerospace Engineering	8/67
WILLIAMS, JOHN S. G.	M.S.	Mechanical Engineering	8/67
BRAYTON, DONALD BRUCE	M.S.	Engineering Science	12/67

CHILDERS, JIM CLYDE	M.S.	Mechanical Engineering	12/67
CHRISS, DONALD EDWARD,	M.S.	Mechanical Engineering	12/67
DZAKOWIC, GERALD S.	Ph.D.	Mechanical Engineering	12/67
KOCH, KENNETH EDWIN	M.S.	Aerospace Engineering	12/67
LANIER, WILLIAM STEVEN	M.S.	Aerospace Engineering	12/67
STARR, ROGERS FLEMING	M.S.	Aerospace Engineering	12/67
WHETSEL, ROGER GIRTON	M.S.	Electrical Engineering	12/67
WHITEHEAD, GURLEY LAVELL	M.S.	Engineering Science	12/67
BONTRAGER, PAUL J.	M.S.	Mechanical Engineering	3/68
DANIEL, THOMAS W.	M.S.	Electrical Engineering	3/68
DELASHMIT, WALTER H. JR.	M.S.	Electrical Engineering	3/68
EAVES, RAYMOND H. JR.	M.S.	Mechanical Engineering	3/68
KITOWSKI, JOHN V.	Ph.D.	Aerospace Engineering	3/68
LITTLE, RANKIN R.	M.S.	Electrical Engineering	3/68
MATHEWS, ALBERT J.	M.S.	Electrical Engineering	3/68
WELCH, JACK L.	M.S.	Electrical Engineering	3/68
WRIGHT, L. E.	M.S.	Mechanical Engineering	3/68
BELZ, RONALD A.	M.S.	Electrical Engineering	6/68
BERTRAND, WILLIAM T.	M.S.	Physics	6/68
BOATMAN, ROGER M.	M.S.	Physics	6/68
DAVIS, MONTIE G.	Ph.D.	Physics	6/68
DEWITT, JACK R.	M.S.	Mechanical Engineering	6/68
FINLEY, BOBBY G.	M.S.	Mechanical Engineering	6/68
LACKEY, ROBERT GRAYSON	M.S.	Aerospace Engineering	6/68
LAZARUS, ALLAN R.	M.S.	Mechanical Engineering	6/68
LEWIS, CLARK H.	Ph.D.	Engineering Science	6/68
MADAGAN, AUBREY N. JR.	M.S.	Aerospace Engineering	6/68
PERKINS, THOMAS M.	M.S.	Aerospace Engineering	6/68
STUDWELL, VICTOR E.	M.S.	Aerospace Engineering	6/68
TODD, DONALD C.	M.S.	Mathematics	6/68
WHITFIELD, DAVID L.	M.S.	Aerospace Engineering	6/68
WILLIAMS, GRADY L.	M.S.	Electrical Engineering	6/68
ANDERSON, JAMES B.	M.S.	Aerospace Engineering	8/68
BLANKS, JAMES R.	M.S.	Mechanical Engineering	8/68
BRAHMAVAR, SUBHASH M.	M.S.	Electrical Engineering	8/68
BRICE, TRAVIS R.	M.S.	Aerospace Engineering	8/68
CARLSON, RANDOLPH L.	M.S.	Electrical Engineering	8/68
CARLSON, THOMAS B.	M.S.	Electrical Engineering	8/68
CHEN, GUANG-JIH	M.S.	Mechanical Engineering	8/68
CHISM, SAMUEL B. JR.	M.S.	Electrical Engineering	8/68
CUNNINGHAM, THOMAS H. M.	M.S.	Aerospace Engineering	8/68
DOTSON, WILLIAM P. JR.	M.S.	Electrical Engineering	8/68
FARMER, WILLIAM MICHAEL	M.S.	Physics	8/68
FOX, JOHN H.	M.S.	Aerospace Engineering	8/68
GERNSTEIN, TERRY M.	M.S.	Mechanical Engineering	8/68
HUBE, FREDERICK K	M.S.	Aerospace Engineering	8/68
HUTCHESON, LEX C. III	M.S.	Mechanical Engineering	8/68
JACKSON, FRANCIS M.	M.S.	Mechanical Engineering	8/68
McALLISTER, JACK D.	Ph.D.	Aerospace Engineering	8/68
McILLWAIN, JOHN T.	M.S.	Engineering Science	8/68
ROBERGE, KENNETH K.	M.S.	Mechanical Engineering	8/68
ROUX, JEFFREY A.	M.S.	Mechanical Engineering	8/68
RUSSELL, THOMAS G. JR.	M.S.	Electrical Engineering	8/68
SHUCKSTRES, DAVID V.	M.S.	Mechanical Engineering	8/68
WILHELM, LUTHER R.	M.S.	Mechanical Engineering	8/68
WILLIAMS, WILLIAM D.	M.S.	Physics	8/68
WINCHENBACH, GERALD L.	M.S.	Aerospace Engineering	8/68
WONG, WILFORD F.	Ph.D.	Aerospace Engineering	8/68
ARNOLD, FREDERICK	M.S.	Mechanical Engineering	12/68
BRYSON, ROBERT JOEL	M.S.	Engineering Science	12/68
BURNSIDE, BELVIN R. JR.	M.S.	Aerospace Engineering	12/68
CARTER, PERCY BRINING JR.	M.S.	Mechanical Engineering	12/68
HOLADAY, BURTON HOWARD	M.S.	Aerospace Engineering	12/68
LEHNER, PAUL	M.S.	Aerospace Engineering	12/68
MADERIA, PATRICK F. JR.	M.S.	Aerospace Engineering	12/68
MENZEL, REINHARD WOLFGANG	M.S.	Physics	12/68
MITCHELL, JAMES GUY	M.S.	Mechanical Engineering	12/68
NENNINGER, RICHARD LEE	M.S.	Aerospace Engineering	12/68
PEELE, DONALD GENE	M.S.	Aerospace Engineering	12/68
SOUKUP, STEPHEN McKAY	M.S.	Aerospace Engineering	12/68
URBANIK, RONALD THOMAS	M.S.	Aerospace Engineering	12/68
WOLF, PETER	Ph.D.	Aerospace Engineering	12/68
AOYAMA, KINYA	M.S.	Aerospace Engineering	3/69
BERG, ANDREW LLOYD	M.S.	Aerospace Engineering	3/69
BRAY, CHARLES WILSON	Ph.D.	Electrical Engineering	3/69
BUCHANAN, WILLIAM THOMAS	M.S.	Mechanical Engineering	3/69
CALDWELL, ROBERT LEO	M.S.	Mechanical Engineering	3/69
CLAY, THOMAS HENRY	M.S.	Electrical Engineering	3/69
EVANS, PAUL ALFRED	M.S.	Mechanical Engineering	3/69
HART, SHERMAN WOLF	M.S.	Mechanical Engineering	3/69
HARTLEY, MICHAEL SCROGGS	M.S.	Mechanical Engineering	3/69
McCLURE, JAMES AUBREY	M.S.	Electrical Engineering	3/69
RHODES, KENNETH EUGENE	M.S.	Electrical Engineering	3/69
ROBSON, GEORGE DAVID	M.S.	Mechanical Engineering	3/69

FORD, JAMES MARLIN	M.S.	Electrical Engineering	6/69
HUNTER, LOUIS G. JR.	Ph.D.	Aerospace Engineering	6/69
JACOB, DIETER	Ph.D.	Aerospace Engineering	6/69
JETT, EDWARD STEPHEN	M.S.	Aerospace Engineering	6/69
LAW, DONALD KEITH	M.S.	Mechanical Engineering	6/69
LIMBAUGH, CHARLES CAMERON	M.S.	Physics	6/69
LITTLE, HERBERT RANDOLPH	M.S.	Aerospace Engineering	6/69
MULLER, PETER	Ph.D.	Mechanical Engineering	6/69
McGREGOR, WHEELER K. JR.	Ph.D.	Physics	6/69
PAULK, ROBERT AUSTIN	M.S.	Aerospace Engineering	6/69
REMERS, RICHARD T.	M.S.	Aerospace Engineering	6/69
RITTENHOUSE, LEWIS E.	M.S.	Aerospace Engineering	6/69
ROBINSON, CHARLES E.	M.S.	Mechanical Engineering	6/69
TAYLOR, ROBERT E.	M.S.	Aerospace Engineering	6/69
WEBB, RONALD O.	M.S.	Electrical Engineering	6/69
BERGT, WOLFGANG	Ph.D.	Aerospace Engineering	8/69
BOUDREAU, ALBERT H.	M.S.	Aerospace Engineering	8/69
BURT, GLEN E.	M.S.	Aerospace Engineering	8/69
CARTER, ROBERT F.	M.S.	Electrical Engineering	8/69
COATS, JACK D.	M.S.	Mechanical Engineering	8/69
OLIVER, ROBERT H.	M.S.	Aerospace Engineering	8/69
SHERRELL, FRED G.	M.S.	Engineering Science	8/69
WRIGHT, WILLARD A.	M.S.	Mechanical Engineering	8/69
APRIL, PAUL KENNETH	M.S.	Aerospace Engineering	12/69
COTT, DONALD WING	Ph.D.	Aerospace Engineering	12/69
FRAZINE, DONALD F.	M.S.	Electrical Engineering	12/69
KEMP, WILLIAM HAROLD	M.S.	Electrical Engineering	12/69
KNESE, PAUL B.	M.S.	Aerospace Engineering	12/69
MILLER, JOHN C. JR.	M.S.	Electrical Engineering	12/69
MORGAN, LAWRENCE A.	M.S.	Mechanical Engineering	12/69
RUBY, EARL CAMPBELL JR.	M.S.	Mechanical Engineering	12/69
SIEBER, CHARLES E.	Ph.D.	Mechanical Engineering	12/69
YODER, LEE O.	M.S.	Aerospace Engineering	12/69
BENEK, JOHN ADDISON	M.S.	Aerospace Engineering	3/70
DEAN, CHESTER FRANCIS	Ph.D.	Aerospace Engineering	3/70
EDWARDS, LARRY WORTH	M.S.	Mechanical Engineering	3/70
JONES, HENRY TURNER	M.S.	Electrical Engineering	3/70
TEMPELMEYER, KENNETH E.	Ph.D.	Engineering Science	3/70
WALKER, WILLIAM LEE	M.S.	Physics	3/70
WURST, CLIFFORD WILLIAM	M.S.	Mechanical Engineering	3/70
CYRUS, MARION SAMUEL	M.S.	Physics	6/70
FLECK, ALVIN	M.S.	Mechanical Engineering	6/70
SALVAGE, JOHN WALTZ	M.S.	Engineering Science	6/70
SU, MICHAEL WEN-SHEAN	Ph.D.	Mechanical Engineering	6/70
BUTKEWICZ, PETER JR.	Ph.D.	Aerospace Engineering	8/70
DOUGHERTY, NATHAN S.	M.S.	Mechanical Engineering	8/70
LASTER, MARION L.	Ph.D.	Aerospace Engineering	8/70
MARTIN, JAMES F.	M.S.	Engineering Science	8/70
SCOTT, MARY H.	M.S.	Electrical Engineering	8/70
AMIN, SATISH M.	M.S.	Electrical Engineering	12/70
BREVIG, OLA	Ph.D.	Mechanical Engineering	12/70
BUSBY, MICHAEL R.	Ph.D.	Aerospace Engineering	12/70
DAVIS, RONALD E.	M.S.	Mechanical Engineering	12/70
HARRIS, RONALD W.	M.S.	Electrical Engineering	12/70
HARSHA, PHILIP T.	Ph.D.	Aerospace Engineering	12/70
HESS, MICHAEL F.	M.S.	Aerospace Engineering	12/70
LENTS, JAMES M.	Ph.D.	Physics	12/70
MARSHALL, DOUGLAS W.	M.S.	Aerospace Engineering	12/70
PATEL, DHIRUBHAI M.	M.S.	Engineering Science	12/70
PENDER, CHARLES W.	M.S.	Engineering Science	12/70
ROCCO, VINCENT A.	M.S.	Mechanical Engineering	12/70
ROUX, JEFFREY A.	Ph.D.	Mechanical Engineering	12/70
SCHULZ, ROY J.	M.S.	Mechanical Engineering	12/70
STEPHENS, JAMES W.	M.S.	Physics	12/70
TOKEN, KENNETH H.	Ph.D.	Aerospace Engineering	12/70
BINION, TRAVIS W.	M.S.	Mechanical Engineering	3/71
FRANCIS, DAVID G.	M.S.	Mechanical Engineering	3/71
HARTSFIELD, HENRY W.	M.S.	Engineering Science	3/71
INTWALA, BIHARILAL D.	M.S.	Mechanical Engineering	3/71
MARTINDALE, WILLIAM R.	M.S.	Aerospace Engineering	3/71
McGEE, BARRY J.	M.S.	Physics	3/71
PELTON, JOHN M.	M.S.	Mechanical Engineering	3/71
RAKOWSKI, WALTER J.	M.S.	Mechanical Engineering	3/71
SMITH, GARY D.	M.S.	Mechanical Engineering	3/71
ALLEN, CHARLES W.	M.S.	Electrical Engineering	6/71
AOYAMA, KINYA	Ph.D.	Aerospace Engineering	6/71
BOYLAN, DAVID E.	M.S.	Aerospace Engineering	6/71
CODY, ROBERT L.	M.S.	Electrical Engineering	6/71
DUNCAN, BILLY J.	M.S.	Electrical Engineering	6/71
HARPER, WALTER L.	M.S.	Mechanical Engineering	6/71
HEIFNER, ROY L. JR.	M.S.	Electrical Engineering	6/71
KUKAINIS, JANIS	MS.	Aerospace Engineering	6/71
LITTON, CHARLES D.	M.S.	Physics	6/71
PEREZ, ERNEST R.	M.S.	Electrical Engineering	6/71

TURRENTINE, WILLIAM A. JR.	M.S.	Mechanical Engineering	6/71
VON RETH, ROLF D.	Ph.D.	Aerospace Engineering	6/71
BELZ, RONALD A.	Ph.D.	Electrical Engineering	8/71
COBLE, JERRY G.	M.S.	Aerospace Engineering	8/71
CRAWFORD, ROGER A.	Ph.D.	Aerospace Engineering	8/71
LINTON, WILLIAM JR.	M.S.	Electrical Engineering	8/71
LOPEZ, JOSE M.	M.S.	Aerospace Engineering	8/71
MOLNAR, DARREL G.	M.S.	Mechanical Engineering	8/71
REYNOLDS, CLIFFORD C.	M.S.	Electrical Engineering	8/71
WALKER, JAMES T.	M.S.	Mechanical Engineering	8/71
CROWL, ROBERT W.	M.S.	Mechanical Engineering	12/71
FRADENBURG, RONALD L.	M.S.	Electrical Engineering	12/71
LIMBAUGH, CHARLES C.	Ph.D.	Physics	12/71
LUTTRELL, ALVIN L.	M.S.	Engineering Science	12/71
McCLURE, DONALD R.	M.S.	Mechanical Engineering	12/71
MURRAY, EDWARD M.	M.S.	Physics	12/71
RICHARDSON, PETER W.	M.S.	Engineering Science	12/71
SANDERS, MARVIN E.	M.S.	Aerospace Engineering	12/71
SHANKLIN, RICHARD V.	Ph.D.	Aerospace Engineering	12/71
SOTOMAYER, WILLIAM A.	M.S.	Aerospace Engineering	12/71
WHITFIELD, DAVID L.	Ph.D.	Aerospace Engineering	12/71
WILHELM, LUTHER R.	Ph.D.	Mechanical Engineering	12/71
WILSON, FREDDIE W.	M.S.	Mechanical Engineering	12/71
DEVERS, ALVA DARRELL	Ph.D.	Mechanical Engineering	3/72
BROWN, JAMES E. JR.	M.S.	Electrical Engineering	6/72
GARNER, JACKIE E.	Ph.D.	Aerospace Engineering	6/72
MAYNE, ARLOE W.	Ph.D.	Mechanical Engineering	6/72
SAHA, HRISHIKESH	Ph.D.	Aerospace Engineering	6/72
SHELNUTT, RILEY C.	M.S.	Physics	6/72
BALDRIDGE, RONALD L.	M.S.	Electrical Engineering	8/72
BUTLER, ROY G.	M.S.	Electrical Engineering	8/72
GLASSMAN, HOWARD N.	M.S.	Mathematics	8/72
KINSLOW, MAX	Ph.D.	Aerospace Engineering	8/72
KRAFT, EDWARD M.	M.S.	Aerospace Engineering	8/72
NICHOLAS, MARIASINGAM T.	M.S.	Mechanical Engineering	8/72
NICHOLSON, MURRAY K.	M.S.	Aerospace Engineering	8/72
SIMPSON, WILLIAM R.	M.S.	Mechanical Engineering	8/72
WILLS, JAMES L.	M.S.	Aerospace Engineering	8/72
VARNER, MICHAEL O.	Ph.D.	Aerospace Engineering	8/72
BUTLER, RICHARD W.	M.S.	Aerospace Engineering	12/72
CHOW, LIN CHEN	Ph.D.	Engineering Science	12/72
KORNSTETT, KENNETH W.	M.S.	Physics	12/72
LADD, THOMAS G.	M.S.	Mathematics	12/72
ROSE, JACK M.	M.S.	Mathematics	12/72
SCHRECKER, GUNTER O.	Ph.D.	Aerospace Engineering	12/72
FARMER, WILLIAM M.	Ph.D.	Physics	3/73
MILLER, JOHN T.	Ph.D.	Aerospace Engineering	3/73
TRIPLETT, MILTON J.	M.S.	Mechanical Engineering	3/73
BOMAR, BRUCE WILLIAM	M.S.	Electrical Engineering	6/73
DAVIDSON, DONALD LEE	M.S.	Engineering Science	6/73
LAZALIER, GLENDON RAY	M.S.	Engineering Admin.	6/73
MANSFIELD, ARTHUR CONRAD	M.S.	Aerospace Engineering	6/73
MOULDEN, TERVOR HOLMES	Ph.D.	Aerospace Engineering	6/73
NADJI, BEHROUZ	M.S.	Engineering Science	6/73
PULLEY, DONALD CHARLES	M.S.	Electrical Engineering	6/73
RICHARDSON, JOHN WILBURN	M.S.	Engineering Science	6/73
ROCHELLE, JAMES K.	M.S.	Engineering Mechanics	6/73
CHANG, KUANG-YEH	M.S.	Electrical Engineering	8/73
PRAHARAJ, SARAT CHANDRA	Ph.D.	Aerospace Engineering	8/73
WICKMAN, JOHN H.	M.S.	Mechanical Engineering	8/73
KEENEY, FRANCES J.	M.S.	Aerospace Engineering	12/73
LOCK, ROBERT C.	M.S.	Aerospace Engineering	12/73
WANG, KONG TENG	Ph.D.	Mechanical Engineering	12/73
BISSINGER, NORBERT C.	Ph.D.	Aerospace Engineering	3/74
MUEHLHAUSER, JOEL W.	Ph.D.	Physics	3/74
SHEN, LIN	M.S.	Mechanical Engineering	3/74
SMITH, DALE K.	M.S.	Aerospace Engineering	3/74
STUDWELL, VICTOR E.	Ph.D.	Aerospace Engineering	3/74
DUANGUDOM, SAHATEP	M.S.	Aerospace Engineering	6/74
IANNUZZI, FRANCESCO A.	Ph.D.	Engineering Science	6/74
KLAUTSCH, RALPH E.	M.S.	Electrical Engineering	6/74
VENGHAUS, HEINZ H.	Ph.D.	Aerospace Engineering	6/74
WALKER, CLARK M.	M.S.	Aerospace Engineering	6/74
WALKER, JAMES T.	M.S.	Engineering Admin.	6/74
BUCKOWSKY, THOMAS G.	M.S.	Avaition Systems	8/74
HASTY, DOYLE E.	M.S.	Engineering Admin.	8/74
MIKKILINENI, RAJENDRA P.	M.S.	Aerospace Engineering	8/74
NADJI, BEHROUZ	M.S.	Computer Science	8/74
PIGOTT, JOSEPH C.	M.S.	Engineering Admin.	8/74
RAPER, RICHARD M.	M.S.	Engineering Admin.	8/74
SMITH, RICHARD H.	M.S.	Engineering Admin.	8/74
DOTTER, ARTHUR MARTIN JR.	M.S.	Engineering Admin.	12/74
MARCHAND, ERNEST O.	M.S.	Mechanical Engineering	12/74
MERZ, GLENN H.	M.S.	Engineering Admin.	12/74

MUTHUSWAMY, JAYARAMAN	M.S.	Mechanical Engineering	12/74
RHUDY, JAMES P.	M.S.	Mechanical Engineering	12/74
SHELTON, JAMES L.	M.S.	Engineering Admin.	12/74
TAYLOR, JAMES LEBRON	M.S.	Mathematics	12/74
TORRES, JOAO-PAULO T.	M.S.	Aerospace Engineering	12/74
WEAVER, HIRAM M.	M.S.	Engineering Admin.	12/74
YEE, PHILIP MING	M.S.	Aerospace Engineering	12/74
CARTER, LARRY DRAYTON	M.S.	Mechanical Engineering	3/75
CHEN, CHWAN-HAW	Ph.D.	Aerospace Engineering	3/75
DORRELL, EDWARD W. JR.	M.S.	Mathematics	3/75
GOETHERT, WINFRIED H.	M.S.	Physics	3/75
GRAY, J. DON	M.S.	Mechanical Engineering	3/75
LI, TZU PANG	M.S.	Electrical Engineering	3/75
LOHR, ALBERT D.	M.S.	Computer Science	3/75
MALONE, LYNN H.	M.S.	Physics	3/75
PHELPS, JOHN A.	M.S.	Aerospace Engineering	3/75
BARNETT, LARRY ROGER	M.S.	Electrical Engineering	6/75
JAI-UA, THAVATCHAI	M.S.	Electrical Engineering	6/75
JOYCE, CHARLES ROBERT	M.S.	Computer Science	6/75
KIMBERLIN, RALPH DACE	M.S.	Aerospace Engineering	6/75
SICKLES, WILLIAM LOWELL	M.S.	Mathematics	6/75
WANG, PU TIEN	M.S.	Electrical Engineering	6/75
CRAWFORD, MARGARET A.	M.S.	Mathematics	8/75
CUNNINGHAM, JAMES W.	Ph.D.	Electrical Engineering	8/75
GERMAN, RICHARD C.	M.S.	Engineering Admin.	8/75
GREEN, FRED M.	M.S.	Computer Science	8/75
JOHNSON, MARTIN R.	M.S.	Physics	8/75
WHITE, JAMES R.	M.S.	Computer Science	8/75
BENEK, JOHN A.	Ph.D.	Aerospace Engineering	12/75
BITTE, JUERGEN	Ph.D.	Aerospace Engineering	12/75
BRIDGES, JACOB H. JR.	M.S.	Computer Science	12/75
HOLBERT, ROY K. JR.	M.S.	Engineering Science	12/75
KAUL, UPENDER KRISHEN	M.S.	Mechanical Engineering	12/75
KRAFT, EDWARD M.	Ph.D.	Aerospace Engineering	12/75
MENZEL, REINHARD W.	Ph.D.	Physics	12/75
MIDGETT, DON C.	M.S.	Computer Science	12/75
ROBERDS, DONALD W.	Ph.D.	Electrical Engineering	12/75
SHAHABI, ALIREZA M.	M.S.	Mechanical Engineering	12/75
BORCHERS, INGO	Ph.D.	Aerospace Engineering	3/76
BURGESS, BRYAN E.	Ph.D.	Engineering Science	3/76
CABLE, ARTHUR J.	M.S.	Engineering Admin.	3/76
CALLENS, EARL E. JR.	Ph.D.	Aerospace Engineering	3/76
HARPER, DAVID CLAIR	M.S.	Aerospace Engineering	3/76
SINCLAIR, DARYL WARREN	M.S.	Mechanical Engineering	3/76
WANG, SHOW TIEN	M.S.	Engineering Science	3/76
HICKEY, ROBERT FRANKLIN	M.S.	Computer Science	6/76
KROEGER, GLEN ARTHUR	M.S.	Physics	6/76
KULKARNI, HARIHAR TRIMBAK	Ph.D.	Mechanical Engineering	6/76
KUWANO, HIDEKI	M.S.	Aerospace Engineering	6/76
PARTIN, ROBERT L. JR.	M.S.	Physics	6/76
SCHULZ, ROY JAMES	Ph.D.	Mechanical Engineering	6/76
SCOTT, MARY HALL	Ph.D.	Engineering Science	6/76
SHELBY, REX THOMAS	M.S.	Mathematics	6/76
ARORA, BALBIR S.	Ph.D.	Mechanical Engineering	8/76
BAKER, WILLIAM B. JR.	Ph.D.	Aerospace Engineering	8/76
CHOU, YOU-LI	Ph.D.	Mechanical Engineering	8/76
MIKKILINENI, PRASAD R.	Ph.D.	Aerospace Engineering	8/76
NOWAK, DIETER KURT	Ph.D.	Aerospace Engineering	8/76
PERKINS, PATRICK E.	M.S.	Computer Science	8/76
PERLMUTTER, MORRIS	Ph.D.	Engineering Science	8/76
TURNER, ROBERT E.	Ph.D.	Engineering Science	8/76
VAKILI-DASTJERD, AHMAD	M.S.	Aerospace Engineering	8/76
WHITE, MARVIS, K.	M.S.	Engineering Science	8/76
WILSON, FRANCES C.	M.S.	Mathematics	8/76
WU, CHUNG-I	M.S.	Mechanical Engineering	8/76
CURTIS, RANDALL S.	M.S.	Aviation Systems	12/76
EVANS, RANDALL L.	M.S.	Computer Science	12/76
FLAHERTY, MICHAEL I.	M.S.	Physics	12/76
GILLEN, ROBERT L.	M.S.	Engineering Admin.	12/76
JACKSON, FRANCIS M. III	M.S.	Aviation Systems	12/76
JACOCKS, JAMES L.	Ph.D.	Aerospace Engineering	12/76
KAR, MUKTA LAL	M.S.	Electrical Engineering	12/76
RICCO, RAYMOND J. JR.	M.S.	Computer Science	12/76
SHIEH, CHIH-FANG	M.S.	Mechanical Engineering	12/76
VAUGHAN, WILLIAM W.	Ph.D.	Engineering Science	12/76
BECKER, ROBERT S.	Ph.D.	Engineering Science	3/77
CLOUSE, RONALD L.	M.S.	Mechanical Engineering	3/77
HUEBSCHMAN, DONALD L.	M.S.	Computer Science	3/77
JOSHI, MAHENDRAKUMAR C.	Ph.D.	Mechanical Engineering	3/77
KESSEL, PHILIP A.	Ph.D.	Mechanical Engineering	3/77
LOWRY, HEARD S. III	M.S.	Physics	3/77
LUCCI, CARMEN A.	M.S.	Aerospace Engineering	3/77
PETERSON, JANET S.	M.S.	Mathematics	3/77
ROY, GABRIEL D.	Ph.D.	Engineering Science	3/77

ZERINGUE, KENNETH R.	M.S.	Physics	3/77
BECKER, LATIKA S.R.	Ph.D.	Engineering Science	6/77
CAMP, DENNIS W.	M.S.	Engineering Science	6/77
CHARBONEAU, GEORGE W.	M.S.	Engineering Science	6/77
JONES, NANCY L.	M.S.	Engineering Science	6/77
KAEWPLOY, PREEDA	M.S.	Aerospace Engineering	6/77
KIMZEY, WILLIAM F.	Ph.D.	Aerospace Engineering	6/77
LIU, BAW-LIN	Ph.D.	Mechanical Engineering	6/77
MUTHUSWAMY, JAYARAMAN	M.S.	Engineering Admin.	6/77
MY, TRAN	M.S.	Electrical Engineering	6/77
PECHINTHORN, SARUN	M.S.	Aerospace Engineering	6/77
PENNINGTON, MICHAEL	M.S.	Computer Science	6/77
SANDERS, EUGENE J.	M.S.	Engineering Admin.	6/77
SHARBER, LESLIE A.	M.S.	Engineering Science	6/77
SUSKO, MICHAEL	M.S.	Engineering Science	6/77
WELCH, DONALD E.	M.S.	Engineering Science	6/77
EATON, ALFRED F. JR.	M.S.	Aviation Systems	8/77
REDDY, KAPULURU RAVIKUMAR	M.S.	Engineering Science	8/77
YAROS, STEVEN FRANCIS	Ph.D.	Aerospace Engineering	8/77
CHAPMAN, JAMES N.	Ph.D.	Engineering Science	12/77
CHEVASUWAN, PAIROJ	M.S.	Aerospace Engineering	12/77
DOWGWILLO, ROBERT M.	M.S.	Aerospace Engineering	12/77
FOWLER, RILEY B.	M.S.	Electrical Engineering	12/77
HARRIS, JOHN AUGUSTUS	M.S.	Computer Science	12/77
PARRIOTT, DAVID LEE	M.S.	Engineering Science	12/77
TATUM, KENNETH E.	M.S.	Aerospace Engineering	12/77
THONGINTRA, CHESTAWEE	M.S.	Aerospace Engineering	12/77
UCHIYAMA, NOBUYOSHI	Ph.D.	Aerospace Engineering	12/77
VOSBURGH, ROBERT J.	M.S.	Aerospace Engineering	12/77
WEINBERG, NORMAN S.	M.S.	Mechanical Engineering	12/77
ZIADY, ZIAD	M.S.	Engineering Science	12/77
BATTE, ZENI	M.S.	Computer Science	3/78
FITZGERALD, JAMES N.	M.S.	Engineering Admin.	3/78
LALLY, LEONARD A. JR.	M.S.	Physics	3/78
LUTZ, RONALD GEORGE	M.S.	Mechanical Engineering	3/78
McLAWHON, GEORGE B. JR.	M.S.	Computer Science	3/78
ABDULRAHMAN, NAZIH N.	M.S.	Mechanical Engineering	6/78
AKERS, DENNIS TEX	M.S.	Mechanical Engineering	6/78
BRANTLEY, DAVID MARK	M.S.	Computer Science	6/78
CROSBY, WILLIAM ARNOLD	M.S.	Engineering Science	6/78
CROUCH, WILLIAM MAURICE	M.S.	Computer Science	6/78
DATHE, INGO	M.S.	Aerospace Engineering	6/78
GOULD, SAMUEL SHULE	M.S.	Electrical Engineering	6/78
PEPLINSKI, STANLEY Z.	M.S.	Physics	6/78
WISE, LILLIAN JUNG	M.S.	Engineering Science	6/78
DURHAM, DON E.	M.S.	Aviation Systems	8/78
EQUERE, IMEH AKPAN	M.S.	Mechanical Engineering	8/78
EY, LILLIAN KIMBROUGH	M.S.	Physics	8/78
GREEN, MICHAEL CLARK	M.S.	Computer Science	8/78
LYONS, SANDRA CARTWRIGHT	M.S.	Computer Science	8/78
NIWA, YOSHIYUKI	M.S.	Aerospace Engineering	8/78
BRYANT, DENISE MAXINE	M.S.	Computer Science	12/78
CALLICUTT, FORREST WADE	M.S.	Mathematics	12/78
CARLSON, PETER K.	M.S.	Physics	12/78
FIKES, DAVID HOWELL	M.S.	Mechanical Engineering	12/78
LYONS, ANTHONY THOMAS	M.S.	Engineering Science	12/78
MATTASITS, GARY RICHARD	M.S.	Aerospace Engineering	12/78
MORRIS, ROGER DALE	M.S.	Electrical Engineering	12/78
VAKILI-DASTJERD, AHMAD	Ph.D.	Aerospace Engineering	12/78
WHORTON, RANDALL D.	M.S.	Mechanical Engineering	12/78
BELEW, PAUL WAYNE	M.S.	Electrical Engineering	3/79
BROWN, JOHNNIE WAYNE	M.S.	Computer Science	3/79
FOX, JOHN HOWARD	Ph.D.	Mechanical Engineering	3/79
HILL, JOHN LOGAN	M.S.	Computer Science	3/79
HUBBLE, JERRY DEAN	M.S.	Aerospace Engineering	3/79
JOLLAY, JAMES P.	MS.	Engineering Science	3/79
POWER, WILLIAM H.	M.S.	Mechanical Engineering	3/79
CHU, JOSEPH Q.	M.S.	Engineering Science	6/79
JONES, JOHN H.	M.S.	Computer Science	6/79
REECE, ROBERT S.	M.S.	Aviation Systems	6/79
SEYMOUR, MARSHALL L.	M.S.	Engineering Admin.	6/79
ARGIALAS, DEMETRE	M.S.	Engineering Science	8/79
BARNES, HOWARD B.	M.S.	Physics	8/79
BRERA, AHMED M.	Ph.D.	Engineering Science	8/79
CHAMBLEE, CLAUDE E.	M.S.	Mechanical Engineering	8/79
McGEE, PATRICK M..	M.S.	Engineering Science	8/79
SCHEIDLER, MICHAEL J.	M.S.	Mathematics	8/79
ABUJELALA, MUFTAH TAHER	M.S.	Aerospace Engineering	12/79
BECKER, BRYAN RUSSELL	Ph.D.	Engineering Science	12/79
KAUFMAN, JOHN W.	M.S.	Engineering Science	12/79
PROVINCE, GARY T.	M.S.	Electrical Engineering	12/79
SRINIVAS, VIJAY K.	M.S.	Engineering Science	12/79
TARABZOUNI, MOHAMED A.	M.S.	Engineering Science	12/79
AGARWAL, ANANT K.	M.S.	Electrical Engineering	3/80

KOEPF, DAVID C.	M.S.	Engineering Science	3/80
MURPHY, PATRICK J.	M.S.	Physics	3/80
PASTEL, ROBERT L.	M.S.	Engineering Science	3/80
PERKINS, PATRICK E.	Ph.D.	Physics	3/80
ROWAN, CHARLES D.V.	M.S.	Engineering Science	3/80
STEELY, SIDNEY L.	M.S.	Engineering Science	3/80
WANG, SHOW-TIEN	Ph.D.	Engineering Science	3/80
DODD, SUSAN G.	M.S.	Mathematics	6/80
GUTHRIE, MICHAEL J.	M.S.	Physics	6/80
LOWE, CHARLES R.	M.S.	Computer Science	6/80
SARKAR, KAMALAKSHA	Ph.D.	Engineering Science	6/80
SHIEH, CHIH FANG	Ph.D.	Mechanical Engineering	6/80
TROUPE, DONALD E.	M.S.	Engineering Admin.	6/80
VICKREY, SUSAN C.	M.S.	Computer Science	6/80
COLLIER, MICHAEL R.	M.S.	Aerospace Engineering	8/80
CYRAN, FREDERIC B.	M.S.	Mechanical Engineering	8/80
GONZALAS, DORA E.	M.S.	Aerospace Engineering	8/80
HAMLETT, EDWARD L.	M.S.	Mechanical Engineering	8/80
MACARAEG, MICHELE G.	M.S.	Engineering Science	8/80
MARQUART, EDWARD J.	M.S.	Engineering Science	8/80
MONTGOMERY, WILLARD W.	Ph.D.	Physics	8/80
ROOKE, KENNY A.	M.S.	Aerospace Engineering	8/80
TURKEL, BARRY S.	M.S.	Aviation Systems	8/80
TYLE, NAVIN	M.S.	Electrical Engineering	8/80
VEERASAMY, VENUGOPAL	Ph.D.	Mechanical Engineering	8/80
WINKLEMAN, BRADLEY C.	M.S.	Physics	8/80
YEAKLEY, PHILLIP L.	M.S.	Mechanical Engineering	8/80
CAMPBELL, LARRY	M.S.	Engineering Admin.	12/80
CARNAL, CHARLES L.	M.S.	Electrical Engineering	12/80
HUANG, KAO-HUA	M.S.	Aerospace Engineering	12/80
PARKER, JAMES Y.	M.S.	Engineering Admin.	12/80
SLOAN, GENE H.	M.S.	Aviation Systems	12/80
DAHM, WERNER J.A.	M.S.	Mechanical Engineering	3/81
ESKRIDGE, RICHARD H.	M.S.	Engineering Science	3/81
HOLT, WILLIAM L.	Ph.D.	Mechanical Engineering	3/81
McCREARY, WILSON E.	M.S.	Electrical Engineering	3/81
TSUI, YEN-YUNG	M.S.	Aerospace Engineering	3/81
AL HAIDEY, KHALAF ALI	M.S.	Engineering Science	6/81
BALASUBRAMANYAN, C.	Ph.D.	Aerospace Engineering	6/81
HESKETH, ANDREW A.	M.S.	Engineering Science	6/81
HUNT, JERRY DALE	M.S.	Engineering Admin.	6/81
KAR, MUKTA LAL	Ph.D.	Engineering Science	6/81
KHAN, MUHAMMAD ASIF	M.S.	Engineering Science	6/81
KIM, BUMBAE	M.S.	Mechanical Engineering	6/81
KURODA, HIROMOTO	M.S.	Engineering Science	6/81
LUKE, GREGORY MICHAEL	M.S.	Engineering Science	6/81
NEUFELD, HERMAN	M.S.	Electrical Engineering	6/81
SIMMONS, MARTHA ANN	M.S.	Engineering Science	6/81
TRIMBACH, JOHN MICHAEL	M.S.	Aerospace Engineering	6/81
WINCHESTER, JAMES EDWARD	M.S.	Engineering Science	6/81
DAVIS, MILTON W. JR.	M.S.	Mechanical Engineering	8/81
JAIN, SUNIL KUMAR	M.S.	Aerospace Engineering	8/81
KUMARAN, ARIC R.	M.S.	Aerospace Engineering	8/81
PRASAD, JAWAHARLAL	M.S.	Chemical Engineering	8/81
SKILES, TERRY W.	M.S.	Mechanical Engineering	8/81
DEE, JYH-REN	M.S.	Mechanical Engineering	12/81
DILL, RICHARD A.	M.S.	Engineering Science	12/81
HARTVIGSEN, THOMAS EARL	M.S.	Mechanical Engineering	12/81
LUCHUK, STEVEN WALLACE	M.S.	Computer Science	12/81
McGEE, JAMES M.	M.S.	Aerospace Engineering	12/81
SHAW, JOSEPH J. JR.	M.S.	Computer Science	12/81
SHEIKHOLESLAMI, AMIR H.	M.S.	Mechanical Engineering	12/81
TARABZOUNI, MOHAMED AHMED	Ph.D.	Engineering Science	12/81
VENKITARAMA, SUNDARAM	Ph.D.	Mechanical Engineering	12/81
VIDYASAGARAN, N.S.	M.S.	Mechanical Engineering	12/81
LANKFORD, JEFFREY PAUL	M.S.	Engineering Science	3/82
LOWRY, HEARD S. III	Ph.D.	Physics	3/82
SU-JYH-CHYUAN	M.S.	Electrical Engineering	3/82
AKKARI, SAFWAN H.	M.S.	Engineering Science	6/82
CHANG, HO-PEN	M.S.	Aerospace Engineering	6/82
DUESTERHAUS, DAVID A.	M.S.	Mechanical Engineering	6/82
HINDS, HARRY S. JR.	M.S.	Aerospace Engineering	6/82
JOHNSON, DALE L.	M.S.	Engineering Science	6/82
KWONG, KIN CHUN	M.S.	Electrical Engineering	6/82
MEYER, ROBERT D.	M.S.	Computer Science	6/82
MOORE, JAMES R.	M.S.	Engineering Science	6/82
PENDLETON, JONES DAVID	Ph.D.	Physics	6/82
POLCE, RONALD L.	M.S.	Mechanical Engineering	6/82
ROSE, CHARLES D.	M.S.	Electrical Engineering	6/82
SON, JUNG-YOUNG	M.S.	Electrical Engineering	6/82
WANTLAND, EDGAR C. JR.	M.S.	Aerospace Engineering	6/82
CAHILL, DAVID MILBURN	M.S.	Aerospace Engineering	8/82
GOSWAMI, RAM KISHORE	Ph.D.	Engineering Science	8/82
NICHOLS, ROBERT HIRAM	M.S.	Aerospace Engineering	8/82

WITT, SANDRA LYNNE	M.S.	Engineering Science	8/82
CHU, JOSEPH Q.	Ph.D.	Engineering Science	12/82
DREHER, JOHN L. III	M.S.	Engineering Admin.	12/82
FOOTE, JOHN PAUL	M.S.	Mechanical Engineering	12/82
NOURINEJAD, ABDOLHOSSEIN	M.S.	Aerospace Engineering	12/82
CROWDER, HERBERT L.	M.S.	Physics	3/83
LIAO, YI-FU	Ph.D.	Mechanical Engineering	3/83
WRIGHT, L.E.	Ph.D.	Engineering Science	3/83
CHANG, JUDY CHIEN-HUI	M.S.	Computer Science	6/83
LAROSILIERE, LOUIS	M.S.	Aerospace Engineering	6/83
LeMASTER, ROBERT A.	Ph.D.	Engineering Science	6/83
LIN, MING-CHANG	M.S.	Mechanical Engineering	6/83
YU, FAN-MING	M.S.	Aerospace Engineering	6/83
BATES, LESTER BRENT	M.S.	Engineering Science	8/83
GILLIAM, FRED T. JR.	Ph.D.	Aerospace Engineering	8/83
STEWART, VAN WILSON III	M.S.	Mechanical Engineering	8/83
DONEGAN, TRACY LEE	M.S.	Aerospace Engineering	12/83
KURODA, HIROMOTO	Ph.D.	Engineering Science	12/83
LEE, JIUNN-JIU	Ph.D.	Engineering Science	12/83
NEWTON, JOHN POLLARD	M.S.	Engineering Science	12/83
PETERS, WILLIAM LEE	M.S.	Aerospace Engineering	12/83
SELMAN, JAMES D. JR.	M.S.	Physics	12/83
CARRINGTON, KENNETH	M.S.	Computer Science	3/84
HUANG, KAO-HUAH	Ph.D.	Mechanical Engineering	3/84
NELSON, CHARLES ROBERT II	M.S.	Industrial Engineering	3/84
STICH, PHILIP BERNARD	M.S.	Mechanical Engineering	3/84
CLAY, THOMAS H.	M.S.	Industrial Engineering	6/84
CRAWFORD, MARGARET ROSS	M.S.	Industrial Engineering	6/84
CROSSWY, FRANK LESLIE	Ph.D.	Engineering Science	6/84
HULL, JAMES DALE	M.S.	Aerospace Engineering	6/84
LAYNE, THOMAS CHRIS	M.S.	Electrical Engineering	6/84
MACARAEG, MICHELE	Ph.D.	Engineering Science	6/84
MATTY, JERE JOSEPH	M.S.	Mechanical Engineering	6/84
SINHA, ATIN KUMAR	Ph.D.	Aerospace Engineering	6/84
SMITH, L. MONTGOMERY	M.S.	Electrical Engineering	6/84
WELCH, JACK	M.S.	Industrial Engineering	6/84
ANSPACH, KEITH MARLIN	M.S.	Computer Science	8/84
AUMALIS, ANSIS EDGARS	M.S.	Engineering Science	8/84
CHI, SHAN	M.S.	Computer Science	8/84
DAVIS, MARTIN HENRY	M.S.	Engineering Science	8/84
HOFFMAN, PETER L.	M.S.	Mechanical Engineering	8/84
HWANG, CHII-JONG	Ph.D.	Mechanical Engineering	8/84
KAUL, RAJ KRISHAN	Ph.D.	Engineering Science	8/84
KIESSLING, EDWARD H. III	M.S.	Mechanical Engineering	8/84
SMITH, MICHAEL STEVEN	M.S.	Physics	8/84
CHAMLEE, JOSEPH BROOKS	M.S.	Engineering Science	12/84
EVANS, JEFFREY AUSTIN	M.S.	Engineering Science	12/84
FOWLEY, MARK DAVID	M.S.	Mechanical Engineering	12/84
GALLAGHER, JOSEPH MICHAEL	M.S.	Computer Science	12/84
GHOSH, ASHOKE	Ph.D.	Mechanical Engineering	12/84
HILL, CHARLES KELLY	M.S.	Engineering Science	12/84
PELHAM, ROGER DALE	M.S.	Computer Science	12/84
RINGNES, ERIK AMUND	M.S.	Aerospace Engineering	12/84
GAMBLE, RICK ALAN	M.S.	Mechanical Engineering	3/85
GRAY, WILLIAM GORDON	M.S.	Mechanical Engineering	3/85
HUGHES, LISA JAN	M.S.	Mechanical Engineering	3/85
HURST, MITCHELL JAMES	M.S.	Computer Science	3/85
MILAM, SAMUEL SCOTT	M.S.	Engineering Science	3/85
MOREHEAD, R. KEITH	M.S.	Mechanical Engineering	3/85
SAP, JAMES LARRY	M.S.	Industrial Engineering	3/85
SUTTON, MICHAEL LEWIS	M.S.	Mechanical Engineering	3/85
THEON, JOHN SPERIDON	Ph.D.	Engineering Science	3/85
WILHITE, LARRY DWIGHT	M.S.	Electrical Engineering	3/85
WILLIAMS, WILLIAM A.	M.S.	Industrial Engineering	3/85
CHANG, HO-PEN	Ph.D.	Aerospace Engineering	6/85
CHO-JOO-YUN (DENNY)	M.S.	Mechanical Engineering	6/85
DEAN, DAVID EDWARDS	M.S.	Mechanical Engineering	6/85
HARIHARAN, JEANETTE	M.S.	Electrical Engineering	6/85
HOLT, CATHY ANN	M.S.	Computer Science	6/85
HOLT, JEFFREY KEVIN	M.S.	Chemical Engineering	6/85
LIENEMANN, KENNETH A.	M.S.	Engineering Science	6/85
PICKENS, ANDREW LEWIS JR.	M.S.	Computer Science	6/85
SON, JUNG-YOUNG	Ph.D.	Engineering Science	6/85
WOOD, CHARLES WADE	M.S.	Aerospace Engineering	6/85
BISHOP, RICHARD M.	M.S.	Electrical Engineering	8/85
BLANNER, JAMES R.	M.S.	Mechanical Engineering	8/85
DUGAS, ROGER M.	M.S.	Mechanical Engineering	8/85
GARDNER, WANDA	M.S.	Engineering Science	8/85
HARTMAN, ANTHONY	M.S.	Mechanical Engineering	8/85
NICHOLSON, LYNN ALAN	M.S.	Mechanical Engineering	8/85
PRATHER, CALVIN	M.S.	Engineering Science	8/85
RINER, WILLIAM C.	M.S.	Mechanical Engineering	8/85
TAPPEN, JOHN	Ph.D.	Engineering Science	8/85
ANDREW, PHILIP LYNN	M.S.	Mechanical Engineering	12/85

BEITEL, GREGG RICHARD	M.S.	Mechanical Engineering	12/85
DAVIS, STEPHAN ROWAN	M.S.	Engineering Science	12/85
GUYTON, FREDERICK CLYDE	M.S.	Aerospace Engineering	12/85
HOEFER, RODNEY WILLIAM	M.S.	Mechanical Engineering	12/85
JONES, MARY WATTS	M.S.	Computer Science	12/85
KNISLEY, DENNIS RAY	M.S.	Chemical Engineering	12/85
LANE, KENNETH GREGORY	M.S.	Electrical Engineering	12/85
MENARD, DAVID GEORGE	M.S.	Physics	12/85
PAYNE, KAREN LYNN	M.S.	Computer Science	12/85
SOLIES, UWE PETER	Ph.D.	Aerospace Engineering	12/85
WANNEWETSCH, GREG DAVID	M.S.	Aerospace Engineering	12/85
YOUNGBLOOD, DOUGLAS BRUCE	M.S.	Aerospace Engineering	12/85
ANDERSON, WILLIAM HOMER	M.S.	Industrial Engineering	3/86
GUYTON, FREDERICK CLYDE	M.S.	Industrial Engineering	3/86
LYNCH, TILDEN PATRICK	M.S.	Engineering Science	3/86
PARKER, ROBERT LEE JR.	M.S.	Industrial Engineering	3/86
PENDER, CHARLES WILLIAM	Ph.D.	Engineering Science	3/86
PHILLIPS, WILLIAM JOE	Ph.D.	Physics	3/86
SHI, ZHI	M.S.	Aerospace Engineering	3/86
TIETZ, THOMAS ANDREAS	M.S.	Aerospace Engineering	3/86
WATSON, WILLIAM THOMAS JR.	M.S.	Industrial Engineering	3/86
BAILEY, ANDREW GRIFFITH	M.S.	Computer Science	6/86
BROOKS, DEBORAH SUSAN	M.S.	Physics	6/86
FREIMARK, ERIC SCOTT	M.S.	Industrial Engineering	6/86
GIBSON, CLYDE PAISLEY	M.S.	Computer Science	6/86
JOLLY, RANDY EUGENE	M.S.	Computer Science	6/86
NICHOLS, ROBERT H.	Ph.D.	Aerospace Engineering	6/86
RUFF, GARY ALLEN	M.S.	Aerospace Engineering	6/86
BARRET, CHRIS	M.S.	Aerospace Engineering	8/86
DANG, HANH QUANG	M.S.	Aerospace Engineering	8/86
DUNN, STEVEN CHARLES	M.S.	Mechanical Engineering	8/86
HOWARD, STEVEN	M.S.	Electrical Engineering	8/86
KEEN, KENNETH SCOTT	M.S.	Aerospace Engineering	8/86
KOLBE, RONALD LYNN	Ph.D.	Mechanical Engineering	8/86
McDANIELS, DAVID MARTIN	M.S.	Aerospace Engineering	8/86
McFEE, JOHN	M.S.	Computer Science	8/86
PAIK, HEUNG CHEON	M.S.	Physics	8/86
SOROOSH, EBRAHIM G.	M.S.	Electrical Engineering	8/86
SUDHARSANAN, SUBRAMANIA	M.S.	Electrical Engineering	8/86
WADDELL, VOREATA SANDERS	M.S.	Physics	8/86
WELLE, RICHARD P.	M.S.	Physics	8/86
ALEXANDER, CHARLES BRYAN	M.S.	Computer Science	12/86
ARMAN, ESSANOLLAH FARSHAD	Ph.D.	Engineering Science	12/86
BUTLER, WILLIAM ALLEN	M.S.	Chemcial Engineering	12/86
BYRD, SHANNON INEZ	M.S.	Engineering Science	12/86
BENNETT, BRIAN KENDALL	M.S.	Mechanical Engineering	12/86
HOBBS, RANDY WAYNE	M.S.	Aerospace Engineering	12/86
KINCAID, WILLIAM KEITH	M.S.	Mechanical Engineering	12/86
MEADOW, JOHN STANTON	M.S.	Electrical Engineering	12/86
SEISER, KAREN M.	M.S.	Engineering Science	12/86
SHAVER, TROY CALVIN	M.S.	Industrial Engineering	12/86
SHIRILLA, KATHLEEN	M.S.	Computer Science	12/86
ANGEL, PAUL ROLAND	M.S.	Mechanical Engineering	3/87
BRIMER, ROBERT ANTHONY	M.S.	Industrial Engineering	3/87
DeSTEFANO, THOMAS DANIEL	M.S.	Mechanical Engineering	3/87
DOBSON, CHRISTOPHER C.	M.S.	Physics	3/87
FRAZIER, JACKSON W. JR.	M.S.	Electrical Engineering	3/87
HALE, MAURICE G.	M.S.	Industrial Engineering	3/87
McKAMEY, MARY CAROLINE	M.S.	Engineering Science	3/87
PARKER, DAVID A.	M.S.	Industrial Engineering	3/87
SUNDARAM, P.	Ph.D.	Aerospace Engineering	3/87
CHIANG, ANDREW SIEW	M.S.	Aerospace Engineering	6/87
CONLEY, ANITA POLK	M.S.	Engineering Science	6/87
CROSS, PAUL TERRY JR.	M.S.	Aerospace Engineering	6/87
DeBUSK, CHARLES RICHARD	M.S.	Industrial Engineering	6/87
DRAKES, JAMES ALAN	M.S.	Physics	6/87
LEE, CALVIN	M.S.	Industrial Engineering	6/87
LIN, KUANHONG HENRY	M.S.	Computer Science	6/87
McAMIS, ROBERT WOOD	M.S.	Mechanical Engineering	6/87
McCARTHY, BARRY GERARD	M.S.	Aviation Systems	6/87
PATEL, MUKESH T.	M.S.	Computer Science	6/87
VONGTANAANEK, TOADSAK	M.S.	Electrical Engineering	6/87
YU, FAN-MING	Ph.D.	Aerospace Engineering	6/87
GROFF, RUSSELL	M.S.	Aerospace Engineering	8/87
HOLCOMB, GREG	M.S.	Electrical Engineering	8/87
NOURINEJAD, ABDOLHOSSEIN	Ph.D.	Aerospace Engineering	8/87
SANFORD, CHARLES STEPHEN	M.S.	Aviation Systems	8/87
SAUNDERS, JAMES L.	M.S.	Aerospace Engineering	8/87
WICHMANN, DOUGLAS	M.S.	Industrial Engineering	8/87
WU, CHIEH	Ph.D.	Engineering Science	8/87
ASKINS, MICHAEL LANCE	M.S.	Computer Science	12/87
BENTLEY, HARRY THOMAS III	M.S.	Industrial Engineering	12/87
CARSON, DAVID LEE	M.S.	Electrical Engineering	12/87
CAYSE, ROBERT WILLIAM	M.S.	Engineering Science	12/87

FLANIGAN, DANIEL PATRICK	M.S.	Industrial Engineering	12/87
GUPTA, UDAY KUMAR	M.S.	Computer Science	12/87
JORDAN, JOHN L.	M.S.	Mechanical Engineering	12/87
LIVER, PETER AUGUST	Ph.D.	Aerospace Engineering	12/87
NEESE, DONALD WAYNE	M.S.	Computer Science	12/87
PACE, LONNIE DeWAYNE	M.S.	Industrial Engineering	12/87
RAMACHANDRAN, K.	Ph.D.	Aerospace Engineering	12/87
SCHARNHORST, DEAN ALLEN	M.S.	Computer Science	12/87
SHUTTLEWORTH, JOHN GEORGE	M.S.	Electrical Engineering	12/87
TYLER, VICTOR CHRISTOPHER	M.S.	Industrial Engineering	12/87
BAKER, RAYMOND NEIL	M.S.	Industrial Engineering	3/88
BOULWARE, ROBERT HENRY	M.S.	Industrial Engineering	3/88
BRIDGES, JACOB HAL II	M.S.	Electrical Engineering	3/88
DEIBNER, GUS W.	M.S.	Mechanical Engineering	3/88
DILL, KEITH MITCHELL	M.S.	Mechanical Engineering	3/88
GROFF, MARY BISKNER	M.S.	Mechanical Engineering	3/88
NICHOLSON, RANDY ALAN	M.S.	Mechanical Engineering	3/88
RAVIPRAKASH, G.K.	M.S.	Mechanical Engineering	3/88
STAMPS, MICHAEL EDWARD	M.S.	Computer Science	3/88
TANJI, FRANK TOSHIO	M.S.	Industrial Engineering	3/88
TIRRES, LIZET	M.S.	Aerospace Engineering	3/88
WALEN, JAMES VERNON	M.S.	Chemical Engineering	3/88
WILLIAMS, JACK L.	M.S.	Industrial Engineering	3/88
WOHLMAN, RICHARD ARTHUR	M.S.	Engineering Science	3/88
BASS, ROBERT EDWIN JR.	M.S.	Computer Science	6/88
BEJNOOD, MOHAMMAD SAEED	M.S.	Mathematics	6/88
BREWER, TIMOTHY LEIGH	M.S.	Computer Science	6/88
DATE', SANJAY S.	M.S.	Engineering Science	6/88
DOTSON, RANDY GREY	M.S.	Computer Science	6/88
GRISSOM, JAMES LLOYD	M.S.	Industrial Engineering	6/88
HANEY, RONALD WILLIAM	M.S.	Aviation Systems	6/88
HOOD, DAVID WESLEY	M.S.	Computer Science	6/88
HORTON, BRADLEY EUGENE	M.S.	Industrial Engineering	6/88
JOHNSON, RABON WARD JR.	M.S.	Industrial Engineering	6/88
McMINN, WESLEY THOMAS	M.S.	Mechanical Engineering	6/88
MEULMANS, RICHARD MICHAEL	M.S.	Industrial Engineering	6/88
MILLER, JOSEPH ANTHONY	M.S.	Aerospace Engineering	6/88
NAYANI, SUDHEER NATH	Ph.D.	Mechanical Engineering	6/88
QUINN, RANDALL WAYNE	M.S.	Aerospace Engineering	6/88
RICHARDSON, JOHN LARRY	M.S.	Industrial Engineering	6/88
WESTERHOF, HENRY ALLEN	M.S.	Industrial Engineering	6/88
WHITE, JAMES R. JR.	M.S.	Computer Science	6/88
ABEL, RALPH JOSEPH	M.S.	Aerospace Engineering	8/88
DE SHELTER, WILLIAM	M.S.	Electrical Engineering	8/88
EVANS, SCOTT	M.S.	Computer Science	8/88
FELCH, MARILYN	M.S.	Industrial Engineering	8/88
HASTINGS, JOSEPH	M.S.	Mechanical Engineering	8/88
HOLLADY, ALTON MARK	M.S.	Industrial Engineering	8/88
JENNINGS, BURT	M.S.	Mechanical Engineering	8/88
LINNEWEBER, CATHY	M.S.	Industrial Engineering	8/88
LITTLE, JEFREY	M.S.	Mechanical Engineering	8/88
Mc CORMAC, JOSEPH	M.S.	Computer Science	8/88
PENG, XIAOHANG	M.S.	Mechanical Engineering	8/88
SCHALLER, DAVID	M.S.	Industrial Engineering	8/88
SHERRELL, FRED	M.S.	Computer Science	8/88
SIMS, JOSEPH PAUL	M.S.	Aviation Systems	8/88
TAYLOR, WILLIAM E.	M.S.	Mathematics	8/88
ZAPATA, JOSE	M.S.	Industrial Engineering	8/88
AGHANAJAFI, CYRUS	Ph.D.	Mechanical Engineering	12/88
BHAT, MAHARAJ KRISHEN	Ph.D.	Aerospace Engineering	12/88
BOWLING, RICHARD F.	M.S.	Industrial Engineering	12/88
BRYANT, MICHAEL RAY	M.S.	Industrial Engineering	12/88
BURNETT, GERALD CLAYTON	M.S.	Computer Science	12/88
BYERS, MICHAEL T.	M.S.	Mechanical Engineering	12/88
CLOYD, JOHN DAVID	M.S.	Aerospace Engineering	12/88
ELROD, DAVID PARKER II	M.S.	Industrial Engineering	12/88
FERBER, HARRY JOSEPH II	M.S.	Computer Science	12/88
FUGERER, ROBERT HENRY	M.S.	Electrical Engineering	12/88
GOPATHI, KUMAR	M.S.	Mechanical Engineering	12/88
HODGES, DOUGLAS ALAN	M.S.	Aerospace Engineering	12/88
JOSEPH, EDWARD STEVEN	M.S.	Aviation Systems	12/88
LOWREY, GORDON ALAN	M.S.	Mechanical Engineering	12/88
MAYBERRY, BILLY RAY	M.S.	Industrial Engineering	12/88
MODESTE, DAVID C.	M.S.	Chemical Engineering	12/88
NASSER, ASAF LORENZ	M.S.	Engineering Science	12/88
O'DONNELL, KEVIN MICAHEL	M.S.	Aerospace Engineering	12/88
PATIL, ABHAY TRIMBAK	M.S.	Chemical Engineering	12/88
PRUFERT, MATT BRIAN	M.S.	Mechanical Engineering	12/88
QIAN, XIAOQING	M.S.	Aerospace Engineering	12/88
RIGNEY, SHARON J.	M.S.	Mechanical Engineering	12/88
SMITH, LEWIS MONTGOMERY	Ph.D.	Electrical Engineering	12/88
STEVENS, DEBORAH DeBERRY	M.S.	Industrial Engineering	12/88
TRUESDALE, ROBERT CARLTON	M.S.	Industrial Engineering	12/88
UTLEY, DAWN ROBESON	M.S.	Industrial Engineering	12/88

WANG, JYH-CHYANG	Ph.D.	Aerospace Engineering	12/88
YORK, CHARLIE EUGENE	M.S.	Industrial Engineering	12/88
ARY, CHARLES STEPHEN	M.S.	Industrial Engineering	5/89
BERRYMAN, NEIL	M.S.	Aviation Systems	5/89
BURGER, RICKY LYNN	M.S.	Computer Science	5/89
BUTLER, ROBERT JOHN	M.S.	Aerospace Engineering	5/89
CLIFTON, MICHAEL JAMES	M.S.	Avaition Systems	5/89
GRAY, PERRY ALVERS	M.S.	Engineering Science	5/89
GULDI, FREDERICK XAVIER	M.S.	Avaition Systems	5/89
IRBY, KAY FRANCES	M.S.	Engineering Science	5/89
MO, JIADA	Ph.D.	Aerospace Engineering	5/89
RASNAKE, DARRYLL GLEN	M.S.	Chemical Engineering	5/89
ROLLINS, DAVID	M.S.	Computer Science	5/89
SEISER, KAREN M.	Ph.D.	Engineering Science	5/89
SHARRAR, LARRY LOGAN	M.S.	Aviation Systems	5/89
SHERMAN, JEFFERSON	M.S.	Aviation Systems	5/89
SILVER, DEAN	M.S.	Industrial Engineering	5/89
SISTERMAN, STEVEN	M.S.	Aerospace Engineering	5/89
SUBRAMANIAN, CHIDAMBARAM	M.S.	Computer Science	5/89
TOONE, LORI DIANE	M.S.	Computer Science	5/89
TOWNE, MATTHEW C.	M.S.	Aerospace Engineering	5/89
TRUSSELL, PHILLIP	M.S.	Industrial Engineering	5/89
WELLS, MICHAEL JOSEPH	M.S.	Computer Science	5/89
WOODWARD, MELISSA	M.S.	Industrial Engineering	5/89
ZYSK, KEVIN	M.S.	Mechanical Engineering	5/89
CARSON, RANDALL E.	M.S.	Industrial Engineering	8/89
DYKHOFF, DAVID CRAIG	M.S.	Aviation Systems	8/89
FOOTE, JOHN PAUL	Ph.D.	Mechanical Engineering	8/89
HALL, CARROL R. JR.	M.S.	Mechanical Engineering	8/89
HAYMOND, JEFFREY	M.S.	Mechanical Engineering	8/89
HOPKINS, JOHN A.	M.S.	Mechanical Engineering	8/89
KRACINOVICH, STEPHEN	M.S.	Aviation Systems	8/89
LANG, MARK THOMAS	M.S.	Industrial Engineering	8/89
LONG, CALVIN WAYNE	Ph.D.	Engineering Science	8/89
McCARTHY, MICHAEL JOSEPH	M.S.	Industrial Engineering	8/89
MELESCUE, JOHN JAMES	M.S.	Aerospace Engineering	8/89
NELSON, KIM DAVID	M.S.	Engineering Science	8/89
NIKHADE, EKANATH	M.S.	Engineering Science	8/89
OLIVER, SUSAN CAROL	M.S.	Computer Science	8/89
RUDZIS, JOHN D.	M.S.	Aviation Systems	8/89
RUYTEN, WILHELMUS M.	Ph.D.	Physics	8/89
SUTLIFF, DANIEL	M.S.	Aerospace Engineering	8/89
WHITBY, DAVID G.	M.S.	Aerospace Engineering	8/89
BRENNAN, PHILIP	M.S.	Aviation Systems	12/89
CROPPER, THOMAS	M.S.	Aviation Systems	12/89
EDWARDS, ROBERT CHARLES	M.S.	Environmental Eng.	12/89
EDWARDS, SAMUEL C.	M.S.	Mechanical Engineering	12/89
ERAMO, ROBERT	M.S.	Mechanical Engineering	12/89
EVERETT, CHARLES E. JR.	M.S.	Aviation Systems	12/89
FAN, XIAOLING	M.S.	Mechanical Engineering	12/89
FORD, JAMES H.	M.S.	Industrial Engineering	12/89
GATZ, CHRISTOPHER	M.S.	Industrial Engineering	12/89
HAWRANEK, WALTER	M.S.	Industrial Engineering	12/89
HINMAN, ELAINE	M.S.	Aerospace Engineering	12/89
HUDDLESTON, DAVID	Ph.D.	Engineering Science	12/89
JACKSON, DOUGLAS	M.S.	Industrial Engineering	12/89
KOOISTRA, ARTHUR	M.S.	Aviation Systems	12/89
LITCHFORD, RONALD	M.S.	Mechanical Engineering	12/89
LUERS, JAMES	Ph.D.	Engineering Science	12/89
MARTIN, DARREN	M.S.	Industrial Engineering	12/89
McBRIDE, MARK	M.S.	Aviation Systems	12/89
OBLEN, RONALD	M.S.	Aviation Systems	12/89
ORR, JOSEPH	M.S.	Mechanical Engineering	12/89
PAINTER, ROGER	M.S.	Chemical Engineering	12/89
PEPPER, ROGER	M.S.	Industrial Engineering	12/89
POTTER, PEGGY	M.S.	Engineering Science	12/89
PURVIS, ROYCE	M.S.	Mechanical Engineering	12/89
SALADINO, ANTHONY	Ph.D.	Aerospace Engineering	12/89
SMITH, CHARLES E.	M.S.	Aviation Systems	12/89
TAYLOR, MICHAEL	M.S.	Mechanical Engineering	12/89
VANARIA, ANTHONY	M.S.	Aviation Systems	12/89

APPENDIX C

RESEARCH SOURCES

Dr. Robert L. Young's memoirs (taped)
Interview Robert W. Kamm (taped)
Interview Dewey Vincent (taped)
Interview Charlie Burton (taped)
Interview Morris L. Simon (taped)
Interview Dr. Frank Bowyer (taped)
Interview Ewing J. Threet
Interview Dr. Marcel K. Newman (taped)
Interview Dr. Fletcher Donaldson (taped)
Interview Sandra Shankle
Support Council Minutes
Minutes of UT Board of Trustees meetings
Dr. B.H. Goethert's unpublished **"From Bomb-Riddled Berlin to America After World War II"/Memories of a German Scientist-Engineer"**
"The Long Walk West" by George Barker, in **The Nashville Tennessean Magazine** July 14, 1957
"UTSI's Dr. Goethert Is Aviation Pioneer," by Dave Stephens, **TEMPO,** January, 1983
Copy of Goethert's remarks at Greater Nashville Chamber breakfast, January 30, 1963
The End of Arcadia by Dr. William Majors, Memphis University State Press
Memorandum File at UTSI
Article written by F. H. Richardson for **USAF News Review,** November 1964
"AEDC'S first 25 years," **High Mach** June 25, 1976
"Bob Kamm: UTSI's Space Age Pioneer" by Gary D. Smith, **TEMPO,** March, 1988
USAF/AEDC history files
UTSI brochures
Letters from former students
Micro-film and UTSI file of news clips of numerous publications including **The Tullahoma News, Manchester Times, Winchester's Herald-Chronicle, The Tennessean, The Banner, The Chattanooga Times, The Chatttanooga Free-Press, and the Knoxville Journal**